P9-EDU-742

Reconstructing Scientific Revolutions

TRANSLATED BY
ALEXANDER T. LEVINE

WITH A FOREWORD BY
THOMAS S. KUHN

THE UNIVERSITY OF CHICAGO PRESS
Chicago & London

PAUL HOYNINGEN-HUENE

Thomas S. Kuhn's Philosophy of Science

PAUL HOYNINGEN-HUENE graduated in physics at the University of Munich and earned a doctorate in theoretical physics from the University of Zurich. As a visiting scholar in the mid-1980s, he spent a year with Kuhn at MIT. He is currently professor of philosophy at the University of Constance, Germany.

The University of Chicago Press, Chicago 60637
The University of Chicago Press, Ltd., London

Printed in the United States of America
02 01 00 99 98 97 96 95 94 93   1 2 3 4 5

ISBN (cloth): 0-226-35550-0
ISBN (paper): 0-226-35551-9

An earlier version was published as *Die Wissenschaftsphilosophie Thomas S. Kuhns: Rekonstruktion und Grundlagenprobleme,* © Friedr. Vieweg & Sohn Verlagsgesellschaft mbH, Braunschweig 1989.

Library of Congress Cataloging-in-Publication Data

Hoyningen-Huene, Paul, 1946–
[Wissenschaftsphilosophie Thomas S. Kuhns. English]
Reconstructing scientific revolutions : Thomas S. Kuhn's philosophy of science / Paul Hoyningen-Huene ; translated by Alexander T. Levine ; with a foreword by Thomas S. Kuhn.
p. cm.
Includes bibliographical references and index.
1. Kuhn, Thomas S. Structure of scientific revolutions. 2. Science—Philosophy. 3. Paradigm (Theory of knowledge) I. Title.
Q175.H88313 1993

501—dc20   92-34288
CIP

⊗ The paper used in this publication meets the minimum requirements of the American National Standard for Information Sciences—Permanence of Paper for Printed Library Materials, ANSI Z39.48-1984.

CONTENTS

TRANSLATOR'S NOTE

ONE CAN HARDLY ATTEMPT a translation of a book on Thomas Kuhn's philosophy of science without recalling Kuhn's eloquent summary of the translator's dilemma:

> [The translator] must find the best available compromises between incompatible objectives. Nuances must be preserved but not at the price of sentences so long that communication breaks down. Literalness is desirable but not if it demands introducing too many foreign words which must be separately discussed in a glossary or appendix. People deeply committed both to accuracy and to felicity of expression find translation painful. (Kuhn 1970b, p. 267)

The line thus described is not one that may be negotiated without any awkwardness, for which I can only apologize.

This book is important to those interested in the Kuhnian project for two reasons: first, it provides the most comprehensive critical overview to date of Kuhn's philosophy of science; and second, it offers a European philosophical perspective on Kuhn's thought otherwise inaccessible to many English-speaking readers. Some of the terms in which the resulting insights are couched will thus be familiar to students of Hegel and Husserl (for example) but not to others schooled in the Anglo-American tradition. It is my hope that the "translator's notes," which I have tried to keep to an absolute minimum, provide some assistance.

Paul Hoyningen-Huene's sensitivity both to the literal content of his own work and to the nuances of the English philosophical idiom has, through his generous corrections and suggestions on earlier drafts, immeasurably improved the quality of this translation. Thanks are also due to Herbert Levine and Sonja Lucky for providing the environment which made preliminary work on the translation possible, and to Ellen Simer and Gary Klungness for their hospitality in Zurich. Jason Frye, Robert Gillis, Louis Matz, Jon Stewart, and Robert Pippin offered helpful suggestions on terminology. Adriana Novoa, Perry Deess, Mariana Ortega, Camilla Serck-Hanssen, and Georg Schwarz also made contributions, too diverse and numerous to list. Neither these friends and colleagues, nor those I've neglected to list, are responsible for my straying to one or another side of Kuhn's line.

FOREWORD

I FIRST MET PAUL HOYNINGEN in mid-August 1984, when he arrived in Boston to spend a year at the Massachusetts Institute of Technology. Both of us were appropriately apprehensive. The visit was an experiment, and neither of us could be confident of its success. His project was a book about my philosophical work, most centrally *The Structure of Scientific Revolutions*. Having studied my writings for several years, he wished now to supplement his results by extended discussions with me. His initiative was accompanied by persuasive recommendations; I welcomed it and had promised help. But I was known to be both busy and irascible, especially with people—to me they seemed very numerous—who persisted in retrieving from my work ideas that had no place there, some of them ideas that I deplored. If I reacted to his views in the same way, trouble for us both lay ahead. Our commitments for the coming year were irretrievably in place.

Within a few days our relief was visible, both to ourselves and to those around us. I rapidly discovered that Hoyningen knew my work better than I and understood it very nearly as well. More important, where I did think his understanding deficient, I found him both uncommonly able to listen and also appropriately stubborn in defense of his views. Our discussions often grew passionate, and it was not always Hoyningen who changed his interpretation of what I had meant. I could not have asked for an interlocutor more patient, more independent, or more concerned to get both detail and overall direction right. Readers who care about resolving the puzzles to be found in my writings will be in his debt for a long time to come. No one, myself included, speaks with as much authority about the nature and development of my ideas.

As a philosopher, Hoyningen is concerned not so much with the development of ideas as with the ideas developed. But his inquiry has required the close comparison of texts written over an interval of more than thirty years. Those texts seldom present ideas in the same words, and do not even regularly present precisely the same ideas. Retrieving a position from them has posed formidable problems of interpretation, few of which I had to face myself before my work with Hoyningen began. Watching his book take form has forced me to rehearse the story of my own development, an experience that has occasionally proved as uncomfortable as it has been enlightening.

What do I take to be the principal shifts in my viewpoint since the publication in 1962 of *The Structure of Scientific Revolutions*? At the head of the list comes a considerable narrowing of focus. Though I thought of that book as addressed primarily to philosophers, it turned out to be pertinent also to sociology and historiography of science. Asked what field it dealt with, I was often at a loss for response. My subsequent attempts to develop the viewpoint have, however, been directed exclusively to the book's philosophical aspects. Its concern with history has gradually been transmuted to a concern with developmental or evolutionary processes in general. Its sociological component survives mainly in the insistence that the vehicle for such processes must be a self-replicating population or group. Such groups I increasingly conceive as language- or discourse-communities, sets of individuals bound together by the shared vocabulary which simultaneously makes professional communication possible and restricts that communication to the profession.

Mention of discourse-communities points toward a series of more specific developments of my viewpoint. *Structure* included many references to the changes in word meanings that accompany scientific revolutions, but it spoke more often of changes of visual gestalt, changes in ways of seeing. Of the two approaches, meaning change was the more fundamental, for the central concepts of incommensurability and partial communication were based primarily upon it. But that basis was far from firm. Neither traditional theories of word meaning nor the newer theories that reduced meaning to reference were suited to the articulation of these concepts. Allusions to altered ways of seeing could at best disguise the deficiency.

Since the publication of *Structure* my most persistent philosophical preoccupation has been the underpinnings of incommensurability: problems about what it is for words to have meanings and about the ways in which words with meanings are fitted to the world they describe. These are the concerns which, a decade after *Structure* appeared, led me to emphasize the role played, at all levels of research, by primitive similarity/difference relations acquired during professional education. Such relations supply what I have more recently come to describe as the taxonomy shared by a field's practitioners, their professional ontology. Some of what a group knows about the world at any time is embodied in its taxonomy, and changes in one or another region of the taxonomy are central to the episodes I have called scientific revolutions. After such changes, many generalizations that invoked the names of older categories are no longer fully expressible.

Another development—the last I shall mention here—is more recent, very much still underway. In *Structure* the argument repeatedly moves back and forth between generalizations about individuals and generalizations about groups, apparently taking for granted that the same concepts are applicable to both, that a group is somehow an individual writ large. The most obvious example is my recourse to gestalt switches like the duck-rabbit. In fact, like other visual experiences, gestalt switches happen to individuals, and there is ample evidence that some members of a scientific community have such experiences during a revolution. But in *Structure* the gestalt switch is repeatedly used also as a model for what happens to a group, and that use now seems to me mistaken. Groups do not have experiences except insofar as all their members do. And there are no experiences, gestalt switches or other, that all the members of a scientific community must share in the course of a revolution. Revolutions should be described not in terms of group experience but in terms of the varied experiences of individual group members. Indeed, that variety itself turns out to play an essential role in the evolution of scientific knowledge.

The separation of concepts applicable to groups from those applicable to individuals is a powerful tool for eliminating the solipsism characteristic of traditional methodologies. Science becomes intrinsically a group activity, no longer even idealizable as a one-person game. The same analytic separation proves crucial also for questions of word meaning. Different individuals may pick out the referents of terms in different ways: what all must share, if communication is to succeed, is not the criteria by which members of a category are identified but rather the pattern of similarity/difference relations which those criteria provide. It is the latter, the shared taxonomic structure, that binds members of the community together, and it does not require that individuals give the same answer to the question: similar with respect to what?

Essential guidance on these and other matters is provided by the book now before you. I recommend it warmly.

Boston, Massachusetts *Thomas S. Kuhn*
August 1988

PREFACE

THE GOAL OF THIS WORK IS, as suggested by its title, to reconstruct Kuhn's philosophy of science, discussing along the way such fundamental problems as appear in the course of reconstruction. We are concerned with the theory first proposed in its entirety by the American historian and philosopher of science Thomas S. Kuhn in this 1962 book *The Structure of Scientific Revolutions.*[1] This work is among the most influential academic books of the past quarter-century[2] and has given rise to what is now an unmanageably vast secondary literature. The terms here coined, "paradigm," "paradigm change," and "scientific revolution" are now commonplace not only in the study of science but also within individual scientific fields, and even in many less scientific domains. Why, then, is a reconstruction of Kuhnian philosophy of science needed?

Three responses suggest themselves. First of all, Kuhn's readers are far from agreeing even on the precise content of his central theses, let alone on their validity. On the contrary, interpretations of Kuhn—hence criticisms, praises, and applications of his work—are extraordinarily varied. Kuhn has himself admitted that weaknesses, obscurities, unclarity, vagueness, confusion, real difficulties, ambiguities, misunderstandings, substantive errors, and provisional formulations grounded in metaphor and intuition are all to be found both in the exposition of *SSR* and in later work, leading to a certain plasticity in the articulation of his position.[3]

Second, as Kuhn himself has often complained,[4] the reception of

1. Hereafter cited as *SSR*. Page references are to the only slightly revised second edition, published in 1970. The 1977 collection of Kuhn's essays, *The Essential Tension: Selected Studies in Scientific Tradition and Change,* will be cited as *ET*. Other works by Kuhn will be cited by year of publication; see bibliography.

2. Sales of the English edition of *SSR* had, by June 30, 1990, exceeded 740,000; *SSR* is also available in at least nineteen translations: German (1967), Polish (1968), Italian (1969), Japanese (1971), Spanish (1971), French (1972), Dutch (1972), Danish (1973), Serbo-Croation (1974), Portuguese (1975), Russian (1975), Romanian (1976), Hebrew (1977), Swedish (1979), Chinese (1980), Greek (1981), Korean (1981), Czech (1981), and Hungarian (1984).

3. 1969c, *ET,* p. 350; 1970b, pp. 234, 249, 250, 252, 259–260, 266; 1970c, *SSR,* pp. 174, 180, 181, 185, 193, 196, 197; 1971a, pp. 139, 145, 146; 1974a, *ET,* pp. 293, 294, 319; 1974b, p. 506; *ET,* pp. xv, xix–xx; 1983a, p. 669. Cf. Holcomb 1989.

4. 1963b, pp. 386, 387; 1970b, pp. 231–232, 236, 259–260, 263 n. 3, 266; 1970c,

his theory has, despite being so widespread, failed by far to do it justice. But neither his expository style, which seems so easily misunderstood, nor the superficiality of its reception can entirely explain the distortions and misunderstandings typical of the popular image of Kuhn.[5] Indeed, such explanation can't be attempted from a completely neutral standpoint, unbiased with respect to both Kuhn and his critics. For, from Kuhn's perspective, we are here dealing with misunderstandings and distortions of a kind typical of situations in which, for the time being, only partial communication is possible.[6] Similar situations also occur, according to Kuhn (though *not* according to some of his critics), in the history of science; and the heart of Kuhn's work, in both the philosophy and the historiography of science, has been the analysis of precisely these situations and their wide-ranging consequences. The "arbitrary" or "ridiculous" appearance of elements of an as yet alien theory, taken "in isolation" from one another, is characteristic of these situations, though the *interplay* of these same elements allows them to "lend each other mutual authority and support."[7] Accordingly, Kuhn's theory, too, can really be understood only when its individual components are likewise considered in their relation to one another.

Third, Kuhn has expanded his theory in several respects since its initial articulation and continues to develop it even now. But, like *SSR* itself, the precise import of published expansions of the theory, the motivation behind them, and their relation to the formulations of *SSR* are subject to the most diverse interpretations. The view is widespread that Kuhn's reflections on *SSR* and on subsequent criticisms, found in his widely read works of the second half of the sixties,[8] constitute a substantive modification of his theory, directed at weakening his original "revolutionary" theses.[9] Others claim that Kuhn "radicalized" his

---

SSR, pp. 175, 198–199; *ET,* p. xxi; 1977c, *ET,* p. 321; 1983a, p. 669; 1983b, p. 712; 1983d, p. 563. Some extreme misunderstandings and distortions are found, for example, in Watkins 1970 and Laudan 1984 (regarding the latter, cf. Hoyningen-Huene 1985). Especially accurate readings are to be found, on Kuhn's view, in Lakatos (Kuhn 1971a, p. 137), and above all in Stegmüller's 1973 reconstruction of Kuhn's theory by means of Sneedian formalism (Kuhn 1976b, p. 179). But even here Kuhn has serious reservations.

5. Stegmüller 1973, chap. 9, § 2, has pointedly exposed several typical misunderstandings of the 1960s and early 1970s. See also Hoyningen-Huene 1988; Kitcher 1983, p. 698 n. 2; Lugg 1987, pp. 181–182; Stegmüller 1986, pp. 79, 340 ff.

6. See especially 1970a, *ET,* pp. 266–269, and 1970b, pp. 231–232, but also earlier, *SSR,* p. x. See also § 7.5.c.

7. 1981, p. 11; Kuhn is referring to Aristotelian physics.

8. In particular, 1970a, 1970b, 1970c, and 1974a.

9. Musgrave 1971 and Shapere 1971 are frequently cited and representative examples of this view; q.v. Curd 1984, p. 4; Newton-Smith 1981, pp. 9, 103, 113–114; Putnam 1981, p. 126; Toulmin 1971; Watanabe 1975, p. 132 n. 1.

"original position" in the essays published in The Essential Tension.[10] Kuhn, by contrast, asserts that all modifications undertaken over the years have left the core of his theory essentially untouched.[11]

Given this decidedly confused state of current discussion, an attempt to clarify fundamentals seems called for. In order to succeed, this clarification must clearly distinguish itself from previous readings of Kuhn. The approach I've chosen consists, for one, in a hermeneutic approach to Kuhnian texts. This approach assumes, first of all, that even texts branded by other readers as incoherent, confused, or inconsistent have some discernible meaning, until the contrary is proved.[12] Furthermore, all of Kuhn's texts must be considered and related to one another, under the assumption that they might thus both illuminate each other's obscurities and allow us to see and understand the evolution of Kuhnian thought.[13] The product of these hermeneutic efforts is a text steeped in footnotes to Kuhn to an extent unusual in studies of science. This thoroughness is justified by the search for a high degree of both precision and tractability in our reconstruction. The adoption of such "philological" standards, though it may expose me to the charge of pedantry, strikes me as appropriate, given the unsatisfactory course of the Kuhnian debate to date. For, if the terms of Kuhn's theory have become household words,[14] the demand for clarity with regard to what this theory does and does not claim, whether and how its claims are justified, what its main problems consist in, and in what respects they call for criticism and expansion is certainly timely.

As a second and equally important method for securing an understanding of Kuhnian theory, I've chosen to establish contact with its author. I spent the 1984–85 academic year at the Department of Linguistics and Philosophy at the Massachusetts Institute of Technology in Cambridge, Massachusetts, where Kuhn has been active since 1979. This one-year sojourn was followed by a number of shorter stays at MIT, each a few weeks in duration. Kuhn read several drafts of this work, and we had the opportunity to discuss many aspects of them

10. Peterson-Falshöft 1980, p. 105.

11. E.g. 1970c, *SSR*, p. 174; 1971a, p. 146; 1974a, *ET*, pp. 318–319; 1983a, p. 671.

12. See, among many others, Laitko 1981; Laudan 1977, p. 231 n. 1; Meiland 1974; Scheffler 1972; Shapere 1964, 1971; Siegel 1976, 1980; Watanabe 1975; Wittich 1981.

13. Even an isolated reading of *SSR* requires that, for a given theoretical issue, many passages—frequently widely dispersed—be considered. The reason for this is that the organization of *SSR* recapitulates a developmental process. Consequently, passages relevant to a given philosophical question are usually scattered throughout the text.

14. It should be noted, by way of qualification, that many citations appear more ritual than substantive in nature. See, for example, the account of how Kuhn's work is used in the psychological periodical literature given in Colemann and Salamon 1988, especially pp. 435–436.

in detail. In addition, I was allowed to peruse Kuhn's unpublished manuscripts, ranging from earlier efforts to work currently in progress. On several occasions, our discussions proved especially helpful in inspiring new ways of reading his texts, revealing hitherto unnoticed connections between bits of theory or written passages and providing me with constructive criticism of my approach. The fact that Kuhn has never laid claim to an interpretive monopoly on his earlier writings helped make my association with him extraordinarily fruitful and equally pleasant. But, for the most part,[15] I will, in this book, make no explicit references either to my discussions with Kuhn or to his unpublished manuscripts, as such reference would undermine the reader's ability to verify my reconstruction.

Some readers may find three things lacking in this book: a historical chapter, sketching the manner in which Kuhnian theory draws on and departs from earlier tradition in the philosophy of science; a separate chapter reviewing criticisms of Kuhn to date; and, finally, a chapter on Kuhn's influence and on the parallels between Kuhn and other authors that constitute, or fail to constitute, evidence for influence in either direction. I will treat these gaps in order.

As regards historical connections with earlier tradition in the philosophy of science in the narrow sense, that is, logical positivism and critical rationalism, a number of expositions are already available;[16] while I do not always agree with their reconstructions of the Kuhnian position, they will suffice as a first approximation. Some connections with other traditions have, furthermore, been acknowledged on several occasions by Kuhn himself. These include debts to Conant School historiography;[17] the historiography of Koyré and his school,[18] which in turn hearkens back to neo-Kantianism; Piaget's developmental psychology;[19] gestalt psychology;[20] Fleck's sociology of science;[21] Whorf's linguistic theory;[22] Wittgenstein's later philosophy;[23] and the philosophy of Quine.[24] These connections have also been confirmed, and to

15. The sole exception is in § 2.2.c.

16. E.g. Suppe 1974; Brown 1977; Bayertz 1980.

17. *SSR*, pp. v, xi; 1979c, p. viii; 1983c, p. 26; 1984a, pp. 30–31.

18. *SSR*, pp. v–vi; 1968a, *ET*, pp. 107–109, 121; 1970e; 1971c, *ET*, pp. 149–150; 1977b, *ET*, p. 11; 1979a, p. 125; 1983c, pp. 27, 29; 1984, p. 243; 1984a, p. 30.

19. *SSR*, p. vi; 1964; 1979c, p. viii.

20. *SSR*, p. vi; *ET*, p. xiii; 1979c, p. ix.

21. *SSR*, p. vi–vii; 1979c.

22. *SSR*, p. vi; 1964, *ET*, p. 258.

23. *SSR*, pp. 44–45; 1968a, *ET*, p. 121.

24. 1961a, *ET*, p. 186 n. 9; *SSR*, p. vi; 1964, *ET*, p. 258 n. 28; 1970b, pp. 268–269; 1970c, *SSR*, p. 202 n. 17; 1971a, p. 146; 1976b, p. 191; *ET*, p. xxii; 1979a, p. 126.

some extent explored, by other authors,[25], so I will attempt no systematic analysis in this book, though I shall comment on them and others in the occasional footnote. Only the relationship between Kuhn's theory and the historiographic approach inaugurated in its essentials by Koyré will be given any systematic exposition, for Kuhn's theory remains unintelligible without some knowledge of its dependence on this kind of historiography. In addition, I will on several occasions compare Kuhn's theory with Kant's critical philosophy and with Popper's critical rationalism, since Kuhn's parallels to and departures from these authors prove instrumental in understanding his theory.

As far as existing criticism of Kuhn is concerned, any systematic treatment of this corpus is necessarily limited by its sheer volume and extreme heterogeneity. Far too many alleged points of criticism are mere misunderstandings, and the enumeration of misunderstandings is of only subordinate interest in the reconstruction of a theory. Criticisms of Kuhn are summarized and discussed now and then in the course of my reconstruction, but usually only in footnotes, for the business of reconstructing the theory in its development from text to text and of confronting the fundamental problems faced by the reconstructed theory seemed the most important task at hand. Of course, I hardly claim to have discussed *all* of these fundamental problems; even disregarding the essentially dubious nature of all such claims to completeness, that would seem an unreasonable goal to set for a single book, especially for a work dealing with such a comprehensive theory.

This book has waived the systematic exposition both of Kuhn's influence and of the parallels between his work and that of other authors, whether or not there is evidence for influence in either direction. Such an exposition would doubtless touch on similarities with the philosophy of Heisenberg and that of Dürr,[26] on the (re)constructive theory of science articulated by Kamlah and Lorenzen,[27] on Feyerabend's work,[28] on Kuhn's influence on Lakatos,[29] on the assimilation of Kuhnian ideas in structuralist theories of science,[30] on parallels to

25. E.g. Barker 1988; Bayertz 1980; Burr/Brown 1988; Cedarbaum 1983; Mandelbaum 1977, pp. 449–450; Maudgil 1989, Merton 1977, pp. 71–109; Poldrack 1983; Stegmüller 1973, chap. 9, § 4; Stock 1980; Wittich 1978a, 1978b, and 1981; Wright 1986; Wuchterl/Hübner 1979, p. 129.

26. See, above all, Heisenberg 1942 (along with Dürr 1988, pp. 131–142); Dürr 1988, especially pp. 26–49.

27. See Kamlah and Lorenzen 1967.

28. See, for example, Feyerabend 1976, 1978a, and 1981a.

29. See, for example, Lakatos 1978.

30. See Sneed 1971, chap. 8, as well as Stegmüller 1973, 1974, 1980, and 1986, along with their citations.

Hübner's philosophy[31] and to Rescher's,[32] on (radical) constructivism,[33] and on recent work by Putnam[34] and Goodman.[35] I will also waive discussion of Kuhn's enormous impact in individual disciplines, especially the humanities.[36] The neglect of these topics is motivated not by lack of interest but rather by the concentration necessary in pursuit of the reconstructive and critical goals of this work. I will occasionally footnote particularly striking parallels to other authors, but with no pretensions to systematicity.

This project has profited from many sources of support. A grant from the Canton of Zurich Fund for the Promotion of Academic Posterity,* a stipend for living expenses from the Swiss National Science Foundation,** and an MIT grant for travel expenses provided the financial assistance that made this effort possible. Victor Gorgé (Bern), Dieter Groh (Konstanz), Trude Hirsch (Zurich), Stefan Niggli (Zurich), Ulrich Röseberg (East Berlin), and Thomas Übel (Cambridge, MA) provided critical commentary of some chapters and sections. The final draft of this work was produced in the stimulating, creative atmosphere of the University of Pittsburgh's Center for Philosophy of Science, where, thanks to an invitation by Adolf Grünbaum and Nicholas Rescher, I spent the 1987–88 academic year as a Senior Visiting Fellow. Special thanks are due to Paul Feyerabend, who read several versions of the entire manuscript as fast as individual sections could come off the printer. And finally, last but not least, my thanks go out to Thomas Kuhn, whose extraordinary hospitality in Cambridge and Boston fostered both my progress and my enjoyment of this project to a degree I would never have expected.

* Kredit zur Förderung des akademischen Nachwuchses des Kanton Zürich

** Schweizerischer Nationalfonds zur Förderung der Wissenschaftlichen Forschung

31. See Hübner 1978.

32. See especially Rescher 1973 and 1982.

33. See e.g. Gumin and Mohler 1985 and Schmidt 1987, with further references.

34. See, above all, Putnam 1978, 1981, and 1983; Kuhn remarks on these parallels in his 1989a, p. 25 n. 26, and in 1990, p. 317 n. 23. An overview of Putnam's recent work which exhibits these parallels clearly (if not explicitly) is provided by Franzen 1985; compare also Stegmüller 1979, chap. 3.3.

35. See Goodman 1978 and 1984, chap. 2.

36. For a treatment of this topic, see e.g. Hollinger 1973 and other contributions to Gutting's 1980 anthology, all of which, however, serve only to reveal the tip of the iceberg.

PART ONE

Introduction

CHAPTER ONE

The Topic of Kuhn's Philosophy of Science

THE TOPIC OF KUHN'S PHILOSOPHY OF SCIENCE is scientific development.[1] As straightforward as this definition might seem at first glance, it calls for many points of clarification. To begin with, the concept of "scientific development" itself requires more accurate definition, first of all because the notion of a "science" has been applied with varying connotations and degrees of breadth. In addition, "development" is properly used only to describe a change over time, and the question of how this change is best conceived remains open (§ 1.1). Next, we must ask in what manner this scientific development becomes accessible; for whatever we understand by scientific development, it is surely not an immediate given. What *are* "given" are testaments to and traces of the past, on the basis of which a reconstruction of scientific development can occur. The question thus amounts to a demand for the principles according to which scientific development is reconstructed from these traces—the principles that turn our unavoidable constructions into an adequate reconstruction (§ 1.2). Third, we must explore the particular focus which guides Kuhn's project, for an interest in scientific development can arise from many different perspectives. Kuhn is concerned with a general theory of scientific development that captures the structure of that process; it is thus natural to ask what we are to understand by the structure of scientific development in general and by the structure of individual scientific revolutions in particular (§ 1.3).

1. This feature is attested at many places, for example 1959a, *ET* pp. 225, 232; 1961a, *ET* p. 221; *SSR* pp. ix, x, 3, 6, 16, 17, 24, 92, 94, 96, 108, 109, 141, 160, 170; 1968a, *ET* p. 118; 1970a, *ET* p. 266; 1970c, *SSR* p. 180; *ET* pp. ix, xv; 1977c, *ET* p. 338; 1981, p. 7; 1983c, p. 27.

1.1 The Issue: Scientific Development

In defining the central issue of a general theory of scientific development, two questions emerge: First, how should the *total domain* of science to which this theory is meant to apply be defined (§ 1.1.a), and second, how should the *elements* of this total domain be identified—how big are those "pieces" of the total domain that count as the theory's particulars? (§ 1.1.b)

a. The Total Domain of Science

The task of defining the total domain of science can be divided into two subtasks. First, we must specify the *disciplines* belonging to this domain (point 1). Does this range include all fields commonly called "the sciences," or is it rather limited to the "hard" sciences?* Does it include only basic science, or also applied studies? Next, we must state those *aspects* of science we take to be relevant in defining our total domain (point 2). Should we limit our exploration to the epistemic aspect of science, thus defining science only *qua* scientific knowledge, or also make use of social, political, economic, and cultural aspects of scientific activity?

1. It is misleading even to say, as we have, that the issue of Kuhn's theory is "scientific development," for "the sciences" encompass more than the "hard science" with which Kuhn is actually concerned. This domain includes the natural sciences and the systematic social sciences;[2] history and philosophy, including the philosophy of science, are explicitly excluded.[3] Kuhn's theory also claims to address the bio-

* Translator's note: Here, as in the following paragraph, the discussion makes use of the diverging connotations of the English "science," and its German counterpart, "Wissenschaft." I have tried to capture this distinction by rendering "science" as "hard science," and "Wissenschaft" as "the sciences."

2. Kuhn's inclusion of the social sciences can be inferred, for example from *SSR*, pp. 15, 21.

3. The exclusion of history, especially the historiography of science, from the sciences is explicitly stipulated in 1970e, p. 67, where Kuhn states that "*Outside of the sciences,* few fields of scholarship have been so transformed in the past thirty years, as the historical study of scientific development" (Emphasis added). The exclusion of philosophy and history from the domain of Kuhn's theory has the consequence that this theory is not self-applicable—at least not immediately (see Wendel 1990, p. 31 n. 1, or similarly p. 168). Thus arguments against the theory which turn on its unmediated self-application become problematic (see e.g. Briggs and Peat 1984, p. 34; Holcomb 1987, p. 475; Kordig 1970; Radnitzky 1982, p. 71; Scheffler 1967, pp. 21–22, 74; Scheffler 1972, pp. 366–367.). Nonetheless, Kuhn himself occasionally appears to apply the theory to itself; witness his references to a crisis in the "philosophical" or "epistemological" paradigm (*SSR,* p. 121; similarly on p. 78), to the "gestalt switch that still divides us [Kuhn and Popper] deeply"

logical sciences;[4] though these are extremely underrepresented in his examples.

Now it might appear that Kuhn's theory prejudices the demarcation criterion separating science from nonscience by importing a more or less determinate understanding of "science." It is indeed the case that Kuhn starts from, and must start from, a determinate understanding of science. But we can distinguish the problematic from the unproblematic features of this understanding. A presupposed understanding of science is unproblematic insofar as it involves only the extensional classification of *uncontroversial* cases; Newtonian mechanics, Maxwellian electrodynamics, Einsteinian relativity theory, and Darwinian evolutionary theory can all be classified as sciences, though the nature of those common characteristics by virtue of which they are so classified, if there are such, remains open. However we choose to define science, our definition is adequate only if it encompasses these fields. Nonetheless, it is possible that our choice of uncontroversial cases biases us in favor of certain areas of scholarship to the detriment of others, thereby undermining our theory's relevance to the neglected disciplines.[5]

A presupposed understanding of science really becomes problematic when it decides *controversial* cases (for example, astrology, psychoanalysis, "Marxism," Aristotelian dynamics), thus prejudicing any definition of the concept of science by preempting a great deal of needed discussion. As far as contemporary fields of controversial scientific character are concerned, Kuhn avoids the problem by leaving their status open. Where historical fields of study are of questionable scientific status, either because it's not clear if they were science or not, or, if they were science, then "good science" or "bad science," our assessment will depend on the manner in which they are reconstructed from accessible remnants.[6] The complex of problems surrounding the assessment of historical fields will be discussed in § 1.2; for now, however, their scientific status must remain an open question.

(1970a, *ET*, p. 292), and to such talking "at cross-purposes" (rather than mere "difference of opinion") that characterizes his dialogue with some of his critics (1970b, p. 233). In § 3.8 I will return to discuss the central argument from self-application against Kuhn's theory.

4. *SSR*, pp. ix, 21.

5. Mayr 1971, pp. 277, 294, claims that biology has been so neglected. See also Greene 1971, Ruse 1970, 1971.

6. The manner of reconstruction can also have a great bearing on our treatment of more recent fields, as illustrated by the controversy over the early history of quantum theory (see Kuhn 1978; Klein 1979; Pinch 1979; Shimony 1979; Foley 1980; Kuhn 1980b, 1980c; Galison 1981; Nicholas 1982; Kuhn 1984). But what's decided by the manner of reconstruction isn't so much the scientific character of these fields but is rather the importance of individual scientific contributions.

The total domain of science must now, *qua* object of Kuhn's theory, be narrowed further, for only "pure science" (or "basic science"), as opposed to applied science or technological invention, is at issue.[7] Pure science is distinct from applied science in that its research topics are chosen on the basis of reasoning and assessment internal to science, not derived from social, economic, or military interests.[8] To be sure, the distinction between pure and applied research appears problematic in many areas of contemporary science, as, for example, in the study of cancer or nuclear fusion. It remains serviceable, however, as witnessed by the fact that many cases, past and present, can be decided without difficulty. Kuhn himself stresses that the pure/applied distinction requires closer examination.[9] As the choices of representative cases of basic science used in the exposition of Kuhn's theory have, to the best of my knowledge, not sparked any controversy in subsequent debate, I will not pursue any problems arising from the distinction here. We must simply remember that Kuhn's theory addresses only "pure" science and thus proceeds under the assumption that pure and applied sciences can be distinguished. Future science may violate this assumption.

For ease of exposition, I will use "science" to denote the *pure sciences,* where not otherwise indicated.

2. The total domain of science must be determined in one further respect in order to prepare it for treatment in a general theory of scientific development: we must specify which of the many aspects of science are of interest and which aren't.[10] One commonly used, narrowly circumscribed conception of science captures only those aspects pertaining to science *qua* form of knowledge: all those objects with special pretensions to scientific status. Roughly speaking, this conception covers the contents of scientific publications. It can be called the "epistemic" (or "theoretical," or "cognitive") conception of science. Less narrow, or at least different, conceptions of science can be obtained by taking other, nonepistemic aspects into consideration. For example, an exploration of the values and norms guiding scientific action allows us to treat science as a "social system."[11] Or one might further examine quantitative aspects of scientific development, such as numbers of scientists or publications.[12] Finally, science can be described as one of

7. 1959a, *ET* pp. 233, 237.

8. 1959a, *ET* pp. 237–238. See Radnitzky 1983, pp. 238–240, and Radnitzky 1986, pp. 105–106.

9. 1959a, *ET* p. 238.

10. See Diemer 1970 for the following.

11. For example, Merton 1942 and Storer 1966 take this approach.

12. As attempted in Solla-Price 1963.

many species of human culture, involved in the most diverse interactions with economics, politics, art, religion, and other such species. This broadest of all descriptions yields the "sociocultural" conception of science.[13]

Kuhn's interest lies in the development of science in its epistemic sense.[14] This definition of the central issue of a theory of scientific development leaves two of its features entirely open. First of all, it presupposes nothing about the form, locus, or degree of certainty of scientific knowledge. Our definition assumes only that the sciences will offer us objects with claims to special epistemic status, the objects with whose development we are concerned. In addition, nothing has been assumed about the kind of ancillary apparatus we will need to adduce in order to explain the development of scientific knowledge. In particular, it remains open whether the corpus of scientific knowledge itself provides enough for this explanatory task or whether social, economic, political, psychological, and other factors must also be taken into consideration. In other words, we remain neutral with respect to the eventual adoption of an internalist or externalist position, regardless of where the proper boundary between internal and external explanations turns out to lie.

b. Permissible Units of Analysis within the Domain of Science

Having fixed the total domain of science *qua* object domain for a theory of scientific development, we still have not yet adequately determined the issue of this theory. For the manner in which elements of this total domain, the particulars to which the general assertions of the theory apply, should be individuated, remains open. The regimentation of our total domain of science, the relevant aspects and disciplines of which we have already specified, might evidently proceed in a number of ways. Division along three different axes allows us to specify "pieces" of different sizes as potential elements of the total domain. A rough division along the *disciplinary axis*[15] allows us to distinguish three categories of pure science; these are the sciences of inorganic, organic, and social worlds, respectively. A somewhat finer-grained categorization along this same axis would distinguish actual disciplines, such as as-

13. The purpose of this taxonomy of the aspects of science is neither the precise differentiation of these possible conceptions nor any assertion regarding the utility of some particular nonepistemic approach (see 1976a, *ET* p. 34). My aim is rather to draw attention to the variability of the concept of science along this axis, thereby emphasizing the need for further definition.

14. See for example *SSR*, p. 11; 1970a, *ET*, p. 267; 1983c, p. 28 and elsewhere.

15. 1970c, *SSR* p. 177.

tronomy, physics, chemistry, biology, etc. A still finer-grained treatment of the disciplinary axis would yield subdisciplines, as for example those of physics: mechanics, acoustics, thermodynamics, optics, etc. Finally, we would arrive at the special problems within each subdiscipline. Along the *temporal axis,* a given disciplinary determined element of the domain can, for example, be examined with a view toward change over a particular time interval, or alternatively, with an eye to its long-term dynamic. And finally, we might focus our attention on the *epistemic axis,* taking the development either of the whole body of knowledge or of such parts as individual concepts, methods, or theories as of special interest.

Although in *SSR* Kuhn fails to address explicitly the aforementioned three viewpoints from which a theory of scientific development's units of analysis might be identified, in his postscript to *SSR* he acknowledges the deficit and treats it in greatest detail.[16]

For Kuhn, the manner in which the total domain of science is to be divided into particulars for treatment in his theory of scientific development is of more than coincidental interest. Indeed, there exists a criterion for determining whether a choice of some particular set of units of analysis is legitimate. This criterion demands that one or more scientific communities actually take, or took, each disciplinary, temporally, and epistemically determined unit in question as its (their) proper field of study.[17] Thus only those elements of the total domain of science for whose dynamic some responsible agent can be found, in the form of one or more scientific communities, constitute acceptable units of analysis. But how, precisely, does Kuhn circumscribe the notion of a scientific community?

In *SSR,* the concept of a scientific community is introduced along with that of a paradigm;[18] the scientific community is characterized by its possession of a paradigm, and conversely, the paradigm is that which makes the scientific community what it is. In papers written in 1969, Kuhn criticizes this mode of introduction for its "intrinsic

16. 1970c, *SSR* pp. 176–181. On p. 180, Kuhn discusses his "vagueness about the nature and size of the relevant communities" in *SSR*. The reason for my reference here to a passage in which the size of scientific communities is at issue will soon become clear.

17. This criterion emerges, for example, from 1970c, *SSR* pp. 179–180 (similarly from 1970b, pp. 254–255.). Here Kuhn defends himself against the claim, put forth by some critics (Shapere 1964, p. 387; Watkins 1970, p. 34; Popper 1970, pp. 54–55), that the theory of matter, as it existed from ancient to modern times, constitutes a counterexample to his theory of scientific development. His reply is that this theory was never, at least until roughly 1920, the province of any particular scientific community. See also 1968a, *ET* p. 109; 1976a, *ET* pp. 32–33.

18. *SSR,* p. viii.

circularity"[19] and suggests that, were he to produce a new edition of his book, he would take advantage of the opportunity to define the concept of a scientific community without reference to the concept of a paradigm:[20]

> I would now insist that scientific communities must be discovered by examining patterns of education and communication before asking which particular research problems engage each group.[21]

In other words, the set of a scientific community's attributes can be divided into two subsets. The first subset, containing features of community education and communication, is used to *identify* the community. The second, which incorporates other sociological features, as well as scientific contents, can be used, among other things, to *explain* the attributes in the first subset.[22] In what follows, we will attempt to trace the path indicated by Kuhn, though our characterization of scientific communities must remain as provisional as Kuhn's own.

"A scientific community consists," Kuhn tells us, "of the practitioners of a scientific specialty."[23] Members of the community exhibit a suite of common traits:

—"similar educations" and
—similar "professional initiations." They have, over the course of their initiation to science,
—"absorbed the same technical literature" and
—"drawn many of the same lessons from it."

19. 1974a, *ET* pp. 294–295; 1970c, *SSR* p. 176; similarly 1970b, p. 252; *ET* pp. xv–xvi. —Of the "consequences" of this circularity, "the most damaging" is thought by Kuhn to be the manner in which it led him to distinguish the "preparadigm period" from the "so-called postparadigm period" (1974a, *ET* p. 295 n. 4; similarly 1970b, p. 272 n. 1; 1970c, *SSR* pp. 178–179). I have incorporated the thrust of these amendments to *SSR* in my formulation of the legitimacy criterion for units of analysis, by stipulating that candidate fields must have been studied by one *or more* scientific communities.

20. 1970b, pp. 252, 271; 1970c, *SSR* p. 176.

21. *ET,* p. xvi; similarly 1970b, p. 253. One possible misunderstanding should be anticipated here. Kuhn does not mean to suggest that scientific communities must be identified exclusively by means of criteria having nothing to do with scientific content, for given such social indicators alone, distinguishing communities of scientists from communities of philosophers, mystics, religious believers, engineers, and the like might prove problematic (see Musgrave 1971, pp. 287–288). The issue is rather the identification of communities *within the total domain of science*—hence our initial discussion of this domain.

22. This second, explanatorily potent class of attributes is the "disciplinary matrix" (1974a, *ET* p. 297; 1970b, p. 271; 1970c, *SSR* pp. 178, 182). See §4.3.

23. 1974a, *ET* p. 296; 1970b, p. 253; 1970c, *SSR* p. 177.

Furthermore,

—"the boundaries of that standard literature mark the limits of a scientific subject matter."[24]

This list applies to *all* scientific communities. Further traits can be specified only when we have distinguished two kinds of scientific community. In the so-called normal case, the scientific community is the only one exploring that subject matter defined by its particular standard literature.[25] In other cases, we find different communities exploring the same field from mutually incompatible points of view. Such communities, in constant competition with one another, are called "schools." While Kuhn specifies no additional typical traits for such schools, he does suggest plausible candidate attributes for the competitionless communities—plausible when one considers the likely past emergence of such communities out of a victory over competing schools:[26]

—"the members of a scientific community see themselves and are seen by others as the men uniquely responsible for the pursuit of a set of shared goals, including the training of their successors."

—"Within such groups communication is relatively full," and

—"professional judgment relatively unanimous."[27]

There is reason to question whether the 1969 approach to the concept of a scientific community really captures Kuhn's own intentions by articulating "the intuitive notion of community that underlies much in the earlier chapters of [*SSR*]."[28] In particular, two objections might

24. 1970c, *SSR* p. 177. The same list can be found in less detail, and without the distinction which follows, in 1974a, *ET* p. 296, and 1970b, p. 253.

25. Two limiting qualifications must be mentioned here. Even in the preface to *SSR*, Kuhn claims that his distinction between pre- and postparadigm periods is "much too schematic" because, among other reasons, "there are circumstances, though I think them rare, under which two paradigms can coexist peacefully in the later period" (p. ix). In this case the field is worked, not by a single community with no competition, but by two peacefully coexisting communities. In later work, Kuhn allows that even where a community appears to be without competition, "Research would, however, disclose the existence of rival schools as well" (1970b, p. 253; q.v. 1970c, *SSR* p. 209). These qualifications do not, however, appear to undermine the distinction between the two different forms of community, especially insofar as these are distinguished by divergent scientific practice (see chap. 5).

26. These traits do not apply to competitionless communities working new fields of study when these have resulted from the splitting or joining of preexisting competitionless communities (1959a, *ET* p. 231; *SSR* p. 15; 1963a, p. 353). For a detailed discussion of the emergence of competitionless communities, see § 5.5.b.

27. 1970c, *SSR* p. 177; q.v. 1969c, *ET* p. 344; 1970b, p. 253; 1985, p. 24.

28. 1970c, *SSR* p. 176.

be raised. To begin with, this concept of community appears too narrowly contemporary,[29] for such features as "similar education" and "similar professional initiation" (as well as the feeling of responsibility for a given discipline) presuppose that science takes the fairly rigidly institutionalized, specialized, and professionalized form that Kuhn himself claims is only some 150 years old.[30] Furthermore, the concept of the second type of community, the school, has been too little developed in its own right and too closely identified with the first type of community.[31] In *SSR,* Kuhn explains how the initial transition to a situation free of competition "transforms a group previously interested merely in the study of nature into a profession or, at least, a discipline."[32] This transformation is claimed to entail a "more rigid definition of the scientific group."[33] Thus an adequate explication of the notion of a scientific school would have to leave room for the diffuse cohesion characteristic of groups engaged in the study of nature prior to the attainment of a competitionless environment.[34]

Regardless of whether our concept of a scientific community is sufficiently sharp, or sensitive to historical change, we must ask the following question: why should only those disciplinarily, temporally,

29. See 1972a, especially p. 177, where Kuhn criticizes Ben-David for his ahistorical use of the concept of a "scientific role." Kuhn here appears to have committed a similar error.

30. *SSR,* p. 19; also 1963a, p. 351. —Revealingly, the examples of scientific communities by which Kuhn illustrates his definition in 1974a, *ET* p. 296, and 1970c, *SSR* pp. 177–178, are mostly taken from modern, professionalized science: communities of physicists, chemists, astronomers, zoologists, organic chemists, protein chemists, high energy physicists, solid state physicists, and radioastronomers.

31. This is a consequence of Kuhn's departure from *SSR,* begun at 1974a, *ET* p. 295 f. 4, in which his treatments of the research undertaken within a school of the prenormal period and of normal science can be seen to converge (see chap. 5 on normal and prenormal science). The road to this convergence, itself a consequence of the changing function of the notion of a paradigm (see §§4.2.b, 5.1), was already prepared in *SSR* (p. ix). In *SSR,* the characteristics of scientific communities are explicitly deduced from the practice of normal science, under the assumption that it is this activity for which scientists are normally trained (*SSR,* pp. 168–169). Though the text of 1974a fails to distinguish the two types of community (*ET* p. 296), the aforementioned convergence is suggested in Kuhn's lengthy footnote 4. The innovation made explicit in this footnote is worked into the distinction between the two types of community as presented in 1970b, p. 253, and 1970c, p. 177. The same footnote also appears, somewhat modified, as 1970b, p. 272 n. 1, and is finally incorporated into the text of 1970c, *SSR* pp. 178–179, where the consequences of the convergence are brought to fruition. See 1974b, p. 500 n. 2, and *ET* p. xx n. 8 regarding the order of composition of 1974a, 1970b, and 1970c.

32. *SSR,* p. 19.

33. Ibid.

34. See the criticism of Kuhn's treatment of the prenormal phase of scientific development offered in Mastermann 1970, pp. 73–75.

and epistemically determined fields of science that are (were) actually studied by one or more scientific communities be taken as acceptable units of analysis for a theory of scientific development? In order to answer this question, we must first address the broad complex of problems surrounding the manner in which we gain access to scientific development *qua* central issue of Kuhn's theory. As the actual chain of past events in which the development of science consists can never be immediately accessible, it must be constructed out of available reports and other traces. But what transforms this unavoidable construction into an accurate reconstruction of the past? According to what principles should the construction be made?

1.2 The Construction of the Target Issue: The Historiography of Science

Reconstructing the development of particular fields of science insofar as they were practiced by particular scientific communities is, of course, the job of the historiography of science, not that of a general theory of the structure of scientific development. Accordingly, Kuhn's theory is methodologically dependent on the historiography of science in that it must allow the latter to furnish it with its particulars. It *reflects* on the representations of particular scientific developments provided by historiography, thus "making explicit some of . . . historiography's implications";[35] it is "metahistorical."[36] The principles according to which particular scientific developments are constructed should thus be sought not in Kuhn's theory itself but in historiography.

But "the historiography of science," thought of as a single field guided by some specific set of principles, hence as in some sense homogeneous, is a fiction. It remains so even when our primary interest is confined to the internal historiography of science, which addresses science only *qua* form of knowledge.[37] For since the 1920s, according to Kuhn, the historiography of science has been in the midst of a thorough transformation, an "intellectual revolution."[38] This revolution is di-

35. *SSR,* p. 3.

36. In *ET*'s table of contents (*ET,* p. vii), Kuhn uses this term to differentiate his "theoretical" work from his historiographic efforts.

37. See 1968a; 1971c, *ET* pp. 159–160, and 1979a on the distinction between the internal and the external historiography of science.

38. 1970e, p. 67; also *SSR,* p. 3; 1962d, *ET* p. 165; 1984, p. 244. In 1986, p. 30, Kuhn calls this change in the historiography of science a "transformation (I shall not say a revolution)."

rected against a tradition of historical writing on science whose methodological apparatus has resulted in the view of scientific development as a cumulative process.[39] Until the 1960s this view of scientific development was almost always accepted without question—and without any cognizance of its dependence on the methodological apparatus of an underlying historiography of science, either. One product of the historiographic revolution is the "new internal historiography of science," whose methodological repertoire is distinct from that of earlier histories.[40] Kuhn's goal is *to propose a new picture of science and scientific development, in particular of scientific progress, grounded in this new historiography.*[41] Accordingly, we must seek the principles for constructing the target issue of Kuhn's theory in the methodological apparatus of the new internal historiography of science.

But how do we determine the features of a particular brand of historiography's "methodological apparatus," the set of presuppositions this historical approach makes about its subject matter? These features may be gleaned by asking after the *criterion of comparative historical relevance* guiding the historiography in question. All historical reporting needs such a criterion, whether implicit or explicit, conscious or unconscious, in reciprocal interaction with the history's narrative content or otherwise.[42] This criterion is indispensable on the grounds that some decision must be made as to what belongs in a given historical narrative and what doesn't.[43] The criterion must be *comparative* in that it allows us to distinguish *degrees* of importance. Stories may be told more or less comprehensively; shorter versions need not omit essentials, and neither must longer versions incorporate inessentials.

The concept of *historical relevance* can be further defined by distinguishing three of its moments (in the Hegelian sense). Each of these moments selects narrative material, and their selections may, in extension, overlap partially or even entirely. The moment of *factual relevance* selects for material which must be included in order for the history of

39. *SSR,* pp. 1–3, chap. 9.

40. As a historian, Kuhn himself is a member of this new school and was influential in promoting its institutionalization in the United States; see 1984a and 1986.

41. *SSR,* pp. ix, x, 1, 3, 160; 1968a, *ET* p. 121; 1970c, *SSR* pp. 207–209; *ET,* pp. xiv–xv; 1980a, p. 188; 1984, pp. 231, 243, 252.

42. See Kuhn 1980a, pp. 182–183, regarding this issue.

43. See, for example, Hegel's *Lectures on the History of Philosophy,* in his *Works,* vol. 18, p. 16ff; Agassi 1963; Danto 1965, chap. 7. Kuhn 1971a, p. 138 (or similarly pp. 141, 142; 1977b, *ET* pp. 14, 16–17). "No historian, whether of science or some other human activity, can operate without preconceptions about what is essential, what is not" (Laudan 1977, pp. 164–165).

a given thing, of a historical "reference point," to be told at all.[44] The moment of *narrative relevance* selects for material which must be taken into account if the resulting text is to be a proper narrative. Such material includes those facts by which a historical report gains needed narrative continuity, or facts which serve to make plausible what would otherwise be implausible.[45] Finally, the moment of *pragmatic relevance* selects for material without which the pragmatic goal of a historical narrative cannot be realized. Thus the content of a historical narrative is determined in part by the audience to which it is addressed and in part by the effect it is meant to have on this audience.[46]

Let us begin by inquiring into the relevance criterion of the old internal historiography of science (§1.2.a) and into the causes which led to the abandonment of this approach (§1.2.b). Then we will be in a position to characterize the new internal historiography and *its* criterion of historical relevance by contrast (§1.2.c).

a. *The Old Internal Historiography of Science*

The criterion of historical relevance of the old internal historiography of science can be understood by reference to the overarching goal of this form of historiography;

> The objective of these older histories of science was to clarify and deepen an understanding of *contemporary* scientific methods or concepts by displaying their evolution.[47]

Their primary goal was thus of the didactic or pedagogical variety, arising either from the particular aim of science education—showing students the historical path to the current state of knowledge in their respective fields—or from a more general desire to extract useful lessons from the history of scientific thought. As might be expected, most authors of such scientific histories were themselves scientists who

44. See Lübbe 1977 on the concept of a historical reference point [*Referenzsubjekt*].
45. See 1971a, p. 142; 1977b, *ET* p. 17.
46. See 1983a, pp. 677–678.
47. 1968a, *ET* p. 107, original emphasis. Regarding the following discussion, see 1955a, pp. 94–95; *SSR*, pp. 1–3, 98–99, chap. 11 1962d, *ET* pp. 165–166; 1964, *ET* p. 253; 1968a, *ET* pp. 105–108; 1970e, pp. 67–68; 1971c, *ET* pp. 148–149; 1976a, *ET* pp. 32–33; *ET* pp. x–xi; 1977b, *ET* pp. 3–4; 1980a, pp. 184–185; 1984, p. 248; 1986, pp. 30–31, 33; 1989b, pp. 49–50. I have limited myself here to references to Kuhn's own work, as Kuhn's perspective on earlier historiography and his reasons for taking it as insufficient have the most direct bearing on the understanding of his theory.

worked in, or had worked in, the field in question. In accordance with its specifically didactic role, this form of historiography usually, though not invariably, found its literary niche in the scientific textbook or, earlier, the "classic," and in such works was concentrated in introductions and scattered passages.

The preferred topic of this historiographic tradition was, again in accordance with its pedagogical intentions, disciplinary and subdisciplinary history. Those components which remained part of a field at the time of its history's composition, and could thus be taken as the completed, enduring property of science, were considered both of *pragmatic* and of *factual relevance.* For they constituted the matter toward which the history's audience was to be guided. The job of the historiographer was to discover these components in earlier texts, then relate in chronological order the inspired, yet methodical, work leading to their discovery or invention, especially by scientist-heroes. In this manner, the progressive accumulation of elements of contemporary science toward attainment of the field's current position could be revealed. It follows that dating the acquisition of new components and crediting their discoverers or inventors was of great *narrative relevance,* for by this means the elements of contemporary knowledge could be organized into a coherent narrative. In addition, factors external to science might occasionally prove of narrative relevance, as when particular technical developments gave impetus toward innovation, or when periods of stagnation could be explained by the influence of religion or other forms of dogmatism. Other factors external to science could be and had to be omitted, for none of the three moments of historical relevance justified their inclusion. Relatedly, the elements of past science that, from the contemporary perspective, appeared as some variety of error, resulting from mistakes, confusion, idiosyncrasies, myth, or superstition, were of no intrinsic interest. For such elements had, in the course of scientific development, been banned from science as "prescientific" or as "bad science." Only that which history had preserved to the present day could be factually relevant.

The old historiography of science, organized in this way, engendered a view of scientific development with the following two features. First, the history of science is a *history of progress,* of the ongoing triumph of "reason," "scientific rationality," or the "scientific method." Second, this progress is cumulative, a steady accretion of new bits of knowledge, some big, some small. The acquisition of new knowledge never affects the truth of prior knowledge, though perhaps some overly ambitious claims of the past, which never really belonged in science in the first place, are corrected.

b. The Critique of the Old Internal Historiography of Science

So how could doubt be shed on the adequacy of this form of historiography?[48] Kuhn lists a total of five factors which ultimately lead to the "historiographic revolution."[49]

1. This type of historiography regularly causes the historian certain difficulties which, paradoxically, are exacerbated rather than alleviated by more detailed historical research. Such difficulties can be divided into three types.

First, we find that any program that seeks to order discoveries chronologically and determine their respective discoverers becomes problematic because it assumes that scientific discoveries are more or less precisely datable events.[50] Closer examination of such discoveries, however, reveals that those belonging to the class of unexpected discoveries are temporally extended processes consisting of numerous phases. The temporal structure of these discoveries is best explained by the fact that the conceptual system in light of which, at a given time, a given past discovery appears as a newsworthy event was itself developed in part over the course of this same process of discovery.[51]

Second, attempts to brand those propositions of past science no longer found in contemporary science as error, superstition, myth, badly conducted science, or prescientific thought also turn out to be problematic. For here closer historical study shows that such typical examples as Aristotelian dynamics, phlogiston chemistry, and caloric thermodynamics are not less scientific views of nature than contemporary science; on the contrary, they

> can be produced by the same sorts of methods and held for the same sorts of reasons that now lead to scientific knowledge.[52]

Third, portraits of the major figures of science drawn by this form of historiography are sometimes quite implausible. Aristotle, for example, appears to be a terrible physicist if we evaluate his *Physics* according to how closely it approximates Galilean-Newtonian physics. But from

48. This inadequacy is now sometimes recognized within the sciences themselves. Thus we find physicist Richard Feynman referring, in 1985, to "'Physicists' history of physics,' which is never correct." This history is a "sort of conventionalized myth-story that the physicists tell to their students, and those students tell to their students, and is not necessarily related to the actual historical development" (Feynman 1985, p. 6).

49. While *SSR*, pp. 2–3, lists only troubles internal to the historiography of science, 1968a, pp. 107–110, treats the broader complex of factors involved in a more balanced manner. See also 1971c, *ET* pp. 148–150.

50. *SSR*, pp. 2, 7, chap. 6; 1962d.

51. See § 7.2.

52. *SSR*, p. 2.

this perspective, neither his observational and reflective acuity in so many other fields, nor the emergence of so long a tradition from so poor a physical theory, is comprehensible.[53] The same holds, *mutatis mutandis,* for Newton, Galileo,[54] and, albeit far less drastically, for Planck.[55]

2. Another problem with traditional discipline- and subdiscipline-oriented historiography arose out of its confrontation with a competing historiographic enterprise. This latter enterprise is motivated by the demand that the historiography of science should not simply document the history of individual disciplines separately but should rather trace the development of the totality of scientific knowledge.[56] Among the various influential consequences of this demand, those which follow from one particular insight are especially important. This insight is the realization that our current taxonomy of the sciences is, in part, a relatively recent product of the history of science itself. At one time, the totality of knowledge was divided into scientific and extrascientific knowledge, and scientific knowledge in turn was divided into individual disciplines in a different manner from that employed today. It appeared that traditional discipline-oriented historiography of science's criterion of historical relevance allowed the construction of historical traditions which had never existed.

3. Yet another trouble with traditional historiography of science emerged with Pierre Duhem's discovery of a medieval tradition in physics.[57] Although this tradition already contained many components of Galilean physics, both methodological and substantive, it was clearly different nonetheless; it was not yet modern science. Yet one could only understand the transition to modern science, the emergence of its typical characteristics, after first understanding this medieval tradition in its own right. Toward this end the old historiography of science's criterion of historical relevance, interested in the past's accumulation of our present heritage, had to be set aside, thus further eroding the criterion's claims to self-evidence.

4. The abandonment of this criterion of historical relevance was further motivated by influences external to the historical study of science, in particular by the influence of work in the history of philoso-

53. *ET,* p. xi; 1981, pp. 8–9.
54. 1980a, pp. 184–185.
55. 1984, pp. 233–238.
56. 1968a, *ET* p. 109; 1971c, *ET* pp. 148–149; 1976a, *ET* pp. 32–33; also 1983d, pp. 567–568.
57. 1968a, pp. 108–109.

phy,[58] identified by Kuhn as, in all probability, the most important factor. In the nineteenth century, the historicist movement brought about the transformation of a range of academic fields, subsequently called the "historical humanities."[59] Fields emerging from this transformation are characterized by the attempt to avoid examining a given cultural domain primarily from the perspective of contemporary culture, seeking instead to capture the cultural other precisely through its difference from the familiar, in all *its own* particular uniqueness and thus in all its (relative) strangeness. While, in the nineteenth century, historicism had already left an emphatic mark on reporting in the history of philosophy, its influence was only felt in the historiography of science beginning in the 1920s and 1930s, and even then primarily in the work of authors springing from the history of philosophy.[60] From their point of view, earlier historiography of science was highly "unhistorical,"[61] history written "backward,"[62] "Whig history,"[63] or "ethnocentric."[64] The view of science and its development extracted from early historiography of science is, so the criticism goes, about as authentic as a picture of an alien culture formed on the basis of travel brochures or language courses.[65]

The heart of early historiography of science's contemporary bias can be discerned in its criterion of historical relevance, which reflects its substantive interest in the current state of science. Only those products of science deemed permanent, complete by the standard of contemporary science, are factually relevant.[66] This judgment projects three components of the present into history.[67] For one, it describes past science using contemporary scientific *concepts,* the possible historicity of which is thereby forcibly ruled out. Furthermore, it takes the *questions* of contemporary science as constant, thus bracketing the potential contextual dependence of research agendas. Finally, it denies those guiding *standards* of science which regulate the acceptability of answers to scientific questions any possibility of historical change.

58. *SSR,* pp. v–vi; 1968a, *ET* pp. 107–108; 1971c, *ET* pp. 135, 149–150; *ET,* pp. xiv–xv; 1977b, *ET* p. 11; 1979a, p. 125.

59. See, for example, Schulz 1972, part 4: "Historicization" [*Vergeschichtlichung*].

60. Kuhn cites Koyré as by far the most important, followed by Brunschwicg, Burtt, Cassirer, Lovejoy, Anneliese Maier, Emile Meyerson, and Hélène Metzger.

61. *SSR,* pp. 1, 137, 138; 1971c, *ET* p. 154.

62. *SSR,* p. 138.

63. 1971c, *ET* pp. 135, 154; 1983b, p. 712; 1983d, p. 568; 1984, p. 246. See Butterfield 1931; on the more recent debate, see Mayr 1990 and his references.

64. 1983d, p. 568; 1984, p. 246.

65. *SSR,* p. 1—See Conant 1947, p. 41.

66. *SSR,* pp. 1, 136, 137.

67. *SSR,* chap. 11, especially pp. 137–138; 1989b, p. 50.

5. In conclusion, the impact of developments in other fields outside the historiography of science, especially the influence of an emerging general science of history, of Marxist historiography, and of German sociology, should be noted. Such influences gave rise to a growing appreciation of nonintellectual factors in scientific development and, in particular, of institutional and socioeconomic factors.[68] While the substantive content of mature sciences is largely insulated from the influence of external factors, it is not so unaffected as the almost exclusively internalist orientation *entailed by earlier historiography's criterion of historical relevance* would suggest. But since the systematic integration of internal and external factors remains more an ideal than the standard practice even in recent historiography of science,[69] this point of criticism against the earlier historiographic tradition may be disregarded here.[70]

c. The New Internal Historiography of Science

A new historiographic tradition arose in response to the aforementioned difficulties and influences confronting the earlier tradition. From roughly 1955 to 1970, this new tradition rapidly gained momentum, especially in North America, becoming increasingly institutionalized and professionalized.[71] This tradition, too, may be characterized by a statement of its criterion of historical relevance, which in turn becomes plausible in light of the new tradition's overarching goal. This goal consists in achieving for the historiography of science what historicism did for other fields of cultural scholarship as early as the nineteenth century: a departure from the contemporary perspective in favor of historicity. Such a departure requires that more recent scientific knowledge be disregarded as far as possible,[72] for only in this manner can we "display the historical integrity of [past] science in its own time"[73] or

68. 1968a, *ET,* pp. 109–110; 1971c; *ET* pp. 148–149, 159–160; 1986, p. 32.

69. 1968a, *ET* pp. 110, 120; 1979a, p. 123; 1983c, p. 29.

70. It should be noted that Kuhn's *theory* can play a potentially important role in the integration of internal and external factors within *historiography*. For Kuhn's theory identifies a central point of contact between science and society: scientific values (see § 4.3.c). It is at this point of contact that the mutual influence of factors external to science and scientific development, frequently postulated but seldom given any detailed functional analysis, becomes more concretely tangible; see 1968a, *ET* pp. 118–120; *ET* p. xv; 1983c.

71. 1968a, *ET* pp. 110–112; 1979a, pp. 121–122; 1986. —In the discussion that follows I once again limit myself to Kuhn's work. My reason, as before, is that Kuhn's own view of recent historiography remains of primary importance to an understanding of the dependence of Kuhnian theory on this historiography. —See e.g. Cohen 1974; Kragh 1987, especially chap. 9; Lindholm 1981; Shapere 1977, pp. 198–199, and Thackray 1970 for other pronouncements on the new historiography.

72. 1968a, *ET* p. 110; 1980a, p. 183; 1984, p. 244.

73. *SSR,* p. 3; similarly 1970e, p. 68, and 1977b, *ET* p. 11.

"analyze an older science in its own terms."[74] Toward this end the historian's task is

> to climb inside the heads of the members of the group which practices some particular scientific specialty during some particular period; to make sense of the way those people practiced their discipline.[75]

It follows that this historiography's legitimate objects of study are scientific fields in the same disciplinary and epistemic extensions in which they were explored by actual scientific communities. An element of not entirely unproblematic arbitrariness is manifest along the temporal axis; unless the historian begins with the earliest phase of a science's development, he or she must choose some starting point congenial to the narration of his or her targeted interval of scientific development. Usually the narration's conclusion will correspond with some factual endpoint, that is, with some sort of culmination in the appropriate line of scientific development.[76]

Narrative and *factual* relevance[77] are thus to be ascribed, first of all, to problems explored by the historical science under study, precisely as they were actually formulated by the relevant community, and to the context in which such problems took on their problematic character in the first place. The next narrative task is to reconstruct the community's behavior around such problems, along with the particular normative standards that regulated this behavior and that can thus serve in retrospect to make it intelligible. In particular, innovations of all kinds must be interrogated with a view toward determining how they were received in their own time, for here the risk of presentistic[78] distortion is especially great. The historically accurate reconstruction of a set of problems and its management by a community further requires an exact reconstruction of the period's conceptual system. Here the historian must contend with issues of *pragmatic* relevance, which govern the degree of detail called for in presenting a historical conceptual system to a chosen audience.[79]

74. *SSR*, p. 167 n. 3; similarly 1984, p. 250; 1989a, pp. 9, 24, 31–32; 1990, pp. 298, 315.

75. 1979a, p. 122; similarly *SSR*, p. vi; 1983c, p. 27. —In virtue of their metaphorical character, these formulations merit further explication; I will return to them later (see n. 84).

76. 1969b, pp. 972–973.

77. Regarding the following discussion, see 1955a, p. 95; *SSR*, pp. 2–3; 1968a, *ET* p. 110; 1969b, p. 969; 1970e, pp. 67–68; 1971a, pp. 138–143; 1971b, *ET* p. 29; *ET*, pp. xi–xiv; 1977b; 1978, p. viii; 1979a, pp. 122, 126–127; 1980a, pp. 182–185; 1983a, pp. 676–678; 1984, pp. 233, 236, 243–252; 1989b, p. 50.

78. This is a common term among historians; see e.g. Groh and Wirtz 1983, p. 192, or Rudwick 1985, p. 12.

79. 1983a, pp. 677–678.

This last task, the historically accurate reconstruction of an old conceptual system, demands special effort on the part of the historian. To begin with, the list of permissible sources must be confined to texts dating from, or antedating, the period itself, where published sources such as period textbooks and journals must, in general, be supplemented by unpublished sources, such as scientific correspondence and diaries. These texts must be read in a *hermeneutic* manner, which for Kuhn means the following:[80]

1. It must be assumed that a given text can be interpreted in several ways.
2. Differing interpretations of a given text are not all of equal value.
3. The evaluation of alternative interpretations of given texts consists, at least for important texts,[81] in assigning a preference to the interpretation with greatest plausibility and coherence.[82]
4. The older a text is, the more improbable it is, in general, that the best interpretation is the one closest at hand for the modern reader.
5. One maxim for the improvement of interpretations consists in the injunction to seek out passages of the text which appear obviously erroneous, implausible, or even absurd. Such passages should be viewed primarily as indicators of the need for improved interpretation, not as signs of the author's confusion.[83] If interpretations of such passages can be found which eliminate their earlier implausibility, the meaning of portions of the text previously thought to be

80. In several autobiographic passages, Kuhn notes that his own (re)discovery of hermeneutics for the history of science began in 1947, with his studies of Aristotle's *Physics* (see *ET,* pp. xi–xiii; 1981, pp. 8–12). As a tradition emanating from philosophy, with fruitful applications in the history of science, Kuhn first encountered hermeneutics primarily in the work of Alexandre Koyré (see especially 1970e; also *SSR,* pp. v–vi; 1971b, *ET* p. 21; 1971c, *ET* p. 150; *ET* p. xiii n. 3; 1977b, *ET* p. 11; 1983c, pp. 27, 29; 1984, p. 243). However, he only encountered the *term* "hermeneutics," along with the European Continental philosophical tradition associated with it, in the early 1970s, and even then this tradition remained fairly opaque to him and his (primarily Anglo-Saxon) colleagues (see *ET,* p. xv). For hints regarding Kuhn's further reception of the hermeneutic tradition, see 1983a, p. 677 and p. 685 n. 11; 1983b, pp. 715–716. —q.v. 1989a, pp. 9–10, 24; 1989b, p. 49; 1990, pp. 298–299.

81. Kuhn never explicitly addresses what makes a given text "important." However, if we read his report of his encounter with Aristotle's *Physics* (*ET,* p. xi; 1981, pp. 8–9) with this issue in mind, it seems that the following measure of a text's importance can be ascribed to Kuhn: A text is important if it has an influential history, or if other texts by the same author have influential histories. These features of a given text allow us to infer its high quality (see also 1964, *ET* p. 253).

82. In the hermeneutic tradition, this stipulation corresponds with what Gadamer calls the *"Vorgriff der Vollkommenheit"*; Gadamer 1960, pp. 277ff. ("prejudice toward authority," English edition p. 269ff.).

83. The appropriateness of this maxim follows from the (at least provisionally) assumed high quality of the text, in virtue of which it was deemed important.

understood may change, for those conceptual changes whose initial, local effect was to make implausible loci plausible propagate throughout the text.[84]

If sources relevant to the history of science are read in this manner, an account of the development of a given field markedly different from a product of the older historiography may result. A comparison of the new account with the old typically yields the following differences.[85] The field's development exhibits fewer chasms bridged by geniuses possessing either the confidence of sleepwalkers, some prophetic intuition, or a seeming ignorance of what they were doing. Instead we find a greater number of transitional steps; the distance from one to the next is accordingly smaller, and the way better prepared and thereby more plausibly negotiable by some sufficiently well endowed person. Participating individuals thus appear in many respects to be "better" scientists; their efforts are more elegant and more readily reproducible. With regard to available source material, we find that it exhibits fewer anomalies, fewer passages which must be "explained away" by recourse to more or less unsatisfying psychologizing hypotheses.[86] In particular, explanatory recourse to the "confusion" of participating scientists can frequently be avoided, because many passages which might have been taken as evidence of confusion are instead taken, under alternative interpretations, as evidence of a more alien past.

This form of historiography evidently leads us to postulate a more

84. Having thus explicated the historian's hermeneutic task, some of those metaphors which Kuhn (and others) commonly apply in this context can be unpacked. Consider claims that the historian must "climb inside the heads of members [of a given scientific community]" (1979a, p. 122; similarly 1983c, p. 27) or "try to think as they did" (1968a, *ET* p. 110; similarly 1977b, *ET* p. 8). These enjoin the historian, first, to identify and master a scientific (and perhaps, to a certain extent, an extrascientific) conceptual system. Successful accomplishment of this task is a necessary precondition for the historically adequate articulation of earlier scientific assertions. In addition, they prescribe the appropriation of the scientific values and norms of the period under study (in part implicitly contained in the scientific assertions of that time), in light of which scientists' decisions become intelligible.

85. I base this account on Kuhn's 1984 reflections on his 1978 historiographic work on black body radiation and the introduction of the quantum hypothesis. My reason is that Kuhn views this as the most representative and most thoroughly developed illustration of that conception of the history of science in which both his historiographic and philosophical efforts are grounded (1984, p. 231).

86. At the same time, however, a great temptation toward psychologizing or sociologizing reasoning arises when we consider the question, how is it that the past of a given field is so often remembered or represented in a distorted manner, not only by participating scientists, but by later scientists, and historians of science as well? This distortion consists in the historically impermissible assimilation of the past to the present. Kuhn dedicates all of chap. 11 of *SSR,* and almost a third of 1984, to this issue (pp. 246–252).

alien, yet at the same time more reasonably alien, scientific past than the old historiography. At first blush, there seems to be no limit to the potential strangeness. Nonetheless, science, for Kuhn, is not without timeless attributes; past science and contemporary science are connected by more than a chain of intransitive similarities. Indeed, the following may be said about all science, whenever conducted:[87]

—Science studies nature, or the world.
—Science aims at an understanding of nature or the world which captures its order with maximal precision and universality.
—Science's orientation toward this goal demands that it search for a set of propositions exhibiting maximal internal coherence and maximal correspondence with nature or the world.[88]
—Science is mostly detail work; it strives toward its understanding of nature or the world by way of a precise understanding of the individual aspects of nature or the world.
—Science proceeds empirically; in other words, the acceptability of propositions is strongly regulated by observation and experience.
—Therefore, there exists a universal characterization both of the production methods of scientific knowledge and of the type of arguments that may be used in support of claims to such epistemic status.[89]

Still, this timeless, universal characterization of science isn't terribly far-reaching; in particular, it in no way determines actual scientific practice in any given historical situation.[90] A range of historically more or less variable factors plays a role in this practice. This role is permissible notwithstanding the above timeless attributes of science, for the latter are *abstract-universal,*⋆ and may thus be instantiated in various ways in different historical situations.[91] Thus *universality, precision, correspondence with nature,* and other key concepts which figure in the timeless characteristics of science may, at different times, be fleshed

⋆ Translator's note: "abstract-universal" is the usual English translation for Hegel's term, "*abstrakt-allgemein.*" See e.g. Hegel, *Wissenschaft der Logik* (*Hegel's Science of Logic,* trans. A.V. Miller, London: Allen and Unwin, 1969), introduction.

87. *SSR,* pp. 4, 42; 1970b, p. 245. See chap. 2, especially § 2.1.a., on Kuhn's concepts of *nature* and *world.*

88. This is an indirect consequence of the constraint placed on historically accurate reconstructions of past science in *SSR,* p. 3, according to which the views of the group under study must be accorded "the maximum internal coherence and the closest possible fit to nature."

89. This is an indirect consequence of the assertion in *SSR,* p. 2, that earlier science was "produced by the same sorts of methods and held for the same sorts of reasons that now lead to scientific knowledge."

90. *SSR,* pp. 3–5, 40–42.

91. See Feyerabend 1981b, pp. 62–63.

out with very different meanings. The new internal historiography of science is of interest to the philosophy of science not on account of any timeless moments of science it may discover (or presuppose) but rather for its ability to uncover, through hermeneutic efforts, the strangeness of science past.[92]

1.3 The Focus: Structure

Any progress we have made so far toward defining the *central issue* of Kuhn's theory doesn't go so far as to fix sufficiently the *topic* of this theory. For though we have discussed the manner in which this central issue, called "scientific development," becomes accessible to Kuhn's theory, we haven't considered the theoretical perspective, the particular *focus,* in which this issue becomes its topic. The word "structure," which appears in the title of *SSR,* serves as an indicator for the focus guiding Kuhn's theoretical exploration of scientific development. Still, on account of its vagueness, this term merely hints at a definite focus of study rather than actually stating it. Furthermore, Kuhn nowhere explicitly addresses the term's meaning. And so the task of clarifying what Kuhn means by the "structure of scientific development" in general, and by the "structure of scientific revolutions" in particular, presents itself.

Before I begin to clarify the specific sense of "structure" at work in Kuhn's philosophy of science, we should bring ourselves up to date on the word's broader meaning. This exercise seems all the more indispensable given the extent to which our cognizance of this notion's proper application yields to its immense popularity.[93]

The most general current meaning of "structure" derives by abstraction from the original meaning of the Latin "*structura,*" "building," or "façade," still alive in contemporary English. When, today, we discuss something's structure, we address three of its features (parallels to the meaning of the Latin word should be obvious):

1. The thing isn't entirely homogeneous or undifferentiated but rather has "parts," "elements," "components," or "moments."
2. The thing's parts don't stand in arbitrary, random, or chaotically changing relations to one another. The relations of the parts are rather more or less determinate, constant, and constitutive of the thing; we thus may speak of an "ordering" of the parts, elements, and so forth.

92. *SSR,* pp. 3–4, chap. 12; 1968a, *ET* p. 121; *ET,* p. xiv; 1984, p. 244.

93. Regarding this discussion and that which follows, see Bastide 1962, Einem 1973, Kambartel 1973, Naumann 1973, and Schlüchter 1974.

3. Because of this ordering, the thing somehow remains a "whole," despite its internal multiplicity.

This general concept of structure is, in the various sciences, differentiated into disparate special concepts of structure by their diverse specifications and emphases. Since to my knowledge, however, such differentiations, even as they occur within historical scholarship, never lead to a notion of structure congruent with Kuhn's, I will not pursue them here.[94]

"Structure" doesn't occur especially often in *SSR,* despite the book's title, and where it does, it's usually not in contexts relevant to the present discussion.[95] In *SSR,* Kuhn uses the term "structure" in a manner potentially revealing of the focus he means it to lend his project only three times.[96]

In two of these instances, Kuhn speaks of the structure of (revolutionary) scientific discoveries:

> Examining selected discoveries . . . , we shall quickly find that they are not isolated events but *extended* episodes with a *regularly recurrent structure.*[97]

Immediately following this sentence, Kuhn lists three temporally and substantively distinct phases of such processes of discovery, the nature of which need not concern us here.[98] This progression of phases, which

94. In historical scholarship, especially studies in social history, the concept *structure* has become the conventional complement of *event,* especially as it figures in the concept (or expression) *structural history* (compare Koselleck 1973, p. 561ff). These "structures" are historical realities which change only slowly (if not necessarily continuously) as compared with events. Such structures delineate the possibility space of historical events without determining the events themselves; in this sense, they are conditions of and factors in events. This notion of structure has only the ground level meaning of "structure" in common with Kuhn's. —For further discussion of this kind of structure, see e.g. Groh 1971 and Lübbe 1977, p. 69 n. 3.

95. In *SSR* we find talk of a "problem-structure" (pp. vii, 137, 144), of "the structure of postrevolutionary textbooks and research publications" (p. ix), of "the structure of the group that practices the field" (p. 18), of "a rather ramshackle structure with little coherence among its various parts" (p. 49), of attempts to give "structure" to "an admittedly fundamental anomaly in theory" (p. 86, similarly p. 56, twice on p. 89), of "the logical structure of scientific knowledge" (p. 95, similarly p. 137), of "the fundamental structural elements of which the universe . . . is composed" (p. 102), and, finally, of "a relatively sudden and unstructured event" (p. 122).

96. We also find this word used in the targeted sense at 1970b, p. 243 ("structure of scientific development"), and in the subtitle of the German edition (though not of the later English edition) of *ET* ("studies in the structure of the history of science" [*"Studien zur Struktur der Wissenschaftsqeschichte"*]).

97. *SSR,* p. 52, emphasis mine; also on p. 8.

98. The title of 1962d, passages of which parallel our present discussion, is "The Historical Structure of Scientific Discovery," which presupposes that such discoveries are temporally extended (*ET* p. 177). Here, too, we find a discussion of the three phases in a process of discovery.

Kuhn sees as constant over all revolutionary discoveries, is later taken up as a (historical) "pattern"[99] or "historical schema."[100] With regard to their structure, such revolutionary processes of discovery may be likened to those great events commonly termed "scientific revolutions,"[101] suggesting that the latter also follow a definite pattern.

The third instance relevant for our purposes occurs in the last chapter of *SSR*. Here Kuhn describes the preceding contents of his book as a "schematic description of scientific development" whose job was to catch "the essential structure of a science's continuing evolution."[102]

What, then, does he mean by "structure"? Kuhn's claims suggest the following:

1. Structure is a feature only of temporally extended processes; point events are, in this sense, fundamentally structureless.
2. A process has structure (or "is structured") if it is organized in temporally and substantively distinguishable phases.
3. The order of these phases isn't arbitrary but is rather itself a component of the structure of the process.
4. The structure of a given process is a universal feature of that process; that is, every element of a given class of processes has one and the same structure.

Obviously, this notion of structure constitutes a concrete version of that most general meaning for "structure" given above. When Kuhn speaks of the structure of scientific development, he is thus concerned with a *universal phase model of scientific development*.[103] Having introduced the phase model, Kuhn immediately qualifies its claims to universality:

> If the historian traces the scientific knowledge of any selected group of related phenomena backward in time, he is *likely* to encounter some *minor variant* of a pattern here illustrated[104]

99. Pp. 53, 54.

100. P. 64.

101. *SSR*, pp. 7–8; similarly 1963b, p. 388. —See § 6.1. for a discussion of this parallel.

102. *SSR*, p. 160. In other passages, and in other works, Kuhn talks in similar terms of a "developmental pattern" (*SSR*, pp. 12, 137; 1970b, pp. 243 [here an explicit synonym for the "structure of scientific development"], 244; *ET*, p. xxi), and of a "developmental schema" (1984, p. 245).

103. Other terms familiar from treatments of science in German historical scholarship include "models of historical progression" (Schlobach 1978, p. 127 [*"Verlaufsmodelle der Geschichte"*]), "process-model" [*"Ablaufsmodell"*], "phase-model" (Berding 1978, p. 272 [*"Phasenmodell"*], "structure-model" (Polikarov 1981 [*"Strukturmodell"*]), and "process-contour" (Marquard 1978, p. 330 [*"Verlaufsfigur"*]).

104. *SSR*, p. 11, emphasis added. Similarly, p. 12.

I will briefly sketch Kuhn's universal phase model here, for reasons which will soon become apparent. As the key concepts of this model have yet to be introduced, my sketch may not be completely intelligible.[105]

In the development of the "mature" sciences of today we can generally find an initial phase, which I will call "prenormal science."[106] There follows the transition to maturity, one phase of which is referred to as "normal science." After a "scientific revolution" of variable length, there follows another phase of normal science, and so on. Between two periods of normal science the phase called "extraordinary science" may occur.

But the claim that Kuhn's philosophy of science aims at a universal phase model of scientific development is misleading, insofar as this model fails to reflect Kuhn's weighting of its individual elements. Kuhn's primary interest is in "the importance to scientific development of 'revolutions.'"[107] Since scientific revolutions are characterized by the special relationship between two consecutive phases of normal science, an understanding of them must be grounded in an understanding of normal science. By comparison, prenormal and extraordinary science, along with other transitions between the various phases, are of lesser importance to Kuhn's theory.

The title of Kuhn's major work, "The Structure of Scientific Revolutions," thus refers not only to the universal phase model of scientific revolutions. For if we take him at his word, this book deals with a particular "conception of the *nature* of scientific revolutions."[108] The "structure of scientific revolutions" not only is an ordering of temporally distinguishable phases; it also picks out the "essential elements" of revolutions: those elements of a scientific revolution which make it what it is.

105. These concepts are properly introduced in Part III.

106. Kuhn himself calls it "preparadigm science"; in § 5.5.b., I will give my reasons for this suggested change in terminology.

107. 1959a, *ET* p. 226.

108. *SSR*, p. 7, emphasis added; see also the title of chap. 9 of *SSR*, "The Nature and Necessity of Scientific Revolutions," and p. 136.

SUMMARY OF PART I

WE ARE NOW IN A POSITION to delineate, in conclusion, the topic of Kuhn's philosophy of science. The object domain of Kuhn's philosophy of science is composed of the total domain of the basic sciences, taken in its epistemic aspect. Only those elements of the total domain that were explored in given temporal and disciplinary extensions by identifiable communities of scientists will be considered. The development of such disciplinary fields will be reconstructed according to the principles of the new internal historiography of science. These fields, taken as the particulars of Kuhn's theory, then provide the basis for a generalized exploration of scientific development. The Kuhnian perspective then sets the stage for later discussion with a universal phase model of scientific development. This phase model includes, for the mature sciences, consecutive phases of normal science separated by scientific revolutions. These scientific revolutions, together with the normal science which makes them possible, then become the primary issue of Kuhn's theory. They will be examined with a view toward the phases in their execution, their defining components, and their consequences.

However, before I can address myself to these themes from the philosophy of science as it is narrowly conceived (part 3), another exploration is called for. For I must first explain what characterizes scientific knowledge and its object, according to Kuhn (part 2). Only after Kuhn's epistemological position or, on the other side of the same coin, his ontological position on this issue has been reconstructed can his conception of normal science and of scientific revolutions be understood.

PART TWO

Scientific Knowledge and Its Object

THE OBJECT OF SCIENTIFIC KNOWLEDGE is nature, or the world. Though this assertion seems both meaningful and reasonable, it is nonetheless problematic, for the meanings of "nature," "world," and of "being an object" are in urgent need of clarification. Chapter 2 will elucidate the meanings of these expressions as they occur in Kuhn's theory. The epistemically operant world concept in Kuhn's work is the notion of a "phenomenal world," of a world constituted by the activities of knowing subjects. Kuhn's view of this process of constitution will be reconstructed in chapter 3. This chapter will survey the nature of scientific knowledge, in the process laying the foundations for an examination of the ways in which such knowledge may develop. Finally, the notion of a paradigm, and the central role it plays in Kuhn's theory of scientific knowledge, will be analyzed in chapter 4.

CHAPTER TWO

The World Concept

Reconstructing Kuhn's Theory requires that we distinguish between two concepts of the world, or nature. § 2.1 will consider how *SSR* gives rise to this necessity. Kuhn himself first explicitly makes the distinction in works composed in 1969 (§ 2.2). Eventually, he returns to this topic, with somewhat different emphases, in 1979 (§ 2.3).

2.1 The Double Meaning of "World" and "Nature" in *SSR* and the Plurality-of-Phenomenal-Worlds Thesis

a. World-in-Itself and Phenomenal World in SSR

According to Kuhn, science of all ages takes, as we have noted,[1] nature or the world as its object. Kuhn usually uses "nature" and "world" synonymously.[2] Such conventional reference to nature or the world as the *object* of science, as the independent counterpart of the scientist, suggests considerable tension with Kuhn's claim that

> though the world does not change with a change of paradigm, the scientist afterward works in *a different world*.[3]

1. See § 1.2.c.

2. The synonymity of "world" and "nature" may be inferred from *SSR*, p. 77, where the "comparison of that theory with the world" is paraphrased, in the next sentence, by the "comparison of both paradigms with nature." Kuhn's use of these terms in *SSR* is not, however, entirely uniform. While, in the above passage among others (e.g. p. 42), the two words are used synonymously, Kuhn sometimes uses "nature" and "world" differently in order to draw the distinction between the two notions to be discussed in this section. See, for example, p. 125, where he speaks of "the world determined jointly by nature and by the paradigms." Synonymous use, however, predominates, and is preserved in later work, e.g. 1979b, p. 414; 1981, pp. 19–20; 1983a, pp. 680–681.

3. *SSR*, p. 121, emphasis mine; similarly pp. 6, 61, 106, 111, 117, 118, 120, 121, 122, 134, 135, 141, 147–148, 150.

Kuhn is conscious of this tension, for in the following sentence he informs the reader of his conviction "that we must learn to make sense of statements that at least resemble these."

So let us try to make sense of this statement. The key to accomplishing this task is only implicitly contained in *SSR*. It involves distinguishing two meanings of "world" and "nature," and is suggested by Kuhn's awareness that the problem lies somewhere in his world concept:

> In a sense that I am unable to explicate further, the proponents of competing paradigms practice their trades in different worlds.[4]

In several of the papers composed in 1969, Kuhn explicitly admits that his "constant recourse in [his] original text to phrases like 'the world changes'"[5] is evidence of vagueness, which he attempts to remove by means of new terminology. Yet even in *SSR*, it's possible to distinguish the two meanings of "world" and "nature."

In its first meaning, "nature" or "world" refers to something which changes over the course of a revolutionary transformation in science, so that we might say that a given such episode had "qualitatively transformed as well as quantitatively enriched"[6] the world. This world is "the scientist's world,"[7] the world in which scientists "live,"[8] "the world within which scientific work was done."[9] It is a "perceived world,"[10] "a world already perceptually and conceptually subdivided in a certain way."[11] Paradigms "are constitutive"[12] of this world or nature, or more precisely, the world is "determined jointly by nature and the paradigms,"[13] where "nature" should be read in its second sense, to be discussed shortly. Whatever these paradigms are,[14] since they are tools "through which to view nature,"[15] they must have some world-constitutive function.

When these passages are taken together, it appears that "world"

4. *SSR*, p. 150. The important first part of this passage has been practically ignored in Kuhn scholarship, with the exception of Grandy 1983, p. 23.

5. 1970c, *SSR* p. 192; similarly 1974a, *ET* p. 309 n. 18.

6. *SSR*, p. 7, similarly p. 106.

7. *SSR*, pp. 7, 111, and elsewhere.

8. *SSR*, p. 134; similarly p. 117.

9. *SSR*, pp. 6, 121, 147, 150.

10. *SSR*, p. 128.

11. *SSR*, p. 129.

12. *SSR*, p. 110; indirectly p. 106.

13. *SSR*, p. 125; similarly pp. 112 (on this passage, see Brown 1983a, p. 97), 123.

14. See Ch. 4.

15. *SSR*, p. 79; similarly p. 94.

and "nature" here coincide to some extent with what Kant calls "nature in the material sense" (*"natura materialiter spectata"*) or even "world": the "aggregate of appearances,"[16] the "object of all possible experience,"[17] the "sum of the objects of experience."[18] For both Kant and Kuhn, epistemic *subjects* are (albeit in different ways) coconstitutive of this world, for which I will henceforth use the term *"phenomenal world."*[19]

In its other meaning, "nature" or "world" refers in Kuhn's work to something which itself remains untouched and uninfluenced by revolutionary change in science. After a revolution, it is merely seen "in a different way,"[20] covered by a "conceptual web" which shifted over the course of the revolution.[21] It is a "hypothetical fixed nature" to which we have no access.[22] This world or nature may be identified as the independent counterpart of the scientist, toward which the scientist's gaze is directed; "Whatever he may then see, the scientist after a revolution is still looking at the same world."[23] This nature or world is understood to contain no moments originating on the side of the epistemic subject; it is that which remains of a phenomenal world when

16. *Critique of Pure Reason,* A114, B163, A334/B391, A418/B446 fn, A419/B447, A455/B483, A506/B534ff. (English translations of Kant's terms taken from the Norman Kemp Smith translation; see especially p. 140); *Prolegomena,* § 36.

17. *Critique of Pure Reason,* A114.

18. *Critique of Pure Reason,* Bxix; similarly A654/B682; *Prolegomena,* § 16; *Metaphysical Foundations of Natural Science,* "Preface."

19. Buchdahl, in 1969, already clearly recognized the existence of certain parallels between Kuhn and Kant; Buchdahl 1969, p. 511 n. 1; q.v. Lane and Lane 1981, pp. 52–53, including n. 4. Kuhn addresses these parallels himself in 1979b, pp. 418–419; compare § 2.3. —Also see Putnam 1978, p. 138, and Putnam 1981, chap. 3.

20. *SSR,* p. 53; similarly p. 118.

21. *SSR,* p. 149; similarly p. 141.

22. *SSR,* p. 118; similarly pp. 111, 114: "the scientist can have no recourse above or beyond what he sees with his eyes and instruments." We may infer from these assertions what about the "fixed nature" is qualified as "hypothetical." Here "hypothetical" can only mean that the assumption of fixed nature can't be taken as proven or otherwise necessarily applicable; in this context, the opposite of "hypothetical" is "known with certainty." What is emphatically *not* implied is that the assumption of fixed nature constitutes a hypothesis that is empirically testable in principle (see von Weizsäcker 1977, p. 190ff.). In his 1969 work, Kuhn explicitly treats this hypothetically fixed nature as a postulate; see § 2.2.a.

23. *SSR,* p. 129; similarly p. 150. Kuhn's use of "looking at" and "seeing" is the only terminological indicator in *SSR* that reveals his vague awareness of the distinction between the two world-concepts. The important textual occurrences of this indicator can be found at pp. 113, 114, 115, 120, 121, 128, 129, 150; see also 1979c, p. ix. Since, for Kuhn, a phenomenal world is "already perceptually and conceptually subdivided in a certain way" (*SSR,* p. 129), "seeing" has both perceptual and conceptual moments.

all of its subject-sided moments have been removed.* Kuhn stipulates this world to be *spatiotemporal, not undifferentiated,* and in some sense *causally efficacious.* These stipulations remain highly indirect in *SSR* and will become somewhat more explicit only in later work.[24] The *spatiotemporality* of this world or nature may be inferred from Kuhn's treatment of *looking at* as a relation that is in some sense spatiotemporally described.[25] The application of *sameness* and *difference* to parts of this world or nature is implied by suggestions that different people "see different things when looking at *the same* sorts of objects,"[26] while "what they look at has not changed."[27] The *causal efficacy* of this world or nature consists in its ability to codetermine a given phenomenal world.[28] In accordance with these properties, Kuhn understands the status of a given part of this world or nature as the epistemic subject's "counterpart" to be spatiotemporally conditioned; to say that a portion of this world is an epistemic subject's counterpart means, primarily, that over a given span of time the epistemic subject stands at a given spatial distance from this portion of the world.

But beyond the claim that it's in some sense the spatiotemporal counterpart of the scientist, nothing may be said about this world or nature. For any further assertions would run the risk of being changed in the next scientific revolution,[29] revealing themselves as inapplicable to this hypothetical, *fixed* world or nature. The price for the independence of this world or nature from any sculpting by human conceptions is that this world is unknowable, at least by direct channels; only a phenomenal world may present itself for empirical access. But Kuhn takes even *indirect* access to the fixed world or nature, whether by

* Translator's note: Hoyingen-Huene uses "subject-sided" [*subjektseitig*] to avoid the unwanted connotations of "subjective." A subject-sided moment of a phenomenal world is some aspect of the phenomenal world originating within an epistemic subject with access to this phenomenal world. It need not at all, for example, be "subjective" in the sense of "not objective."

24. See § 2.2.b.

25. See references to "looking at" listed in n. 23.

26. *SSR,* p. 120; emphasis mine, Kuhn's emphasis dropped. In § 2.2.d, I will discuss the difficulties that arise from applying the relations *sameness* and *difference* to parts of a world conceived as purely object-sided.

27. *SSR,* p. 150.

28. *SSR,* pp. 112, 123, 125; also p. 150.

29. Kuhn's view that this world or nature may be ascribed no historically invariant attributes beyond spatiotemporality is only (relatively) clearly expressed in 1979b, where he describes his view as approaching Kant's, "but . . . with categories of the mind which could change with time" (pp. 418–419). What is noteworthy about this passage is Kuhn's silence on the Kantian forms of intuition, space and time, which suggests that Kuhn indeed places space and time on the object side.

inference from *one* phenomenal world to the fixed world or nature or by inference from a comparison of *several* phenomenal worlds,[30] for an impossibility. As this is not the place to discuss this latter view in detail, let us be content for now with Kuhn's insistence on the unknowability of a fixed, purely object-sided world or nature.

In accordance with an obvious parallel to the thing-in-itself of Kant's critical philosophy, I will call this world or nature the *world-in-itself*. The most important parallel consists in Kuhn's and Kant's insistence on the pure object-sidedness, hence unknowability, of the world-in-itself and thing-in-itself, respectively. Kant's intended opponent is, primarily, rationalist metaphysics. Kuhn, by contrast, is chiefly warding off the possibility of *empirical* access to the world-in-itself, that is, the naive realist interpretation of science. But in addition, he wishes to reject the more refined realist philosophy of science which sees the scientific process as a progressive "drawing closer to the truth," or "rising degree of verisimilitude."[31] While Kant, in contrast to some later self-ordained transcendental philosophers, saw the thing-in-itself as indispensable on conceptual grounds,[32] Kuhn attempts in 1979 to do without the world-in-itself; I will return to this effort later.[33]

Having thus explicitly distinguished two world concepts, we are now in a position to make sense of the sentence cited at the beginning

30. *SSR*, pp. 170–173. Here Kuhn rejects the view that science "draws constantly nearer" to the truth. The notion of a "one full, objective true account of nature" is branded not only unnecessary, but an outright hindrance to the explanation of scientific development (*SSR*, p. 171). Later treatments of the same topic may be found at 1970a, *ET* pp. 288–289; 1970b, pp. 264–266; 1970c, *SSR* pp. 206–207; 1974b, pp. 508–509; 1979b, pp. 417–418; 1984, p. 244. —I will return to this problem-complex in § 7.6.

31. See n. 30. The chief target here is Popper's school.

32. E.g. *Critique of Pure Reason*, Bxxviff.: "But our further contention must also be duly borne in mind, namely, that though we cannot *know* these objects as things in themselves, we must yet be in position at least to *think* them as things in themselves; otherwise we should be landed in the absurd conclusion that there can be appearance without anything that appears" (English edition, N. K. Smith, 1965 p. 27). Similarly, see A251ff., B308, and elsewhere; see also Martin 1969, § 29. One might criticize Kant for simply endorsing the view, entailed by our conventional use of words, that appearances must be appearances *of something*. Cf. Heidegger 1927, § 7.A.

33. See § 2.3. Kuhn's vacillation has parallels in the Kantian debate since Jacobi; see e.g. Martin 1969, §§ 21ff, or Shaw 1969. Another substantive difference between Kant and Kuhn resides in their criteria for identifying the genetically subject-sided moments of knowledge. For Kuhn, the criterion consists in the *possibility of change* in such moments of the perceived or the known, or equivalently, in the existence of a learning process for such moments (see chap. 3). For Kant, such moments are those which enable the epistemic subject to experience objects, or, what comes to the same thing, to make certain synthetic a priori judgments. The most obvious consequence of this difference in criteria is seen in the differing roles assigned to space and time by Kant and by Kuhn, respectively.

of this section: "though the world does not change with a change of paradigm, the scientist afterward works in a different world."[34] The world-in-itself is independent of all subject-sided moments, hence of any changes taking place within them. By contrast, a phenomenal world is constituted both by the object-sided world-in-itself and by subject-sided moments (which, taken together, are sometimes called a "paradigm"), the (at this point as yet unanalyzed) interaction of which gives rise to the perceptual and conceptual differentiation of this phenomenal world. The way is thus opened for the possibility of a change in phenomenal world, despite the constancy of the world-in-itself; such change occurs precisely when those world-constitutive moments located in the epistemic subject, together called the paradigm, change—not in such a way as to lose their world-constituting function, but rather in such a way as to give rise to a different phenomenal world.

b. The Plurality-of-Phenomenal-Worlds Thesis and Its Justification

Kant's distinction between appearance and thing-in-itself was motivated both by the critical attitude toward traditional metaphysics and by epistemological scruples: with the help of this distinction, rationalist metaphysics could be rejected, transcendental philosophy grounded, and the possibility of the natural science and mathematics (of Kant's time) explained. To accomplish these tasks it was enough to distinguish between the thing-in-itself and *one* world of appearances. Kuhn, however, is primarily concerned with the *difference between distinct phenomenal worlds,* the assumption of which he holds as necessary if we are to understand the structure of scientific development in a historically adequate way. This fundamental assumption of Kuhn's theory I will call the *plurality-of-phenomenal-worlds thesis.*[35] In addition to this thesis, Kuhn also occasionally speaks of the distinction between phenomenal world(s) and world-in-itself; I will consider his reasons in § 2.2.

The plurality-of-phenomenal-worlds thesis, together with the aforementioned claim that the world-in-itself is inaccessible to us, forms the basis for two central doctrines of Kuhn's theory, at which, for the time being, we can only gesture. First, Kuhn's skepticism to-

34. *SSR,* p. 121.

35. See Goodman 1975 and 1984, chap. 2; Putnam 1981, pp. 49, 54, 79, 134–135. Historically, this thesis appears to have been first formulated by Cassirer (O. Schwemmer, personal communication; see Goodman 1975 and Stegmüller 1985, p. 30). The plurality-of-phenomenal-worlds thesis at issue here has nothing to do with either the "plurality of worlds thesis" with which Lewis articulates his "modal realism" (Lewis 1986, esp. p. 2), or with the "plurality of worlds" necessary for the existence of intelligent extraterrestrial life (Dick 1984 and Crowe 1986).

ward attempts to create a "neutral observation language" is explained by this assumption.[36] For such a neutral observation language is supposed to refer to a completely non-theory-laden, immediate given, meaning either (on the "naive realist" interpretation) the world-in-itself, or (on the "phenomenological objectivist" interpretation) the *one* phenomenal world to which we have access. But neither is the world-in-itself accessible, nor is there only one phenomenal world. Second, the plurality-of-phenomenal-worlds thesis entails Kuhn's rejection of any dichotomy between the purely factual and the theoretical, or between observational and theoretical concepts, or, relatedly, between "invention" and "discovery."[37] For again, the purely factual could only belong either to a world-in-itself conceived as accessible or to a unique phenomenal world.

Kuhn diagnoses his theory's plurality-of-phenomenal-worlds thesis (or, alternatively, his views on the nonuniqueness of the relationship between the one world-in-itself and the many phenomenal worlds) as an essential break with an epistemological tradition going back to Descartes and persisting almost without interruption to the present.[38] This tradition is characterized by the assumption that, in a given perceptual situation, "observations . . . themselves are fixed by the nature of the environment and of the perceptual apparatus,"[39] where the perceptual "apparatus" is taken to be uniform over all humans. In consequence of this assumption, that which, in a particular situation, is given to the senses depends on neither individual psychological nor cultural factors, though it may be subject to differing interpretations.[40] According to the tradition, only *one* phenomenal world corresponds (both locally and globally) to the *one* world-in-itself. For Kuhn, by contrast, a plurality of possible phenomenal worlds, whose differences aren't the result of varying interpretations, at least not in any "ordinary sense of the term 'interpretation,'"[41] corresponds to the one world-in-itself.

This diagnosis of the break with the epistemological tradition raises at least four questions. To begin with, one might ask whether Kuhn's identification and characterization of the epistemological tradition is true to the history of philosophy. As we are not here interested in

36. *SSR,* pp. 126–129, 145–146; 1970a, *ET* p. 267; 1970b, pp. 234–235, 266; 1970c, *SSR* p. 201; 1974a, *ET* p. 308. On this issue, see §§ 3.2, along with 3.6.b and 3.6.c.

37. *SSR,* pp. 7, 17, 52, 53, 66, 141; 1962d, *ET* p. 171; 1970a, *ET* p. 267; 1974a, *ET* pp. 300, 302 n. 11; 1974b, p. 505; 1977c, *ET* p. 338; 1979b, p. 410. On this issue, see § 3.6.a, point 2.

38. *SSR,* pp. 121, 126. Also 1974a, *ET* p. 308; 1970b, p. 276; 1970c, *SSR* p. 195.

39. *SSR,* p. 120.

40. *SSR,* pp. 120–121, 126.

41. *SSR,* p. 123; similarly pp. 121–122.

Kuhn's merits as a historian of philosophy, we may safely ignore this question. A second question concerns the precise sense in which the differences between phenomenal worlds are said not to result from interpretation; and what does "interpretation" mean here, anyway? We will consider this question in § 3.5. Third, is there any connection between this break with epistemological tradition and the break with historiographic tradition described in § 1.2? There is, in fact, some reciprocity between the two breaks with tradition. On the one hand, the hermeneutical approach to the pertinent sources of the history of science uncovers past ways of viewing nature in a way that resonates with the plurality-of-phenomenal-worlds thesis. Conversely, this thesis motivates the kind of approach to sources that can bring to light the alterity of other phenomenal worlds potentially hidden in such texts. Finally, and by far most importantly for any systematic reconstruction of Kuhn's theory, it should be asked how Kuhn justifies his plurality-of-phenomenal-worlds thesis. We will consider this question without further ado.

In a certain sense, the plurality-of-phenomenal-worlds thesis is grounded in the *experience of the historian of science,* in particular of the historian of science who practices his trade according to the principles of the "new internal historiography of science."[42] For this historian of science attempts to reproduce a past scientific community's way of thinking and to present his or her chosen period of scientific history from the viewpoint of participating agents. If this attempt is successful, it *may* produce a different phenomenal world—different, that is, by comparison with the historian's own phenomenal world. Furthermore, it *may* turn out that the phenomenal world discovered by the historian exhibits change over the course of a particular line of scientific development, as actually happens, according to Kuhn, in those periods of scientific development afterwards called "scientific revolutions."[43] In the historian's retracing of these events, it appears

as if the professional community had been suddenly transported

42. Consider the first sentence of the central chapter of *SSR,* chap. 10, "Revolutions as Changes of World View": "Examining the record of past research from the vantage of contemporary historiography, *the historian of science may be tempted to exclaim* that when paradigms change, the world itself changes with them" (*SSR,* p. 111, emphasis mine; similarly p. 117). "Contemporary historiography" refers, of course, to the new internal historiography of science; see § 1.2.c. —Kuhn's reliance on the historian's experience is retained in 1983b, in deliberate consonance with *SSR:* "The historian's discovery of the past repeatedly involves the sudden recognition of new patterns or gestalts. It follows that the historian, at least, does experience revolutions. Those theses were at the heart of my original position, and on them I would still insist" (p. 715). Similarly 1984, pp. 245–246.

43. On this issue, see § 6.2.

> to another planet where familiar objects are seen in a different light and are joined by unfamiliar ones as well.[44]

This kind of experience on the part of the historian serves to ground the plurality-of-phenomenal-worlds thesis to the extent that it familiarizes us with the phenomena for which this thesis is meant to account. Just such an experience proved Kuhn's own point of departure in developing his theory.[45]

But the claim of any one historian or group of historians[46] to such an experience carries no special argumentative weight, so long as there is reason to believe that experiences of this kind can never be authentic. And according to Kuhn, the possible authenticity of such experiences is precisely what is denied by the modern epistemological tradition, which holds the products of perceptual activity to be fixed perceptions, codetermined by environment and perceptual apparatus, subject merely to differing interpretations. It follows that (1) the fundamental *possibility* of perceptual changes occurring, leading to other phenomenal worlds, and (2) their *actual occurrence* in the history of science must be given intersubjectively verifiable demonstration.

Regarding (1): In *SSR*, Kuhn argued the plausibility of perceptual changes that lead to new phenomenal worlds by reference to the visual gestalt shifts explored in perceptual psychology.[47] Still, the results of perceptual psychology can offer no more than plausibility, for, as Kuhn is well aware, the parallels between the kind of perceptual change that gives rise to new phenomenal worlds and that studied in gestalt shift experiments are limited.[48] What perceptual psychology can do is draw our attention to the following three characteristics of perceptual processes.

First, the product of an act of perception is not determined solely

44. *SSR*, p. 111.

45. That is, the (reversible) change in his own phenomenal world which occurred with his success in working in to Aristotelian dynamics; *SSR*, p. v; 1970b, pp. 241–242; *ET*, pp. xi–xii; 1979c, pp. vii–viii; 1981, pp. 8–12; see also 1989b, p. 49.

46. Kuhn is referring to the fact that "Other colleagues have repeatedly noted that history of science would make better sense if one could suppose that scientists occasionally experienced shifts of perception" (*SSR*, p. 113). He has primarily N. R. Hanson in mind.

47. *SSR*, pp. 62–64, 85, 111–114, 126–127.

48. "*Either as a metaphor or because it reflects the nature of the mind,* that psychological experiment provides a wonderfully simple and cogent schema for the process of scientific discovery" (*SSR*, p. 64, emphasis mine); "That parallel [between gestalt shift experiments and paradigm shift] can be misleading" (*SSR*, p. 85); "though psychological experiments are suggestive, they cannot, in the nature of the case, be more than that. They do display characteristics of perception that *could* be central to scientific development, but they do not demonstrate that the careful and controlled observation exercised by the research scientist at all partakes of those characteristics" (*SSR*, p. 113).

by the features of the object toward which the perceiving person directs his or her attention. This claim is not meant to imply merely that the object's physical distance, the duration of exposure to it, or similar considerations play a role in the clarity of object perception. The products of perceptual acts are rather codetermined by "conceptual categories,"[49] "conceptual parameters,"[50] or "perceptual categories,"[51] in the sense that such parameters determine *as what* a given object of perception will be identified. Perception is thus not a purely passive taking-in; its product results, rather, from the cooperation of factors properly located both on the object-side *and on the subject-side.* Second, these perceptual categories should not be assumed constant over all humans, for they depend to an important degree on the perceiver's learning history;

> What a man sees depends on what he looks at and also upon what his previous visual-conceptual experience has taught him to see.[52]

Third, these categories, though coconstitutive of perception, can nonetheless be changed in certain perceptual situations. Experimentally, such situations are those in which the perceiving subject becomes confused, as a result either of a physical change in his or her perceptual organs or of exposure to an object not present in his or her usual perceptual world. Only with the passage of time, the needed amount of which varying from individual to individual, can perceptual categories adapt to a confusing situation in such a way that the confusion vanishes and "the initially anomalous has become the anticipated."[53]

Regarding (2): According to Kuhn, the demonstration of the *actuality* of perceptual changes in the history of science now required runs into the difficulty that "if perceptual switches accompany paradigm changes, we may not expect scientists to attest to these changes directly."[54] For in psychological experiments, the experimental subject can become aware of subject-sided perceptual switches in two ways. First of all, the subject might voluntarily bring about a perceptual switch, at the same time recognizing that nothing in the external world has changed. Second, the subject, relieved by his or her ability to bring about the switch, might also see the drawing itself as permitting different visual gestalts without his or her particular subject-sided perceptual molding. Neither of these possible roads to the recognition of subject-

49. *SSR*, pp. 63, 64, 123.
50. *SSR*, p. 124.
51. *SSR*, p. 116.
52. *SSR*, p. 113; similarly p. 63.
53. *SSR*, p. 64.
54. *SSR*, pp. 114–115.

sided moments of perception, familiar from experiments in perceptual psychology, has any analogues in those perceptual changes in science—if, indeed, there are such—which give rise to new phenomenal worlds.

For one thing, according to Kuhn, "the scientist does not preserve the gestalt subject's freedom to switch back and forth between ways of seeing."[55] This ability is not denied the scientist in principle,[56] however, since the historian of science who diagnoses such perceptual changes must also be able to reproduce them. It is rather the scientist's interests which prevent him or her from returning, after a perceptual change he or she believes to be advantageous, to the mode of perception left behind. If the scientist considers the old mold at all, it is reflectively to reject it, not to reproduce it in all its immediacy.[57] Second, there is no scientific analogue of the drawings, familiar from gestalt switch experiments, which may be seen either as a collection of lines and grey-scales, or, alternatively, as different pictures separated by perceptual switches; "The scientist can have no recourse above or beyond what he sees with his eyes and instruments."[58] For such "recourse" could come only from that which remains of a phenomenal world when all contributions by the perceiving subject have been subtracted: the world-in-itself.

So if the history of science does contain changes of phenomenal world brought about by perceptual switches, we can't count on finding any direct evidence of them. We must thus rely on "indirect and behavioral evidence" in determining whether such changes occur in the history of science.[59] And indeed, Kuhn has at his disposal many examples from the history of science which might be *interpreted* as changes in perceptual mode.[60] But precisely because they require interpretation and offer no direct proof, such examples in fact provide only limited argumentative support for the plurality-of-phenomenal-worlds thesis; such interpretations will only be accepted by those who deem perceptual switches both *possible* and *to be expected* in the history of science. Those who lack these convictions will rather seek alternative interpretations of the examples proffered.[61] And this seems to be the reason why

55. *SSR*, p. 85; similarly p. 114.

56. Although, taken literally, Kuhn's claims might also be read in this way.

57. *SSR*, pp. 114–115, 125. —In 1984, pp. 246–252, Kuhn takes up the issue of scientists' interests in looking back on scientific revolutions with a slightly different emphasis.

58. *SSR*, p. 114.

59. *SSR*, p. 115.

60. *SSR*, pp. 115–120, 123–125.

61. Kuhn shows full awareness of this possibility in *SSR*. There his epistemological reflections are motivated by the existence of an alternative interpretation of his historical

Kuhn, in the papers composed in 1969, in which significant portions of the theory presented in *SSR* are reformulated, doesn't try to expand his arsenal of examples of perceptual shifts in the history of science in support of the plurality-of-phenomenal-worlds thesis. Instead, this thesis is given a new formulation and further theoretical development.

2.2 Stimulus and Sensation in the 1969 Papers

In some of the papers written in 1969, Kuhn undertakes thorough reformulations and further explications of the central theses of *SSR*, responding to the most important criticisms.[62] Whether, in the process, he weakens or dilutes his theory, as some critics have suggested,[63] obviously can't be determined without a detailed comparison of the respective versions of the theory's components. As far as our present topic, the distinction between world-in-itself and phenomenal world, and the plurality-of-phenomenal-worlds thesis, is concerned, we find a *reformulation* of his position couched in terms used, albeit in passing, in almost the same way in *SSR* (§ 2.2.a). But *stimulus,* the central concept of this new presentation, is not entirely unequivocal (§ 2.2.b). Kuhn's reasons for introducing what we shall call the stimulus ontology are discussed in § 2.2.c. The stimulus ontology turns out to be fraught with difficulties so severe (§ 2.2.d) that none of the modifications close at hand will restore its consistency with other important parts of Kuhn's theory (§ 2.2.e).

a. The Transition from SSR

In the revised presentation of 1969, the concepts *stimulus, sensation,* and *data* serve to differentiate and clarify the somewhat obscure world concept of *SSR*.[64] Instead of saying, as he did in *SSR*,

---

examples: "there is obviously another and far more usual way to describe all of the historical examples outlined above. Many readers will surely want to say that what changes with a paradigm is only the scientist's interpretations of observations that themselves are fixed once and for all by the nature of the environment and of the perceptual apparatus" (*SSR*, p. 120). See Shapere 1964.

62. Their order of composition is 1974a, 1970b, 1970c, and 1974b; see 1974b, p. 500 n. 2 and *ET*, p. xx n. 8 concerning the order.

63. E.g. Laitko 1981, p. 185; Laudan 1977, p. 231 n. 1; Newton-Smith 1981, pp. 9, 103, 113–114; Shapere 1971, pp. 707, 708; Toulmin 1967, p. 471 n. 8; Toulmin 1971, p. 60, among others.

64. 1974a, *ET* pp. 308–309; 1970b, p. 276; 1970c, *SSR* pp. 192–193; 1974b, pp. 509, 511. Kuhn talks of "data" and discusses their fixed nature at several points in *SSR* (pp. 121–122, 125–126); this concept clears the way for the introduction of the distinction

> that members of different scientific communities live in different worlds and that scientific revolutions change the world in which a scientist works,

Kuhn now wants

> to say that members of different communities are presented with different data by the same stimuli. Notice, however, that that change does not make phrases like "a different world" inappropriate,[65]

for the members of different groups "do *in some sense* live in different worlds."[66] They do so in the sense that "the given world, whether everyday or scientific, is not a world of stimuli,"[67] rather, "our world is populated in the first instance not by stimuli but by the objects of our sensations."[68] "Objects of our sensations" are just those objects which appear to us through our sensations.[69] "Object" should here be taken as having a very broad meaning, potentially referring both to a perceived thing[70] as well as, for example, to a perceived color.[71]

By contrast to the objects of sensation, "the given in experience,"[72] stimuli belong to that which is not given:

> We have no recourse to stimuli as given, but are always—by the time we can see or talk or do science—already initiated to a data world that the community has divided in a certain way.[73]

---

between "stimulus" and "sensation" in 1974a, *ET* p. 308. "Retinal imprints" or "retinal impressions" make their appearance in *SSR*, pp. 125, 127, 128, and 129. What in 1969 will be called "sensation" appears in *SSR* primarily as "immediate experience" (pp. 128, 129).

65. 1974a, *ET* p. 309 n. 18.

66. 1970c, *SSR* p. 193.

67. 1974a, *ET* p. 309 n. 18.

68. 1970c, *SSR* p. 193.

69. This follows indirectly from 1970c, *SSR* p. 194, where "to have a sensation" is paraphrased by "to perceive something."

70. This follows from Kuhn's paraphrasing of "their sensations are the same" by "they . . . see things . . . in much the same ways" at 1970c, *SSR* p. 193.

71. This becomes evident at 1974a, *ET* p. 308, where Kuhn takes "the perception of a given color" as an example of sensation, not the perception of a particularly colored thing. This broad use of the concept of an object ignores Kant's problem: how do we synthesize *objective perception* (for Kant, "experience") of things from the manifold given to sensibility? This is also evident from the facility with which Kuhn moves from "sensations" to "compounds" and "elements" of sensations: 1974a, *ET* p. 308. One reason for this lacuna is that Kuhn is interested not in the constitution of thinghood *per se* (see Heidegger 1962) but rather in the possible perception of *different things* given the *same stimuli*.

72. 1974a, *ET* p. 308.

73. 1974b, p. 509.

But not only are stimuli not given, there is also "no way to get outside [this data world]—back to stimuli."[74] Accordingly, the world of stimuli must be treated as a *postulate,* assumed out of philosophical motives:

> We posit the existence of stimuli to explain our perceptions of the world, and we posit their immutability to avoid both individual and social solipsism.[75]

The cited passages from Kuhn's 1969 work give every appearance of knitting seamlessly with *SSR*: *SSR*'s hypothetical world-in-itself is now the posited world of stimuli; *SSR*'s phenomenal world is now the world of objects revealed to sensation. The motive for distinguishing between stimuli and sensations is identical with the corresponding motive in *SSR,* the desire to speak of the different worlds of different scientific communities despite our conviction that all humans live, in a certain sense, in *one* world. And even the strategy for reconciling two apparently irreconcilable doctrines is the same as in *SSR;* it proceeds by distinguishing a purely object-sided realm of stimuli from the world constituted by epistemic subjects. This process of constitution produces different worlds according to the particulars of its concrete execution.[76] *SSR*'s plurality-of-phenomenal-worlds thesis now becomes the doctrine of the *non-uniqueness of the relation between stimulus and sensation*:

> very different stimuli can produce the same sensations; . . . the same stimulus can produce very different sensations.[77]

But the position thus articulated in the 1969 papers raises a series of questions and problems which merit discussion.

b. The Ambiguity of the Stimulus-Concept

Confronted with the 1969 papers, some of Kuhn's readers were confused by his use of "stimuli" in two different meanings, without any textual hint on the transition from one to the other.[78]

74. Ibid.

75. 1970c, *SSR* p. 193.

76. See chap. 3 for an analysis of the process of constituting a phenomenal world.

77. 1970c, *SSR* p. 193; similarly 1974a, *ET* p. 308; 1970b, p. 276; 1974b, p. 509. —The same is said of "retinal impressions" and "to see a thing" in *SSR; SSR,* pp. 126–127.

78. The most gripping example of this equivocation may be found in 1970c, *SSR* pp. 192–193; other, less obvious examples are 1974a, *ET* pp. 308–310, and 1970b, p. 276. One critic of the equivocation is Shapere, in his 1971, p. 708. —Compare the problems discussed in this and other subsections of 2.2 with the analogous problems found in Kant and the Kantian debate; here we find them treated under the heading "transcendent

The first meaning of "stimuli" is found, above all, in Kuhn's remarks on the metaphysical-epistemological problem of solipsism:

> If two people stand at the same place and gaze in the same direction, we must, under pain of solipsism, conclude that they receive closely similar stimuli. (If both could put their eyes at the same place, the stimuli would be identical.). . .
>
> . . . We posit the existence of stimuli to explain our perceptions of the world, and we posit their immutability to avoid both individual and social solipsism. About neither posit have I the slightest reservation.[79]

From this passage, we may extract the following features of the first meaning of "stimuli":

—Reference to stimuli in this sense is allowed by a posit, not by any empirically determinable fact.
—This posit is motivated by its ability to explain our perceptions of the world, and to avoid individual and social solipsism.[80]
—Stimuli are granted their own being; that is, their existence is independent of their reception or nonreception by epistemic subjects.
—This independent being amounts to spatiotemporal existence.
—Stimuli are granted their own determinacy; that is, their attributes are independent of their reception or nonreception, and of their particular mode of reception by epistemic subjects.[81]
—The determinacy of stimuli is causally relevant in perception, though it doesn't fully determine object perception.[82]

In this meaning of "stimuli," stimuli are purely object-sided, and have independent spatiotemporal existence and independent, determinate characteristics. But nothing more can be said about stimuli within the compass of Kuhn's theory, for the theory insists that the purely object-sided is inaccessible.[83] Stimuli are thus *determinate* but *cannot be*

---

versus empirical affection." See Vaihinger 1922, vol. 2, p. 35–55, especially p. 52ff., along with p. 363ff., p. 366 n., and p. 463. Kuhn finds himself, as we shall see, in an "unfortunate position" similar to that which Vaihinger sees Kant as occupying (p. 363).

79. 1970c, *SSR* pp. 192–193; similarly 1970b, p. 276.

80. I will return for a more precise discussion of these motives in § 2.2.c.

81. This determinacy is contained within the "immutability" posit; q.v. 1974b, pp. 508–509.

82. This last claim follows from the doctrine of the nonuniqueness of the relation between stimulus and sensation; see § 2.2.a.

83. In contrast to naive realism, on which the object-sided is more or less directly accessible, and to the somewhat more refined version of realism, on which the subject-sided aspects of our knowledge may somehow be subtracted.

determined by us.[84] Stimuli in this sense are neutral with respect to different phenomenal worlds: they are just that which the object-side contributes to all phenomenal worlds, and thus that which, for Kuhn, can explain object perception and ward off solipsism.[85]

The second meaning of "stimuli" emerges from Kuhn's opposition between stimuli and sensations. While sensations (and their component elements) are directly "given" us, we know of stimuli in the second sense only by way of theoretical constructions: they are that which empirical science allows us to identify as causally responsible for our sensations, as, for example, sound waves, photons, and the like. In this second sense, too, stimuli have spatiotemporal existence and their own determinate attributes, but here these attributes are accessible by means of scientific theories. Accordingly, these stimuli aren't purely object-sided, for their identification by science has changed historically and might change again. And of course stimuli in this sense aren't neutral with respect to different phenomenal worlds; on the contrary, they are themselves a construction contingent on a given phenomenal world, constituting a kind of theoretical superstructure of this phenomenal world.[86]

The two meanings of "stimuli" correspond, as Husserl might say, to different *standpoints*.[87] From the *natural* standpoint, stimuli are unhesitatingly ascribed absolute reality just as they appear to empirical examination; they are taken, along with all of their characteristics, as purely object-sided. From the particular *critical epistemological* standpoint characteristic of Kuhn's theory, however, the assumption of the pure object-sidedness of such stimuli is no longer tenable. The premise that the world *I* (and other members of my community) take to be real, is, in the same way, *the* real world for *all* humans, which appeared so self-evident from the natural standpoint, is now called into question. For the plurality-of-phenomenal-worlds thesis, or the corresponding doctrine of the nonuniqueness of the relation between stimulus and sensation, prevents us from generalizing the self-evident truths of the natural standpoint; it prevents me from viewing the world I take to be real as *the* real world. Consequently, the critical epistemological standpoint prevents any attributes ascribed to stimuli from the perspective of a given phenomenal world (or in a given phenomenal world)

84. For similar attempts in the Kantian debate to ascribe certain determinacies to the thing-in-itself, see Vaihinger 1922, vol. 2, p. 180–184.

85. I will treat the alleged contributions of the stimulus ontology in greater detail in § 2.2.c.

86. As noted by Kuhn himself in *SSR*, p. 129.

87. See Husserl 1922, §§ 27ff. [See Gibson's translation; "standpoint" is the usual rendering of the German *"Einstellung."* —A.T.L.]

from being retained as attributes of the purely object-sided. In a certain sense, the critical epistemological standpoint makes room for the plurality-of-phenomenal-worlds thesis in a space that, from the natural standpoint, was the exclusive province of a single phenomenal world. The critical epistemological standpoint places a demand on philosophers analogous to the demand on historians of science, that they set aside contemporary scientific knowledge of the world so as to leave themselves open to the potential alterity of earlier world views.[88] Any substantive assumption about the nature of stimuli constitutes a prejudice in favor of some particular phenomenal world (or some particular class of phenomenal worlds) and is thus to all appearances a methodological error. The plurality-of-phenomenal-worlds thesis demands that philosophers examining the structure of scientific development maintain their neutrality toward the multiplicity of phenomenal worlds.

The radical nature of the critical epistemological standpoint thus prohibits us from taking the attributes of purely object-sided stimuli as describable, for all descriptions of empirical science run the risk of being overturned, for good reasons, in the next scientific revolution, thereby revealing themselves never to have been genetically object-sided in the first place. Despite the radical character of this critical epistemological standpoint, Kuhn insists on ascribing to stimuli independent being and independent attributes. His reason lies in the contributions credited to the stimulus ontology, the very reason the ontology was introduced to begin with.

c. Contributions Credited to the Stimulus Ontology

The motives Kuhn cited for positing purely object-sided, spatiotemporal stimuli with determinate attributes indescribable by us, and with a causal impact in perception, involved their (partial) ability to explain our perceptions of the world, and to ward off individual and social solipsism.[89] Let us examine these motives more closely.

To begin with, Kuhn wants to explain why communication about

88. Compare § 1.2.c.

89. See § 2.2.b. —My reconstruction of Kuhn's position in this subsection relies on a number of (occasionally quite heated) discussions I held with Thomas Kuhn in the spring of 1985, and on his written and oral responses to earlier versions of this chapter. The need for these unorthodox sources arises from the fact that, while many of the relevant passages in Kuhn's work are cryptically brief, this aspect of Kuhn's theory nonetheless strikes me as too important simply to pass over. The price I must pay for this is that the accuracy of this portion of my reconstruction becomes only partially verifiable.

many objects of a shared external world is largely unproblematic *within* a given social community, for example, within a given scientific community, and why many of the members' nonverbal reactions to this external world exhibit such group-specific similarity.[90] This behavioral similarity is assumed to follow from the fact that all members of a given community have *the same object perceptions* when they look in the same direction at the same time from (approximately) the same place. Three quite different factors must work together in order to bring about this parity of object perceptions: the perceiving subjects must have the same perceptually relevant *biological* equipment, the same perceptually relevant *cultural* background,[91] and finally, the same incident object-sided *stimuli*. Each of these factors is necessary to explain the parity of perceptions, but only jointly may they be sufficient. In particular, there would be no reason to expect even two biologically and culturally identical perceptual subjects to have the same perceptions in a given situation unless they also received the same stimuli. But this assumes that a given stimulus (or a given set of stimuli) can't be processed into *just any* object perception; instead, the stimulus offers a certain amount of resistance to being processed into an object perception. In other words, the stimulus must possess some proprietary, determinate feature causally relevant in perception.[92]

It is worth insisting that three things *do not* follow from this demand for the proprietary determinacy of causally efficacious features of stimuli. First, the proprietary determinacy of stimuli does not entail that a given stimulus uniquely, exceptionlessly determines the object perception of a perceptual subject with a given cultural background. This failure of determination is illustrated by the "flip-flopping" of pictures used in gestalt shift experiments. Second, this proprietary determinacy of stimuli surely doesn't entail that the determinate features of a stimulus, working together with the perceptual subject's biological endowment, could completely override the perceptual subject's cultural background. The cultural molding of the perceptual subject by his or her respective community is rather a necessary, additional determinant

90. 1970c, *SSR* pp. 193–194; similarly 1970b, p. 276.

91. See § 3.5 for the details of Kuhn's perceptual theory.

92. The motive for Kuhn's position (though not the motive's consequences) is clearly identical with one motive in the Kantian debate, formulated by neo-Kantian Alois Riehl as follows: "The notion of an object independent of the perceiving subject, which he and all other subjects have in common—for this and nothing else is the concept of a "thing-in-itself—isn't all that horrifying. . . . indeed it can be justified as a demand of the economy of thought, as the simplest hypothesis which explains different subjects' arriving at a unanimous judgement by way of their various perceptions" (Riehl 1925, p. 39 [My translation. —A.T.L.]). Also see Schmidt 1985.

for the product of a given perceptual act. And third, it doesn't follow that those proprietary features of stimuli in virtue of which they resist being processed into just any object perceptions must be knowable by us. Stimuli, like other objects in the world, are only accessible to us by way of theory, and we appear to have no guarantee for the finality of theories. Indeed, we can't even assume that successive theories' determinations of stimuli will "draw closer" to the determinate features of the stimuli themselves.[93]

So much for our sketch of the role Kuhn intends stimuli with determinate yet indescribable features to play *within* a given social community. They are an indispensable component of the explanation of the unanimity of object perceptions within the community and thus needed for avoiding individual solipsism. But these stimuli are meant to play a further role in the avoidance of "social solipsism." What does this further role amount to?

Social solipsism, for Kuhn, is the doctrine on which a similarity of object perceptions, and, accordingly, successful communication about the external world, is possible only *within* a given social community. Furthermore, the members of a given community are denied any possibility of learning another community's way of thinking, speaking, or perceiving. The success of ethnologists, historians, linguists, and other social scientists speaks against this doctrine. How should the possibility of success in such disciplines be explained? The chasm sundering us from an alien community may be bridged by the fundamental biological similarity of members of other cultures and by the possibility of sharing stimulus situations with them. If stimuli offered no resistance to the cultural conditioning of perception, members of two different communities might, in the same situation (whatever that might mean, under our present supposition) and within certain biological limitations, see *arbitrarily many* different things when looking in the same direction from close to the same place. In particular, what one speaker takes to be the *same* perceptual situation, and applies the same words in, another speaker might take to be *unpredictably different* situations. In that case, learning the language of another community would be impossible.[94] But the fact that alien languages, and consequently the cultures in which they are spoken, are not, at least in principle, inaccessible to members of other communities, suggests that the cultural conditioning of perception is not without limits—limits which can be ex-

93. See § 7.6.d.

94. This assumes that learning the language of an alien culture depends fundamentally on the successful perceptual identification of the referents of words in the alien language. For Kuhn's views on this issue, see chap. 3, especially § 3.6.

plained by the determinate, proprietary features of the stimuli themselves.

Of course, these proprietary features of stimuli provide no *guarantee* of access to the culture of a given alien community. All that the stimulus ontology allows us to say is this: *as far as the stimuli are concerned, it is not in principle impossible, in the same stimulus situation, to have the same perceptions as members of an alien community.* This claim does nothing to explain the success or failure of any individual effort to understand an alien culture; a success only confirms that success was possible, and a failure can only be consoled by the recognition that the malleability of the stimuli was not to blame. As far as the stimuli themselves are concerned, the stimulus ontology's defense against social solipsism only goes so far as to deny the in-principle impossibility of breaking out of one's own culture. But this contribution by the stimulus ontology is precisely what Kuhn needs for his theory: the ability to ward off the complete relativization of reality to culture, without at the same time ethnocentrically or presentistically projecting one's own conception of reality onto alien cultures. The preemption of a total relativization of reality to the cultural spectrum resonates both with the success of the social sciences and with Kuhn's insistence that science's passing through revolutions results in progress, not mere change.[95] Furthermore, this caution with regard to the projection of one's own conception of reality onto other cultures resonates with our experience with the alterity of past scientific cultures, which subjects nearly all elements of contemporary scientific culture to potential historical change.[96]

To summarize: The determinate, proprietary features of stimuli are meant to have the function of warding off the complete relativization of the concept of reality to individuals or communities, which threatens when we adopt the critical epistemological standpoint characteristic of Kuhn's theory. Although we may never determine their features, stimuli do their job of resisting the impending arbitrariness of perception and theory formation and thus prevent the onset of thoroughgoing relativism, in virtue of their *own* being and their *proprietary* determinacy.

d. Troubles with the Stimulus Ontology

The impressive power of the stimulus ontology at first blush turns out, on closer examination, to be fraught with fundamental difficulties. Let

95. See § 7.6.
96. See § 2.1.b.

us consider the doctrine of the nonuniqueness of the relation between stimulus and sensation once more:

> very different stimuli can produce the same sensations; . . . the same stimulus can produce very different sensations.[97]

Kuhn makes this statement in connection with his reformulation of the plurality-of-phenomenal worlds thesis and his defense against solipsism. It follows that these stimuli should be understood as *at least* stimuli in the purely object-sided sense: determinate but, for us, indescribable stimuli. For Kuhn, the central consequence of the cited doctrine is

> that two groups, the members of which have systematically different sensations on receipt of the same stimuli, do *in some sense* live in different worlds.[98]

But here, the difficulty appears, how can *equivalent* stimuli be factually distinguished from *different* stimuli, if these stimuli are indescribable by us? If the doctrine of the nonuniqueness of the relation between stimulus and sensation is to be at all justifiable, and indeed operationally applicable, we must at least be able to identify equivalent stimulus situations.

The indescribability of stimuli notwithstanding, a special feature of the stimulus ontology seems to allow the identification of equivalent stimulus situations;

> If two people stand at the same place and gaze in the same direction, we must . . . conclude that they receive closely similar stimuli. (If both could put their eyes at the same place, the stimuli would be identical.)[99]

The parenthetical sentence contains the justification for the criterion for the *approximate* equivalence of stimuli which immediately precedes it: the nature of object-sided influence depends exclusively on the given point in space and, we must add, time. The parenthetical sentence is thus simply another formulation of Kuhn's stipulation of the spatiotemporality, determinacy, and causal efficacy of stimuli in perception.[100] The implied *object-sidedness of space and time* seems to permit a special avenue of contact with purely object-sided stimuli which, while

97. 1970c, *SSR* p. 193; compare § 2.2.a.
98. Ibid.
99. 1970c, *SSR* p. 192.
100. Compare §§ 2.1.a and 2.2.b.

it doesn't make them describable by us, still allows us to identify equivalent stimulus situations.

But the spatial extension of the human body and its sensory organs prevents us from ever bringing the sensory organs of two people together in precisely the same place at precisely the same time. Thus, in the above passage, Kuhn sets an operationally inert sentence in parentheses, replacing it with an approximation. But if it is to fulfill its purpose, this approximation must obviously impose a continuity axiom on stimuli conceived as purely object-sided: as small spatial displacements can give rise to only small changes in received stimuli, it follows that stimuli must be relatively evenly distributed in space. But how might this continuity axiom be justified? Even in concrete cases (let alone the general case), *perception* is unsuited to this task. For the doctrine of the nonuniqueness of the relation between stimulus and sensation, as it now stands, prohibits any inference from the (approximate) equivalence of perceptions to the (approximate) equivalence of stimuli. Neither can empirical scientific *theories* make any assertions on the spatiotemporal distribution of stimuli conceived as purely object-sided. For such theories are only concerned with stimuli in the second sense, and these stimuli always contain genetically subject-sided components—they are always contingent on some particular phenomenal world. So the continuity axiom required in order to draw relevant consequences from the stimulus ontology turns out to be an *additional metaphysical postulate*.

It is highly doubtful whether Kuhn actually intended to incorporate this additional metaphysical postulate into his theory, and highly doubtful whether he should. For the aforementioned proximity criterion for the identification of equivalent stimulus situations is inapplicable in many of the situations relevant to Kuhn's efforts toward a theory of the structure of scientific development. This criterion is only applicable in situations in which two observers find themselves *in approximately the same place at the same time*. Scientists engaged in a controversy over theory choice may occasionally be in such a position. The historian whose job it is to reconstruct the history of science, however, is generally not in this position: he or she is not, for example, when the issue is the transition from Aristotelian to Galilean dynamics.

And in fact, Kuhn proceeds in a different way in many of the corresponding passages of *SSR*.[101] Where, in 1969, equivalent *purely object-sided* stimulus situations are to be identified, *SSR* refers to a phenomenal world. With regard to the transition from Aristotelian to Gali-

101. This other procedure can be worked into a modification of the stimulus ontology; see § 2.2.e.

lean dynamics, for example, Kuhn claims, "when Aristotle and Galileo looked at *swinging stones,* the first saw constrained fall, the second a pendulum."[102] Here the shared fixed[103] object of both Aristotle's and Galileo's observations, the swinging stones, is *identified within a phenomenal world,* the phenomenal world which Kuhn assumes himself to share with his readers. No recourse to the shared, *purely object-sided* stimuli of Aristotle and Galileo is even considered here.[104] But *SSR* already tends toward an ambiguous notion of stimulus, as we can see from Kuhn's desire, in some cases, to answer the following question in the affirmative: "Did these men really *see* different things when *looking at* the same sorts of objects?"[105] For, on the one hand, Kuhn's use of the terminological distinction between "to see" and "to look at," to draw the distinction between the *product* of an act of perception and the *as yet undetermined object* of this act, implies that he imputes "the same sorts of objects" to the world-in-itself. But on the other hand, "the same sorts of objects" can't be identified, as such, at least not directly, within the world-in-itself, but rather can only be identified as they are within a phenomenal world. And in accordance with Kuhn's own theory, an inference from the identification of a class of objects in a phenomenal world to the organization of the world-in-itself is impermissible, for such classifications of objects are subject to unpredictable change in scientific revolutions.

At such points it appears that the failure to differentiate sufficiently the two meanings of the stimulus concept takes its revenge.[106] For, on the one hand, Kuhn here makes use of stimuli in the sense in which they are purely object-sided, hence indescribable. And rightly so, for the articulation of the plurality-of-phenomenal-worlds thesis demands that we adopt a critical epistemological standpoint, from which all exclusive claims to knowledge of external reality are suspended in favor of a neutral stance toward all phenomenal worlds. But on the other hand, when it comes to the identification of equivalent stimulus situa-

102. *SSR,* p. 121; emphasis mine.

103. Or perhaps swinging.

104. Even such talk of equivalent stimuli in the second sense, scientifically describable stimuli, is problematic. For "equivalent stimulus situations" might be concretely realized by *very* different stimuli. For example, the swinging stone's color, size, shape, and brightness, the shadow it casts, the length and thickness of the thread, may all be irrelevant, but these features make a substantial contribution to the stimuli impinging on the observer's eye. The possibility of *seeing* a given stimulus situation *as* a swinging stone would thus have to function as a criterion for the "relevant" similarity of stimulus situations—but here, again, is a criterion taken directly from a given phenomenal world. On this issue, cf. Jonas 1973, pp. 239ff.

105. *SSR,* p. 120; original emphasis; similarly p. 122.

106. Compare § 2.2.b.

tions, Kuhn relies on criteria which are legitimate only from the natural standpoint—criteria taken from a particular phenomenal world. The switch from one standpoint to the other is masked by the ambiguity of the stimulus concept, and the fundamental problem of identifying equivalent constellations of purely object-sided stimuli thus appears soluble.

Once the radical character of that critical epistemological standpoint which suspends our own phenomenal world's claims to exclusive reality has been recognized, the explanatory power Kuhn attributes to the stimulus ontology is called into serious question, and not just on account of our highly dubious prospects for identifying equivalent stimulus situations. For once the recourse to a phenomenal world involved in identifying equivalent stimulus situations has been branded illegitimate, talk of the equivalent biological endowment and cultural conditioning of perceptual subjects appears suspect as well.[107] Such talk, central to the exposition of the stimulus ontology's explanatory virtues, involves assertions which, because they were gleaned from theories coined in the *natural* standpoint, can't be neutral with respect to all phenomenal worlds. There seems to be no reason to accept that *physical* theory's claims about stimuli must be suspended in order to allow the plurality-of-phenomenal-worlds thesis to be formulated, while the propositions of *biology* and the *social* sciences are left standing as if *they* contained no prejudice in favor of some particular phenomenal world (or class of phenomenal worlds).

This problem, whose severity should now be clear, affects the "viewpoint" of any analyst who attempts an unbiased examination of the reality presupposed by an alien subjectivity and allows the possibility that this reality is different from the one he or she presupposes. This effort demands, for one, impartiality toward the different possible ways of conceiving reality. In particular, the analyst must suspend his or her own conception of reality. If the analyst does this as radically as Kuhn's stimulus ontology suggests, he or she simultaneously gives up all the tools needed in order actually to carry out the examination. For example, he or she is no longer in any position to identify equivalent stimulus situations, since such identification can only issue from, or occur within, a given phenomenal world. In addition, any possibility of making biological or anthropological assumptions is lost; for such assumptions can only be legitimated in the framework of a particular phenomenal world, and once so legitimated, they become illegitimate as judged by the standard of complete impartiality toward other phenomenal worlds.

107. Compare § 2.2.c.

But all these reflections show, in the end, that the instantaneous and complete fulfillment of the ideal of impartiality toward other phenomenal worlds is an illusion. We can't simply leave our own phenomenal world behind and go trotting off in unbiased pursuit of the (re)construction of other cultures' alien phenomenal worlds. The historian, the ethnologist, and also the philosopher who studies alien phenomenal worlds are bound to their respective phenomenal worlds, and any attempt on their part to construct other phenomenal worlds results in a construction *which issues from their own phenomenal worlds*. The necessity of proceeding from one's own phenomenal world is ineluctable. If (as might be denied) emancipation from this starting point is possible at all, it presumably only occurs step by step.[108]

What has become of Kuhn's stimulus ontology, and of the motives which prompted its introduction?

Stimuli were meant to account for the similarity of object perceptions enjoyed by members of a given community in similar stimulus situations and for the fact that, at least in principle, members of different communities aren't denied the possibility of learning each other's modes of perception.[109] But it turns out that stimulus situations in which the operant purely object-sided stimuli are *approximately* the same can't, on Kuhn's theory, be identified as required in order actually to apply the stimulus ontology to any concrete case.[110] In addition, the desired explanation also requires premises taken from biology and the social sciences that, since achieved from the natural standpoint, stand in stark contrast to those purely object-sided stimuli drawn from the critical epistemological standpoint. And so the stimulus ontology is incapable of actually accomplishing the explanatory tasks for which it was introduced in the first place.

e. *The Modified Stimulus Ontology*

Can the stimulus ontology be so modified as to perform its intended explanatory functions? The modification closest to hand, that implicitly used in *SSR*,[111] consists in the following: the approximate equivalence of stimulus situations *in the second sense*, that is, of incident electromagnetic radiation or sound waves, serves as the approximate equivalence criterion for stimulus situations *in the first sense*, where stimuli are

108. I will return to this entire problem complex in § 3.8.

109. Compare § 2.2.c.

110. Unless one pays the price of assuming the rather questionable additional metaphysical postulate discussed above.

111. Compare § 2.2.d.

purely object-sided. Scientific methods thus become permissible in identifying approximately equivalent, purely object-sided stimuli, and such identification becomes relatively unproblematic. The shift from stimuli in the first sense, conceived as purely object-sided and indescribable, to stimuli in the second sense, conceived as scientifically describable, implies a simultaneous shift from the critical epistemological standpoint to the natural standpoint.[112] This shift appears to have a further advantage in that it allows us to apply theories from biology and the social sciences developed from the natural standpoint, which, from the critical epistemological standpoint, appeared so problematic.[113]

So let us consider this proposed modification of the stimulus ontology. We will suppose that the avoidance of individual and social solipsism is successful if, arguing as in § 2.2.c., we take the scientifically identified approximate equivalence of stimulus situations as our criterion for the approximate equivalence of purely object-sided stimulus situations.[114] One theoretical consequence which now seems unavoidable stands in sharp contrast to Kuhn's original intentions. For we have moved substantially nearer to the form of realism generally taken for granted in empirical science[115] and characteristic of Popperian philosophy. I will call this "Peircean realism."[116] According to this form of realism, science captures reality, albeit neither absolutely nor incorrigibly, still to a fair and, over the years, ever-improving approximation.[117] As we will show, the employment of scientific theories in identifying purely object-sided equivalences raises fewer unpleasant consequences for Kuhn's theory than the fact that empirical, scientific statements will have to be made about perceptual subjects.

The employment of scientific theories in identifying approximately equivalent stimuli actually implies a form of realism subtly weaker than the Peircean variety. For if we allow the march of science to exhibit

112. Compare § 2.2.b.

113. Compare § 2.2.d.

114. I am not now concerned with justifying this criterion. In what follows, I will merely analyze the implications of the position called the "modified stimulus ontology."

115. With the exception of certain interpretations of quantum mechanics.

116. In general, this doctrine is ascribed to Peirce (see e.g. Levi 1985, p. 622; Rescher 1984, chap. 5; Scheffler 1967, pp. 11, 19, and 73). For Popper's presentation and defense of this form of realism, see in particular his 1963, chap. 10; 1972, chap. 2, and 1979, pp. 371–372.

117. The chief problem with this position isn't really its justification but is rather the task of explicating what "fair approximation" actually means. See § 7.6.d for Kuhn's rejection of this position. For a defense of the position against Kuhn see Shimony 1976; q.v. Niiniluoto 1985, 1987, and Oddie 1986, along with their references.

any cognitive progress, as Kuhn does,[118] it follows that an identification of approximately equivalent stimulus situations entailed by a *better* theory must, under our assumptions, be taken as a *better* identification. In what sense can an identification of object-sided equivalence be "better"? Only in the sense in which it better captures the actual object-sided equivalence (or difference). Note that this does not imply any progress toward more closely capturing the *nature* of the purely object-sided (whatever that might mean); all it implies is that we more closely capture an object-sided *equivalence relation.* Scientific progress would thus consist not in a "drawing closer to the truth" in the sense of a progressive approximation of the true nature of things but rather in an ever-improving ability to identify purely object-sided equivalence.

But, in order to apply this modified stimulus ontology, some theoretical statements about perceptual subjects must also be adduced.[119] Among these is the general claim that various perceptual subjects all share the same basic biological endowment. Yet this general claim isn't enough; parameters for the equivalent stimulation of human sense organs must also be specified, implicitly or explicitly. For example, concrete assumptions must be made about excitation thresholds and frequency sensitivities. Such assumptions are needed if we are actually to identify classes of equivalent stimulus situations by means of scientific theories. For the equivalence of two stimulus situations doesn't imply that the two situations are identical with respect to *everything* impinging on the relevant perceptual subjects.[120] Two stimulus situations rather count as equivalent if they agree in those features which, for biological reasons, are *perceptually efficacious,* even if they differ in their perceptually inert aspects. For example, two perceptual situations differing only in the ultraviolet spectrum would count as situations with equivalent visual stimuli. We thus obviously need some empirical scientific theories of human perception in order for the modified stimulus ontology to identify equivalent stimulus situations in the desired way.

But now we must consider the status of these theories of human perception. Do such theories represent human perception at least approximately *as it really is,* or must the distinction between better and worse theories be drawn, here as elsewhere, without any way of gauging their "proximity to the truth"?

To begin with, it is easily shown that interpreting such theories

118. See §§ 5.5 and 7.6.

119. Compare § 2.2.c.

120. On this issue, compare footnote 104 to 2.2.d in which a further aspect of the difference between two stimulus situations taken as equivalent is considered.

according to Peircean realism would grant them a status allowing them to fulfill their intended function. For if human perception indeed works in the manner described by theories of perception, these theories can be taken as approximately true of *all* perceptual subjects, *regardless* of which phenomenal world they hold to be the real world. The proximity to the truth of theories of human perception allows me to apply the claims of such theories to perceptual subjects not just as they appear to *me,* as *objects in my phenomenal world,* but as they actually are—regardless of whether these perceptual subjects share my perceptual theories. Equivalent stimulus situations can thus be identified by means of *realistically interpreted* theories of stimuli and human perception, enabling us consciously to place ourselves in the same stimulus situations as members of an alien culture, whatever the actual stimuli involved. And this ability, we recall, was a necessary precondition for the accessibility of alien cultures.[121]

But this Peircean realist interpretation of proffered perceptual theories is highly unsatisfactory. For there seems no adequate reason for allowing *these* theories, but not the theories about stimuli, to be interpreted as more or less close to the truth. If we wish to remain consistent within the bounds of Kuhn's theory, we will have to interpret theories of human perception as relative to particular phenomenal worlds; and though the possibility of comparing the respective merits of different perceptual theories isn't thereby precluded, the possibility of applying absolute or relative claims of proximity to the truth is.[122] But now social solipsism poses a formidable threat; I can capture the perceptual workings of those perceptual subjects to whose phenomenal world I seek access no longer *even approximately as they really are* but only *as they appear to me as objects in my phenomenal world.* To be sure, I can work out better or worse theories about their perceptual functioning, by my own standards, and can thus identify approximately equivalent stimulus situations for them and for me well or poorly, by my standards. But all this occurs *within my phenomenal world,* and I have absolutely no grounds for claiming that the equivalent stimulus situations thus identified *come even close* to capturing the perceptually relevant object-sided stimuli the aliens and I share. The claim that equivalent situations identified by scientific means could serve as the neutral basis and secure point of departure for access to alien cultures becomes untenable; all such identifications occur from within a given phenomenal world, and can thus make no claims to neutrality toward other phenomenal worlds. The prospects for understanding an alien phenomenal

121. Compare § 2.2.c.

122. See § 7.4 on the comparison of theories.

world by the means sketched in § 2.2.c. now appear highly suspect, at least if our goal is to understand the alien phenomenal world *in itself,* as this phenomenal world is to members of the alien culture. The most I can have is pictures or models of alien phenomenal worlds, *whose proximity to the truth can, in principle, never be evaluated,* as a consequence of the assumption that the proximity to the truth of my perceptual theories can't be evaluated. Social solipsism follows.

It appears that the modification of the stimulus ontology we have been discussing can't do its intended job if the theories of human perception it employs are interpreted in the same critical manner in which Kuhn interprets scientific theories in general, as having no determinable relation to reality *itself. The one* reality, the world-in-itself, or the world of stimuli (in the first sense) is inaccessible, and so a theory's distance from or proximity to this world can be gauged in neither absolute nor relative terms. A realistic interpretation of theories of perception powerful enough to make use of the stimulus ontology appears to be ruled out by considerations of consistency. And so the modified stimulus ontology can't perform the functions which made its candidacy as a prop for the foundations of Kuhn's theory plausible.

Neither in *SSR* nor in his 1969 papers does Kuhn appear to be fully conscious of the troubles faced by both the original and the modified versions of the stimulus ontology. The double meaning of "nature" and "world" in *SSR* and the corresponding ambiguity of the stimulus concept in the 1969 papers are chiefly to blame. In addition, Kuhn appears not to have taken sufficient account of the fact that members of alien cultures can, initially, only appear to us as objects in our own phenomenal worlds; thus the critical skepticism we adopt toward the claimed uniqueness of our own culture's notion of reality also applies to what we perceive as members of alien cultures, if they can count as epistemically accessible at all. The less realistic reading of "nature," "world," and "stimulus" seems in better accord with the goals of Kuhn's theory, as it was molded by the particular critical epistemological standpoint required by this theory. But the critical epistemological standpoint gives up all of the tools we would need to battle individual and social solipsism, for the identification of equivalent stimulus situations is prohibited by the inaccessability of the purely object-sided, and the employment of theories about perceptual subjects illegitimate. The more realistic reading of "nature," "world," and "stimulus," molded by the natural standpoint, leads to the "modified stimulus ontology." On this version, the identification of approximately equivalent stimulus situations is unproblematic and the employment of theories of human perception justified. But now a dilemma arises: if the employed theories of perception are interpreted in accordance with Kuhn's theory, social solipsism can no longer be avoided—yet the defense against solipsism was the most important motive for introducing the stimulus ontology. If, on the other hand, we interpret perceptual theories according to Peircean realism, we stand in opposition to the central thrust of Kuhn's theory, which asserts that *reality itself* is inaccessible to us.

Kuhn's undeclared shifts between the two readings of the central concepts of world or nature and stimulus (especially in his postscript to *SSR*)[123] makes it seem as though the advantages of both variants of the stimulus ontology could somehow be combined. But, in fact, the fundamental difficulties the stimulus ontology causes for Kuhn's theory are thereby masked. The fundamental problem, that of the viewpoint of the analyst examining the objectivity presupposed by alien subjectivity, will reemerge in chapter 3, in our analysis of the construction of phenomenal worlds, and I will return to it in § 3.8. But first we must consider such changes in Kuhn's understanding of the world as the object of scientific inquiry as occur after 1969, and especially after 1979.

2.3 The Phenomenal World after 1969

Kuhn's writings over the decade following 1969 teach us nothing about the epistemological-ontological problem of the phenomenal world. Kuhn returns to this issue only in 1979b, where his position exhibits two important points of change.

First, Kuhn now explicitly does without the world-in itself, the "one real world, still unknown."[124] He describes his position as "also . . . Kantian, but without the 'things in themselves'" (and with temporally mutable categories).[125] These texts leave open whether, and if so, how Kuhn avoids the solipsistic pitfalls of this position which threatened in 1969. Though Kuhn claims his position to be a "realistic" one, the precise meaning of this claim remains, admittedly, unexplained.[126] In any case, Kuhn asserts, though with some reluctance, that a position which is Kantian in this sense "need not . . . make the world less real."[127]

Second, the phenomenal world is now addressed with changed focus. In one of the most important developments in Kuhn's thought,

123. Compare § 2.2.b.
124. 1979b, p. 418.
125. 1979b, pp. 418–419; also compare 1979c, p. xi.
126. 1979b, p. 415.
127. 1979b, p. 419.

the role previously ascribed to perception makes way for an even stronger emphasis on the role of language.[128] In *SSR* and in the 1969 papers, language was already strongly implicated in the phenomenal world,[129] though visual purchase remained the guiding prototype for any encounter with the world. In *SSR,* the phenomenal world is a world that is "seen," in both the literal and the metaphorical senses. In the 1969 papers, our encounter with the world is mediated by stimuli, always conceived as *visual* stimuli. Of course, different ways of seeing the world also have their linguistic consequences, but the dominant role is ascribed to vision itself. Now the weight is shifted in favor of language. In 1982, Kuhn claims "If I were now rewriting *The Structure of Scientific Revolutions,* I would emphasize language change more."[130] *SSR* asserted of the phenomenal world, for example, that it was "already perceptually and conceptually subdivided in a certain way."[131] Now the typical claim is that there are "clusters of interrelated terms" which must "be learned together, and which, when learned, give a structure to some portion of the world of experience;"[132] in short, "language structures the world."[133] The "structure of the world" is here understood as the network of similarities and differences between objects in the world, especially as evidenced by the extensions of concepts and the relations between them.[134]

But this emphasis on language should not be understood as entirely eliminating the role of perception in Kuhn's theory. For one, first language acquisition is tied to perception.[135] In addition, perception plays an important role in the dynamic of (scientific) language:

> alterations in the way scientific terms attach to nature are not—logical empiricism to the contrary—purely formal or purely linguistic. On the contrary, they come about in response to pressures generated by observation or experiment.[136]

128. A second important line of development involves the notion of a paradigm; see § 4.2.

129. On this issue, see especially § 3.6.b.

130. 1983b, p. 715. —This trend toward an increasing emphasis on language continues in works which will appear after 1990, and thus can't be considered in this study. Consider the title of the Thalheimer Lectures, which Kuhn gave at Johns Hopkins University on November 12–19, 1984: "Scientific Development and Lexical Change."

131. *SSR,* p. 129.

132. 1983a, p. 680.

133. Ibid.; similarly 1979b, p. 418; 1981, p. 21; 1983a, pp. 676, 682; 1989a, p. 11; 1990, p. 300.

134. I will return to this issue in greater detail in chap. 3.

135. 1983a, pp. 680–681.

136. 1979b, p. 416; similarly p. 419.

The greater emphasis on language also has an impact on the formulation of the plurality-of-phenomenal-worlds thesis. This thesis becomes the "assertion that different languages impose different structures on the world."[137] The thesis is justified as in earlier works: Kuhn offers examples of different structurings of the world by different languages (or sociolects), or of structural changes brought about by linguistic change.[138] In addition, he offers an analysis of the way in which the concepts at work in such structures might be learned, thus giving a plausible account of the way in which different languages lead to different structurings of the phenomenal world.[139]

But such formulations as "different languages impose different structures on the world" and "terms attach to nature"[140] might be misunderstood as claiming that the structureless, completely object-sided world came first, and was followed by different languages, by which various structures were imposed upon the world. This claim suggests recourse to a world-in-itself, a road which Kuhn now explicitly rejects. He rather insists on the *impossibility of separating the purely object-sided from the genetically subject-sided,* as suggested by the following rhetorical questions:

> Does it obviously make better sense to speak of accommodating language to the world than of accommodating the world to language? Or is the way of talking *which creates that distinction* itself illusory? Is what we refer to as 'the world' perhaps a product of a *mutual* accommodation between experience and language?[141]

The genetically subject-sided moments of the world can't be identified, let alone subtracted. And neither can the genetically object-sided moments of language be identified, let alone subtracted:

> In much of language learning these two sorts of knowledge—knowledge of words and knowledge of nature—are acquired together, not really two sorts of knowledge at all, but two faces of the single coinage that a language provides. . . . If I am right, the central characteristic of scientific revolutions is that they alter the knowledge of nature that is intrinsic to the language itself and

137. 1983a, p. 682; similarly 1979b, p. 414; 1981, p. 21; 1983a, pp. 680, 683. —Also see the earlier 1970b, p. 277.

138. 1979b, pp. 416, 418; 1981, p. 19; 1983a, pp. 679–680.

139. On this issue, see chap. 3, especially § 3.6.d.

140. 1979b, p. 416; 1981, pp. 18, 19, 20, 21; 1983a, pp. 680–681.

141. 1979b, p. 418; emphasis mine; similarly 1983a, pp. 681–682.

that is thus prior to anything quite describable as description or generalization, scientific or everyday.[142]

The genetically subject-sided and genetically object-sided moments in the constructions of language and the world are thus inseparable. This is the reason Kuhn can drop, and must drop, the postulate of a purely object-sided world-in-itself, however conceived. For if this purely object-sided world is to have any of the explanatory power for which it was introduced in the first place, it must have *determinate, determinable* features; nothing indeterminate or indescribable can serve as an *explanans* (for something *determinate*). But we have recourse to nothing which might establish the pure object-sidedness of any attribute, neither an apparently subject-sided language, nor the apparently object-sided world; for, in truth, the two moments are inextricably linked. It now remains an open question in what sense Kuhn's is a realistic position,[143] of the sort which might mandate a rejection of social solipsism.

142. 1981, p. 28; similarly 1983a, p. 682; 1989a, pp. 15–16, 18, 20; 1990, pp. 302, 304, 306.

143. I will return to this question in §3.2.

CHAPTER THREE

The Constitution of a Phenomenal World

THE SCIENTIST'S WORLD, the world that provides the scientist and other members of his or her community with objects for study, is accessible by means of perception, language, instrumentation, and theory. As we learned in chapter 2, this world is a *phenomenal world;* in other words, despite appearances to the contrary, its objects contain genetically subject-sided moments in addition to purely object-sided moments.

Now, as Kuhn has noted so clearly since 1979, the genetically object-sided and genetically subject-sided moments of a phenomenal world are not properly separable. Does this mean that we can never analyze phenomenal worlds with respect to their subject-sided moments? Or, more pointedly, can we never distinguish changes in a phenomenal world that are solely the result of alterations in subject-sided moments from changes not only subject-sided in origin? Obviously this cannot be the case for Kuhn, lest we be forced to deny any sense to his talk of scientific revolutions as changes in the scientist's world.[1] Despite the inseparability of genetically subject-sided from genetically object-sided moments of a phenomenal world, we may judge *some* changes in a phenomenal world to be subject-sided in origin without thereby imposing a *complete* sundering of subject-sided from object-sided. It follows that phenomenal worlds may, up to a point, be analyzed with respect to their genetically subject-sided moments, even though these always occur along with object-sided moments. I will call such an analysis, carried out now not on a particular phenomenal world but for phenomenal worlds in general, the *general analysis of the constitution of phenomenal worlds.*[2]

Kuhn attempts such a general analysis of the constitution of phe-

1. See § 6.2.

2. In recent work Kuhn has also emphasized the importance of contributions by the historiography of science to such analyses; see 1986, p. 33.

nomenal worlds by exploring the process *whereby the individual members of a given (scientific) community gain access to the community's phenomenal world,* a process in which "education, language, experience, and culture" play a role.[3] This formulation of the program of general analysis of the constitution of phenomenal worlds invites three points of clarification.

First, our use of the expression "to gain access to a phenomenal world" should neither suggest that the phenomenal world has an existence independent of and prior to the community, and thus needs only to be discovered by its members, nor be taken to imply that the community creates the phenomenal world entirely according to its own dictates; the constitution of a phenomenal world is neither purely passive reception nor purely active invention. On the contrary, the inseparability of genetically subject-sided from genetically object-sided moments of a phenomenal world implies that gaining access to a phenomenal world is something which lies between the two poles, "discovery" and "invention."[4]

Second, a phenomenal world is specific to a given community, and in a (not entirely unproblematic) sense, this community may be called the constituting agent for the phenomenal world.[5] How, then, can the constitution of phenomenal worlds be analyzed by means of an examination of the process whereby *individual* members of the community "gain access" to them, in the sense described? The answer to this question lies in the peculiar way in which the individual members are conceived when our task is to explore the constitution of phenomenal worlds. For in this case individual members are considered only with regard to their membership in the community; their common nature *qua* members of the community is in view. From this perspective their individuality vanishes, as each member becomes a representative of the whole community. Of course, this point of view presupposes that the diverse individuals do, indeed, have some group-specific common traits; we will have to ask both in what these group-specific traits consist and how they come about.[6]

Third, Kuhn approaches the issue of the constitution of phenomenal worlds by way of the question of individual access. This question embodies the search for a *genesis,* namely the coming about of a phe-

3. 1970c, *SSR,* p. 193.

4. In § 3.2 I will demonstrate at what point the intermediate position of such access becomes palpable.

5. Thus a change in phenomenal world of the kind that occurs in scientific revolutions, resulting from a change in genetically subject-sided moments, is a process attributable to an agent: the scientific community. See § 6.1.

6. See § 3.4.

nomenal world's being-for-someone. But we must sharply distinguish this quest from the demand for the *real genesis* of a phenomenal world, or the real genesis of the objects contained in a phenomenal world. For the question of real genesis is concerned, for example, with the manner in which the components of a given thing must have come together; in what way, perhaps, they might have been transformed, or might have transformed each other, in order to bring about this thing as an effect. And in general,[7] only that which is actually in the world, which has or had a real existence *apart from the analyst,* can count as one of the thing's true "building blocks." The demand for real genesis is, obviously, posed from the "natural standpoint,"[8] according to which primary interest lies in the present and past objects in the (given phenomenal) world.

In seeking the genesis of phenomenal worlds (or of regions of, or objects in, these phenomenal worlds) in the sense of their constitution, the interest lies elsewhere. For then the fact that phenomenal worlds also contain relevant, genetically subject-sided moments must be taken into account, and our task is to analyze these moments with regard to their world-constituting function. An ideally successful analysis would result in an understanding of phenomenal worlds on which two features have become transparent: first, that phenomenal worlds do, indeed, contain genetically subject-sided moments, the nature of which is known just to the extent to which they are separable from genetically object-sided moments. In addition, the overwhelming propensity of any phenomenal world to appear to those living in it as the *only* purely object-sided world, as *the* real world, must also be transparently intelligible. In order to launch any inquiry into the constitution of phenomenal worlds, the natural standpoint, which holds the complete objectivity, and hence the absolute reality, of the given phenomenal world to be self-evident, must clearly be abandoned. The necessity of leaving the natural standpoint in order to carry out a general analysis of the constitution of phenomenal worlds brings with it a number of methodological problems, which I will discuss shortly.

First, however, I would like to address the question of why the issue of the constitution of phenomenal worlds is of such importance for Kuhn's project. Let us assume that Kuhn could satisfactorily meet our demand regarding the constitution of phenomenal worlds, namely, that he could provide a general analysis of phenomenal worlds with respect to certain of their genetically subject-sided moments. Let us

7. I am here avoiding the special complex of problems that appears to arise from the interaction of observer and observed in quantum physics.

8. Compare § 2.2.b.

further assume that Kuhn could demonstrate the existence of a possibility space within which the subject-sided moments of phenomenal worlds might vary, such that variation in subject-sided moments results in a difference in phenomenal worlds. Under these conditions Kuhn would have a devastating argument for his central plurality-of-phenomenal-worlds thesis. But beyond the brute fact of the plurality of possible phenomenal worlds, we might also come to appreciate *how* it is that various genetically subject-sided moments serve to constitute diverse phenomenal worlds. Kuhn's theory would thereby gain real substance, for then the claim that a given community lives in a given phenomenal world would also imply that some specific features of this phenomenal world could be illuminated by recourse to the individual member's mode of access (in the defined sense) to the community's phenomenal world. Furthermore, the preconditions for determining in what way, and by virtue of what factors, the (phenomenal) world of a scientific community can change during the course of a scientific revolution would have been met.[9]

But the general analysis of the constitution of phenomenal worlds faces a methodological difficulty bearing a more-than-coincidental similarity to the difficulty which stymied Kuhn's use of the stimulus ontology.[10] Kuhn attacks the one-sidedness of work in the earlier historiographic and epistemological tradition for having recognized only its own phenomenal world as the real world, thus excluding (implicitly or explicitly) the possibility of alternative phenomenal worlds.[11] In order to overcome this predilection for our own phenomenal world and adopt an impartial attitude toward other possible phenomenal worlds, we must suspend all of our own phenomenal world's exclusive claims to absolute reality. In other words, the general analysis of the constitution of phenomenal worlds must be carried out strictly from the critical epistemological standpoint. But this does not mean that our own phenomenal world's claim to exclusive existence is only, as Husserl might have it, "bracketed"; in that case we could remain completely convinced of the exclusivity claim, though choosing temporarily, for methodological reasons, not to make any use of it.[12] Here our own

9. See § 6.2.

10. Compare §§ 2.2.d–e.

11. Compare §§ 1.2.c, 2.1.b.

12. See Husserl 1922, §§ 31, 32 (compare this with the treatment of certain aspects of the relationship between phenomenology and Kuhnian theory in Embree 1981). —We are here dealing with a "pathway of doubt" vis-à-vis our phenomenal world which might more precisely be called "the way of despair. For what happens on it is not what is ordinarily understood when the word 'doubt' is used: shilly-shallying about this or that presumed truth, followed by a return to that truth again, after the doubt has been

conception of what is empirically real must rather, in accordance with the plurality-of-phenomenal-worlds thesis, actually be *given up* in order to make room for alternative conceptions of empirical reality.

But does this radically critical epistemological standpoint even permit us to attempt a general analysis of the constitution of phenomenal worlds? Aren't we committed to taking the constituting *agents* as existing absolutely, that is, independently of the analyst's phenomenal world? Mustn't we make certain assumptions about these agents, assumptions regarding the faculties in virtue of which they become constituting agents of a phenomenal world? From what sources could such assumptions derive their plausibility?[13] An empirical justification seems unacceptable, on the grounds that it could only take account of the constituting agents of other phenomenal worlds *qua objects* of the analyst's phenomenal world, thus injuring crucial neutrality toward diverse phenomenal worlds. A reflective justification of these assumptions faces a similar difficulty, the implausibility of arriving by reflection at the set of *invariant* world-constituting faculties of agency common to *all* (human) subjects. On this issue, Kuhn himself claims at one point that, at least for scientists, a transcendence of one's *own* phenomenal world complete enough to allow an unbiased survey of *all* possible phenomenal worlds is impossible, though a scientist situated in a given historical context might face a choice between *two* possible phenomenal worlds.[14] We must therefore conclude that Kuhn would deny the possibility of such a completely neutral standpoint for the philosopher or historian of science as well.

It thus seems unavoidable that the general analysis of the constitution of phenomenal worlds must be conducted from the standpoint of our own phenomenal world. Assumptions verifiable only with respect to the analyst's own phenomenal world must, indeed, be made. But how, then, can the general analysis of the constitution of phenomenal worlds ever attain its goal: a reconstruction, *unprejudiced* by our phenomenal world, of the process whereby the members of a given community access a phenomenal world alien to us? This methodological difficulty is never considered in Kuhn's writings, an inattention remi-

---

appropriately dispelled—so that at the end of the process, the matter is taken to be what it was in the first place. On the contrary, this path is the conscious insight into the untruth of phenomenal knowledge" (G.W.F. Hegel, *The Phenomenology of Spirit,* trans. A.V. Miller [Oxford: Clarendon, 1977], pp. 49–50). The parallel between Hegel's *Phenomenology of Spirit* and Kuhn's theory noted here is only one among many—but this is not the proper place for a discussion of this topic.

13. Compare these questions with those posed from a Humean perspective by Feigl 1964, pp. 46–47, as challenges to certain premises of Kantian transcendental philosophy.

14. 1974b, p. 509.

niscent of his undeclared shift between the two meanings of the notion of "stimulus," which obscured troubles with the stimulus ontology.[15] Here as before the problem lies in finding an appropriate viewpoint for the unbiased exploration of a reality *presupposed by alien subjectivity*. It follows that Kuhn's general analysis of the constitution of phenomenal worlds, which seeks its goal along a path not mediated by any reflective treatment of the analyst's viewpoint, will only produce results that are, strictly speaking, undetermined.

In what follows I will nonetheless reconstruct Kuhn's analysis without first reflecting on the analyst's viewpoint. Only after the analysis has been completed will we consider, retrospectively, whether there *is* a viewpoint from which such an analysis is possible, and, if so, what weight we should place on our results.[16]

In one respect, however, we may already limit the demands we place on Kuhn's general analysis of the constitution of phenomenal worlds; Kuhn hardly claims to have treated the constitution of phenomenal worlds exhaustively. He rather selects only *one* element of the constituting process for discussion, albeit one he sees as fundamental to an understanding of the structure of scientific progress.[17] But this element, a specific sort of learning, is not only relevant to the constitution of such phenomenal worlds as characterize the *sciences* but indeed universal in the sense that it plays a role whether we are dealing with the phenomenal world of "an entire culture or a specialists' subcommunity [of experts in a given scientific field] within it."[18] It follows, as a methodological consequence of this universality, that certain, relatively easily analyzed learning processes important in constituting an *ordinary* world might also be taken as representative of the processes involved in constituting the phenomenal worlds of science. Kuhn makes frequent use of this possibility in illustrating the constitution of a scientific world, an "extraordinarily complex enterprise,"[19] by nonscientific examples.[20]

15. Compare §§ 2.2.b, 2.2.d–e.

16. See § 3.8.

17. Kuhn speaks of "one of the fundamental techniques" in 1970c, *SSR* p. 193, and similarly at 1979b, p. 410; 'technique' is used here because the constitution of a phenomenal world is a learning process dependent on technical guidance, as we shall see later.

18. 1970c, *SSR,* p. 193, also 1974a, *ET* p. 313, where the invariance of the process is asserted for the worlds of daily life, taxonomy, and such "more abstract sciences" as Newtonian mechanics; see also 1970b, p. 270, and even *SSR,* p. 127.

19. 1974a, *ET* p. 309.

20. 1974a, *ET* pp. 309–318; 1979b, pp. 412–413; 1983a, p. 682. It follows that there is a great deal of continuity between mundane knowledge and scientific knowledge, which I will not pursue here.

The process of constituting phenomenal worlds, insofar as it is here given a general analysis, is characterized by the following schema: We are dealing with a learning process (§ 3.1) in which, by means of ostension (§ 3.2), a certain type of similarity relation (§ 3.3), which members of a given social community have already mastered (§ 3.4), is learned. These similarity relations are codeterminants for perception (§ 3.5) and for the formation of empirical concepts (§ 3.6). Concepts introduced by means of the similarity relations cannot be precisely explicated, and they contain knowledge of nature (§ 3.7).[21] The transformations which Kuhn's conception of the phenomenal world, and with it his conception of the process of constitution, have undergone, especially those occurring between the late 1960s and the late 1970s, are given special attention in § 3.6. Finally, the problem of viewpoint for the analyst attempting a general analysis of the constitution of phenomenal worlds is treated in § 3.8.

3.1 The Learning Process

Access to the phenomenal world of a given (scientific) community may be gained by way of an at least partially reversible *learning process.*[22] Characterizing the process whereby the learner discloses a phenomenal world in this way suggests three negative features. First, such contributions to the phenomenal world as fall to the side of the subject aren't entirely innate.[23] Second, as far as the subject side is concerned, the disclosure of a world isn't a process consisting merely in the realization of some predetermined possibility.[24] Third, the learning process isn't a completely irreversible imprinting; its product can be, at least in part, unlearned or overlearned.[25] In the following sections, we will discuss

21. The Kuhnian analysis of the constitution of a phenomenal world exhibits notable similarities with "linguistic access to the world," as described by Kamlah and Lorenzen in a book fundamental to (re)constructive philosophy of science (Kamlah and Lorenzen 1967). In what follows, I will note parallels in my footnotes without attempting a systematic comparison. —See Kamlah and Lorenzen 1967, p. 46, on the aforementioned universality of the constitution process and its methodological consequences.

22. *SSR,* p. 113; 1970a, *ET* p. 285 n. 34; 1974a, *ET* pp. 309, 312; 1970b, p. 275; 1970c, *SSR* p. 196; 1979b, pp. 412–413; 1981, p. 20; 1983a, pp. 680–681; 1983d, p. 566; 1989a, pp. 14–15; 1990, p. 302. —Compare Kamlah and Lorenzen 1967, p. 29.

23. 1974a, *ET* pp. 308–309. —Of course, this feature doesn't rule out the possibility that some conditions underlying the capacity for undertaking such a learning process might be innate.

24. As is the case, according to Kant, in the attainment of the categories; compare Vaihinger 1922, vol. 2, pp. 89–101.

25. E.g. 1970a, *ET* p. 285 n. 34; 1983a, p. 677.

what is learned in this learning process and how the process proceeds. Section 3.2 will present what is chiefly learned, § 3.3 will consider the means employed in learning, § 3.4 will discuss who does the learning, and the sections which follow will trace the function of successful learning in, and its consequences for, the phenomenal world and scientific knowledge of that world.

The aforementioned negative features of the learning process entail two attributes of phenomenal worlds central to Kuhn's theory of the structure of scientific development: the phenomenal worlds of different communities may also be different; and the phenomenal world of a given community can change.[26] This latter attribute of phenomenal worlds, in particular, will prove of decisive importance in characterizing scientific revolutions.[27]

3.2 Similarity Relations

Kuhn's general analysis of the constitution of phenomenal worlds doesn't, as we have noted, claim to illuminate all aspects of this process. Instead, his analysis focuses primarily on one aspect of special import to his theory of scientific development. This is the role played by *a particular kind of similarity relation* in the constitution of a phenomenal world.[28] Before I get into the peculiarities of this form of similarity relation, I will state the proximate function of similarity relations in general.

Each similarity relation is constitutive in the formation of a *similarity class,* the set whose elements include everything the relation deems similar.[29] Kuhn distinguishes three separate domains within which similarity classes may be formed:[30]

26. See Kamlah and Lorenzen 1967, pp. 46ff., p. 67.

27. See § 6.2.

28. Although Kuhn's use of the term 'similarity relation' begins only in the 1969 papers, the term's referent is of central importance as early as *SSR*. See the footnotes which follow for textual references. —The idea originates in Wittgenstein's later work; *SSR,* pp. 44–45.

29. 1970c, *SSR* p. 200.

30. The explicit, worked-out differentiation of the three domains is found only in 1970c, *SSR* p. 194. 1974a, *ET* p. 313 moves directly from similarity relations as applied to objects to similarity relations as applied to problem situations; 1970b, pp. 273–274 first alludes to the distinction, which may trace back to Suppe 1974a (or at least to the version presented at the March, 1969 symposium), especially pp. 486–489; see 1974b, p. 504. In 1981, p. 20, and in 1983a, pp. 680–682, Kuhn again lists "objects and situations" as the relata of similarity relations.

—Different sensory perceptions of the same object. The formation of the similarity class of perceptions similar in virtue of belonging to the same perceptual object is constitutive of the recognizable identity of that object.

—Different objects, which in the phenomenal world being constituted belong to the same species. Kuhn also calls such similarity classes "natural families" or "natural kinds."[31]

—Different problem situations,[32] susceptible to treatment by means of the same symbolic generalization.[33]

Learning similarity relations always and necessarily includes *learning dissimilarities*.[34] This is true not only in the sense that every similarity is a similarity only given an indeterminate background of dissimilarity. Similarity classes are, moreover, always learned *as they contrast one from the other,* so that the learning process gives rise to a more or less comprehensive network of similarities and dissimilarities within the given domain.

Kuhn is most interested in a particular kind of similarity relation, the defining characteristic of which sustains important aspects of his theory.[35] These similarity relations are *immediate* in the sense that the

31. *SSR,* p. 45; 1970a, *ET* p. 285; 1974a, *ET* p. 312; 1970c, *SSR* p. 194; 1979b, pp. 411, 413, 414; 1981, pp. 10, 20. The adjective "natural" shouldn't be taken to suggest that only natural objects are at issue; the similarity is rather a "natural" one in the sense suggested by the immediacy of similarity relations, which we will discuss soon.

32. The term "problem situation" is the most precise of those Kuhn uses (e.g. in 1970b, p. 273). Other designations include "problems" (e.g. in 1974a, *ET* p. 306; 1970c, *SSR* p. 189), "situations" (e.g. in 1970c, *SSR* pp. 189, 190; 1990, p. 314, and elsewhere), "research problems" (e.g. *SSR,* p. 45), and "intended applications" (of a theory; 1976b, pp. 193–195).

33. What it means for problem situations to be "susceptible to treatment by means of the same symbolic generalization" will be clarified in § 3.6.e. —Similarity relations in separate domains may be related by dependencies. To begin with, the formation of natural families can depend on similarity relations constitutive of the perceptible identity of individual objects. In addition, the formation of similarity relations which hold between problem situations may exhibit interdependence with the formation of natural families of objects. This occurs when certain objects (or certain aspects of these objects) are identifiable only in a certain class of problem situations, and, conversely, such problem situations obtain only when these objects are given (1983a, p. 686 n. 14; also 1970b, p. 274).

34. Kuhn emphasizes this fact, which is important to his theory, at many points, e.g. 1970a, *ET* p. 285; 1974a, *ET* p. 312; 1970b, p. 274; 1970c, *SSR* pp. 193–94; 1974b, p. 504; 1976b, pp. 195, 199 n. 14; 1979b, p. 413; 1981, p. 20; 1983a, pp. 680, 682, 683; 1989a, p. 16; 1990, p. 303. —See Kamlah and Lorenzen 1967, p. 30.

35. But Kuhn says explicitly that *not all* of a phenomenal world's operant similarity relations need share in this characteristic; *SSR,* p. 47; 1974a, *ET* pp. 312–313, 318.

similarity involved *isn't derived from defining characteristics of the relata.*[36] This characterization of the immediacy of certain similarity relations rejects, in Kuhn's words, all claims that certain "characteristics," "rules," "principles," and "criteria" constitute defining *tertia comparationis* for similarity.[37]

Two consequences follow from the immediacy of this kind of similarity relation. First, elements of a similarity class formed by an immediate similarity relation are tied to one another by "family resem-

36. Kuhn expresses this kind of immediacy for similarity relations in a variety of ways. With regard to similarity relations between ways of viewing objects, or between objects, Kuhn rejects the demand for a similiarity-conferring third ("similar with respect to what?") as illegitimate (e.g. 1974a, *ET* p. 307; 1970b, p. 274; 1970c, *SSR* p. 200). Such similarity classes or the recognition of similarities is called "primitive" in virtue of the illegitimacy of this demand (1974a, *ET* p. 312; 1970b, p. 275; 1970c, *SSR* p. 200). The recognition of members of natural families as such is similarly called "immediate" (1970c, *SSR* p. 197 n. 14), and Kuhn asserts that the "integrity" of this cognitive process must be accepted (1974a, *ET* p. 313, 1970c, *SSR* p. 195). With regard to the immediate similarity relations which hold between problem situations, Kuhn speaks of the "direct inspection of paradigms" (*SSR,* p. 44) or "direct modeling" (*SSR,* p. 47; 1974a, *ET* p. 308), which need not be preceded by an answer to the question "similar with respect to what?" (1974a, *ET* p. 308; 1977b, *ET* p. 17). In his work since 1979, in which the linguistic aspect of world-constitution bears more weight, Kuhn claims that scientific terms ordinarily "are introduced and thereafter deployed . . . [without] acquiring a list of criteria necessary and sufficient to determine the referents of the corresponding terms" (1979b, p. 409). The similarity is "inexplicit" (ibid.), and we have no explicit definition for the implicated terms. The same issue is addressed (though not with the same "reference determination" terminology) in the 1969 papers, e.g. in 1974a, *ET* p. 312. —See Kamlah and Lorenzen 1967, p. 29.

37. *SSR,* pp. 42, 43, 44; 1977b, *ET* p. 17; 1974a, *ET* pp. 307, 313; 1970b, p. 275; 1970c, *SSR* pp. 190–191, 192, 194, 197; 1974b, p. 511; 1979b, p. 409. —We should here draw attention to a possible source of misunderstanding in Kuhn's texts. Kuhn uses the word "criterion" in both narrow and broad senses. In the narrow sense, "criterion" refers to some *defining trait;* in this sense, those similarity relations in which Kuhn is most interested are without recourse to any criterion. In its broad sense, this term refers to some feature of relata whose status as either a defining conceptual feature or an empirical feature is *indeterminate.* In this sense, the perception of a given trait can count as a criterion for placing a problem situation in a given similarity class. See, for example, 1974a, *ET* p. 308; here both narrow and broad uses of "criterion" occur in one sentence: "his basic criterion [in the broad sense —P.H.] is a perception of similarity that is both logically and psychologically prior to any of the numerous criteria [in the narrow sense —P.H.] by which that same identification of similarity might have been made." For uses of "criterion" in the broad sense, see e.g. 1964, *ET* p. 259; 1970c, *SSR* p. 197 n. 14; 1981, pp. 8, 18, 19, 20. Kuhn lays this source of possible misunderstanding to rest only in his 1983a, pp. 685–686 n. 13, where he states "'criteria' is to be understood in a very broad sense, one that embraces whatever techniques, not all of them necessarily conscious, people do use in pinning words to the world. In particular, as used here, 'criteria' can certainly include similarity to paradigmatic examples." For the same, broad use of "criterion," q.v. Shapere 1977, n. 4.

blances," in the late Wittgensteinian sense.[38] In other words, elements of the similarity class are not similar in virtue of their sharing of one general characteristic. We are rather dealing with

> a complicated network of similarities overlapping and criss-crossing: sometimes overall similarities, sometimes similarities of detail.[39]

In addition, such similarity classes turn out to be sets with fuzzy boundaries.[40] In other words, an immediate similarity relation doesn't determine membership in its similarity class for all conceivable cases, for all possible objects or situations.

Since similarity relations hold within a phenomenal world, it might be asked whether they're genetically object-sided or genetically subject-sided. Or, as Shapere puts the question to Kuhn, are similarity relations something that exists, and can be *discovered,* or are they something that has to be *invented*?[41] Where Kuhn's theory should be placed in the field spanned by such labels as "realism," "idealism," "subjectivism," "relativism," "conventionalism," and "nominalism" depends heavily on the answer to this question.

To begin with, it is easy to show that the similarity relations to be learned can't, for Kuhn, be purely genetically object-sided. If they were, they would stand in contradiction with central claims of Kuhn's theory, depriving talk of the subject-sided constitution of phenomenal worlds, and thus the plurality-of-phenomenal-worlds thesis itself, of all meaning. Are similarity relations, then, purely genetically subject-sided?[42] Surely not, in the sense of complete obedience to the whim of a scientific community or even of an individual. Confronted with the disjunction between subject-sided and object-sided origins, Kuhn replies,

> In [a certain] sense, learning a similarity relationship is learning something about nature that there is to be found. . . . But in another sense the group does put them [the similarity relations] there (*or finds them already there*), and what I want most to resist

38. *SSR,* p. 45; 1970a, *ET* pp. 285, 286 n. 35; 1974a, *ET* p. 312; 1970c, *SSR* p. 194; 1979b, pp. 410–415; 1981, pp. 10, 20.

39. Wittgenstein 1953, Part I, no. 66 (trans. G.E.M. Anscombe).

40. This holds for Kuhn as well as for Wittgenstein; Wittgenstein 1953, Part I, no. 71; Kuhn 1970a, *ET* pp. 285, 287 n. 37; 1974a, *ET* pp. 316–318 n. 21; 1983b, p. 715. —Cf. Kamlah and Lorenzen 1967, p. 47.

41. Shapere in Kuhn 1974b, pp. 506–507.

42. This would entail a "subjective-idealist" or "subjectivist" position of the kind attributed to Kuhn by Scheffler 1967, p. 19.

> about the question is the implication that it must have a yes or no answer.[43]

This passage might be taken to imply that scientific communities are, in their free choice of similarity relations, constrained only by previous choices and not by genetically object-sided resistance. This, however, is not Kuhn's view. Kuhn rejects it, without using the term "similarity relations," in a passage of *SSR* we are not yet ready to interpret fully. Here he calls normal science "an attempt to *force* nature into the preformed and relatively inflexible box that the paradigm supplies."[44] But this attempt may fail, for "nature cannot be forced into an arbitrary set of conceptual boxes."[45] This resistance of nature is also emphasized in positing the world-in-itself and in setting forth the stimulus ontology; the world-in-itself codetermines phenomenal worlds in virtue of its own proprietary features, and stimuli are similarly granted their own being and causally efficacious proprietary features.[46] It follows that the immediate similarity relations have *both* a genetically object-sided *and* a genetically subject-sided moment, though the two are inseparable.[47]

In *SSR,* the interaction of genetically object-sided and genetically subject-sided moments in scientific knowledge is described only in very general terms, without reference to similarity relations. Here Kuhn makes use of the opposition between (genetically object-sided) "observation and experience" and a (genetically subject-sided) "element of arbitrariness":

> Observation and experience can and must drastically restrict the range of admissible scientific belief, else there would be no sci-

43. 1974b, p. 509, original emphasis. —Compare Goodman 1975, p. 22, and Putnam 1981, p. 54.

44. *SSR,* p. 24, emphasis mine; similarly pp. 5, 151–152; 1970a, *ET* pp. 270–271; 1970b, pp. 260, 263; also 1961a, *ET* pp. 200–201. The "relatively inflexible box" is the conceptual system imposed by way of similarity relations and paradigm examples; see § 3.6.

45. 1970b, p. 263; the entire passage from which this clause is taken emphasizes the resistance of nature. —In anticipation of the constitutive role of similarity relations in perception, to be discussed in § 3.5, we might also consider the following passage. Scientists sundered by a revolution "see different things when they look from the same point in the same direction. Again, that is not to say that they can see anything they please" (*SSR,* p. 150). Kuhn repeats this claim in his postscript: "To say that the members of different groups may have different stimuli is not to imply that they may have just any perceptions at all" (1970c, *SSR* p. 195).

46. Compare §§ 2.1.a, 2.2.b, and 2.2.c.

47. This view of similarity relations differs from that of the "Edinburgh School" in the sociology of science, for whom similarity relations are genetically entirely subject-sided; see Barnes 1982, especially §§ 2.2. and 2.3. —See also Putnam 1981, p. 54.

> ence. But they cannot alone determine a particular body of such belief. An apparently arbitrary element, compounded of personal and historical accident, is always a formative ingredient of the beliefs espoused by a given scientific community at a given time.[48]

Kuhn's conception of similarity relations as, in part, genetically object-sided is doubtless his reason for classifying his view as a *realist position.* Of both Popper and himself he claims, for example,

> we both insist that scientists may properly aim to invent theories that *explain* observed phenomena and that do so in terms of *real* objects, whatever the latter phrase may mean.[49]

He also claims, analogously, of Boyd and himself that "we are both unregenerate realists,"[50] though adding a qualifying disclaimer to the effect that he has not yet worked out the consequences of his realism. Boyd himself, however, describes Kuhn's view as an antirealistic position, or, more precisely, as "constructivist antirealism,"[51] for it claims that

> the world that scientists study, in some robust sense must be defined or constituted by or "constructed" from the theoretical tradition in which the scientific community in question works.[52]

In his most recent work, Kuhn admits that his position can't be classed as unqualifiedly realist. On the contrary, it constitutes a "threat to realism,"[53] since it insists that similarity relations contain a genetically subject-sided moment and hence that a world constituted by such

48. *SSR,* p. 4; similarly 1963b, p. 393. I take "apparently" in this passage to mean *"obviously,"* not *"seemingly,"* for Kuhn begins the two paragraphs following this passage with references to this "element of arbitrariness," whose presence is now taken for granted. See also *SSR,* p. 76, where Kuhn alludes to the well-known fact "that more than one theoretical construction can always be placed upon a given collection of data."

49. 1970a, *ET* p. 267.

50. 1979b, p. 415.

51. Boyd 1984, pp. 51–58.

52. Boyd 1984, p. 52. —Hacking, by contrast, suggests that Kuhn's position be characterized by means not of the opposition between *idealism* and realism, but rather of the older distinction between *nominalism* and realism, where Kuhn's view is best described as "revolutionary nominalism" (Hacking 1984, p. 117; also Hacking 1983, pp. 108–111; his reasons will become clear when we discuss the constitutive role of immediate similarity relations in concept formation [§ 3.6] and in determining the characteristics of scientific revolutions [chap. 6]). One might reply to Hacking that Kuhn's position ought to be placed *between* nominalism and realism, as the similarity relations of which concepts are constituted contain both genetically subject-sided *and* genetically object-sided moments.

53. 1989a, p. 23; 1990, p. 317 n. 21.

relations can't be, as realism would have it, purely object-sided.[54] Kuhn puts off for a later date confronting the urgent problem that arises for the realist perspective from the presence of genetically subject-sided moments.[55]

3.3 Ostension

The formation of immediate similarity relations, according to Kuhn, is a learning process in which ostension and its subspecies, ascription and exclusion, play a great role, hence a process containing not only linguistic moments but also an important extralinguistic moment.[56] Kuhn explains in detail the role played by ostension in learning similarity relations which hold between *objects*.[57] This learning process features an instructor and a pupil.[58] In the presence of the pupil, the instructor ostends different members of a similarity class already established (for the instructor), ascribes them to the similarity class, and then ostends nonmembers of the similarity class and denies them membership. Ascription to and exclusion from a given similarity class does not proceed by listing defining features of the class, which would allow the pupil to recognize the similarity of individual members by reference to such defining characteristics. Instead, the class is identified by the mere pronouncement of its name. The pupil must thus learn the targeted similarity without any specification of the *defining* features of the similarity class, instead attempting to repeat the instructor's procedure, receiving confirmation for correct ascriptions and exclusions and correction for wrong ascriptions and exclusions. As a matter of empirical fact, learning immediate similarity relations in this way seems possible; that is, the pupil can form the same similarity class as the instructor without the help of mediating criteria to make the similarity explicit or furnish any *definition* of the similarity class.[59]

54. Or, in Kuhn's own words, "the world itself must be somehow lexicon dependent" (1989a, p. 24). I will return to the notion of a lexicon in §§ 3.6.g and 4.4.a.

55. 1989a, p. 24.

56. Kuhn describes the process of ostension as "non-verbal or incompletely verbal," "not fully linguistic," or "non-linguistic" in 1970b, pp. 270, 271; similarly 1970c, *SSR* p. 191. The importance of ostension is emphasized in 1974a, *ET* p. 309, where Kuhn calls it a "primary pedagogic tool"; similarly 1970a, *ET* p. 285; 1970b, p. 274; 1970c, *SSR* p. 193; 1974b, pp. 503–506; 1979b, pp. 411–414; 1983a, p. 680; 1989a, p. 16; 1990, p. 302. —See Kamlah and Lorenzen 1967, p. 27ff.

57. 1974a, *ET* pp. 308–309; 1981, p. 20.

58. I will return to the roles of instructor and pupil in the next section.

59. 1974a, *ET* p. 309; 1979b, pp. 412–413. I will shortly return to discuss the conditions which must be met in order for such learning to occur. —See Kamlah and Lorenzen 1967, p. 29.

Ostension also plays a decisive role in learning those immediate similarity relations that hold between *problem situations,* though here Kuhn's view softens somewhat;

> the same technique [employed in learning immediate similarity relations between objects], if in a less pure form, is essential to the more abstract sciences as well.[60]

The process here is "ostension or some elaboration of it,"[61] for it should be obvious that simply extending one's finger isn't enough to refer to a problem situation.[62] In his latest work, Kuhn remarks that mere description may suffice to refer to a problem situation, without the actual physical presence of that situation.[63]

Which objects or problem situations must be ostended in order for immediate similarity relations to be learned? This question is answered by *paradigms* (in a certain sense),[64] which somehow constitute the fixed points in the net of similarity and difference relations at the heart of world constitution. But the notion of a paradigm and its development in Kuhn's work requires thorough discussion, which would hold up our present analysis of world constitution. The discussion of paradigms as the objects of ostension must thus be put off until chapter 4.

Kuhn operates under the assumption that there are such processes, not exclusively linguistic but rather, as we shall see in § 3.6, in a certain sense language-grounding, of ostension, ascription, and exclusion. These processes can be demonstrated, understood, and imitated.[65] This assumption is accepted by Kuhn as a fact, and it is explored with a view toward its functions and consequences; the fact itself, however, is given no further analysis.[66] But such analysis is of importance to Kuhn's theory in at least three respects.

1. According to Kuhn, immediate similarity relations are formed not only in the domains of objects and problem situations but also in the domain of individual perceptions belonging to a single object.[67]

60. 1974a, *ET* p. 313.

61. 1970b, p. 271; similarly 1974b, p. 504; 1979b, pp. 413–414.

62. I will further discuss such problems as affect even deictic reference to visible material objects later.

63. 1989a, p. 16; 1990, p. 302.

64. 1970a, *ET* pp. 284–285; 1970b, p. 271; 1974a, *ET* pp. 318–319.

65. This point, according to Kuhn, is the locus of contention between his theory of science and Popper's: "If Sir Karl and I have a fundamental philosophic dispute, it is about the relevance of this last mode of language-nature learning [by ostension] to the philosophy of science. . . . I believe he misses a central point, the one which led me to introduce the notion of paradigms in my *Scientific Revolutions*" (1970b, p. 271).

66. This is particularly evident in 1981, p. 20.

67. See § 3.2.

The example Kuhn uses to illustrate this point involves the individual perceptions of a mother by her child, whose world as yet contains no persons, conceived as persisting individuals. According to Kuhn, these individual perceptions are linked by immediate similarity relations to form perceptions of one and the same person.[68] This example addresses ontogenetically very early processes of object construction—in this case, the first construction of a person. But it's hardly clear how ostension could be understood when the one who's supposed to understand lives in a perceptual world containing, as yet, no discrete persons. For, in general, to understand an act of ostension presupposes perception of the ostending person and an understanding of that person's intention. Precisely this condition isn't met in the case at hand, where similarity relations have yet to construct discernable individual persons.[69] The role played by ostension in generating those similarity relations responsible for the ontogenetically early processes of person construction is thus opaque.

2. The case just discussed makes it clear that in order to understand ostension, ascription, and exclusion a pupil must meet certain prerequisites. These divide into three groups.

First, the pupil must be able to understand the instructor's acts of reference as such. So, for example, when an extended finger is used to show the pupil something, the pupil must understand this gesture as pointing. If, instead, the pupil sees only the hand, registering only its physical motion, he or she must necessarily fail to appreciate an act of reference to that at which the finger is pointed.[70]

68. 1970b, p. 274; 1970c, *SSR* p. 194.

69. One tempting reply is that ostension, in this case, should be understood not as literally an act of pointing but rather as drawing the child's attention by sounds and touch. The parallel to the other cases discussed above would then consist in the absence of any verbal presentation of defining characteristics on the part of the instructor. But how could the instructor then ascribe the relevant individual perceptions to the same similarity class? Is this part of the learning process really the product of instruction?

70. See Friedmann 1981, p. 48. —According to Giehl 1969, the ability to point is specifically human; indeed, human beings might even "be defined as the pointing creatures" (p. 53 [My translation. —A.T.L.]). If this is right, Holenstein's explanation for the understanding of pointing is highly unsatisfactory: "It is conceivable, that a mere index finger can be understood immediately . . . by virtue of an indwelling tendency, longitudinal and unidirectionally attenuating, of objects to gesture beyond themselves toward some possible continuation" (Holenstein 1980, p. 23 [My translation. —A.T.L.]). Whatever this "tendency to gesture beyond" might be, if it were something "indwelling" in physical objects, understanding pointing would require no prior treatment of specifically human traits. —Ontogenetically speaking, pointing appears to me to derive from grasping; on this issue, I'm in agreement with Holenstein. It begins with the infant's grasping reflex. Next, things within reach are grasped under visual supervision. Finally, the child attempts to grasp objects beyond its reach, leaning and extending arm and

Second, such acts must be understood not just as reference but as reference to something in particular. This further understanding is clearly problematic even for pointing. At what is the extended finger pointing? What is the "end point" of this pointing?[71] Mere pointing doesn't determine whether what's meant is some object, some view or attribute of the object, some group of objects or the relations between them, or simply the direction of pointing itself. Accordingly, Kuhn, in the example in which the similarity classes of swans, geese, and ducks are learned, presupposes that the pupil has already learned to recognize birds.[72] Pointing must evidently achieve closure by the prior determination of the domain out of which an act of pointing is to select something. But if understanding every act of pointing as an act of pointing to something in particular depends on the prior understanding of a domain specification, then the possibility of intelligible pointing must presuppose a mastery of the most general domain specification, which, for its part, can't be learned by pointing. These most general domain specifications are *categories,* such as *thing, attribute, relation,* and the like. Kuhn must thus presuppose that both instructor and pupil understand a set of categories that in turn make particular acts of pointing a possible means for learning immediate similarity relations.

A third group of prerequisites emerges when we consider that objects and situations must be not only ostended but also ascribed to and excluded from the appropriate similarity classes. An understanding of ascription and exclusion presupposes that the pupil must, in some sense, already know what to look for: the similarity, in which the particulars coincide.[73] In addition, the pupil must also understand negation in order to understand the exclusion of particulars from similarity classes.

3. These prerequisites are interesting not just for what they say about the abilities a pupil must come with in order to learn immediate similarity relations by ostension, ascription, and exclusion. For the same prerequisites imply conditions that must be fulfilled by any phenomenal world, if it is to be constituted by this process of instruction.[74]

---

hand. Intentionality toward the momentarily ungraspable object is now fully developed. If the actual desire to grasp is lacking, pointing remains. —My thanks go to Sarah Hoyningen-Huene for supplying empirical data on this topic.

71. On this issue, see Wittgenstein 1953, no. 28ff.

72. 1970a, *ET* p. 285; 1974a, *ET* p. 309. Learning the vocabulary of Newtonian mechanics presupposes, analogously, familiarity with certain other concepts: 1989a, pp. 15–16, 1990, p. 302.

73. This issue should be familiar from Plato: *Phaedo* 72e–77d, and *Theaetetus* 184b–186e.

74. On this issue, see Kant, *Critique of Pure Reason* A158/B197.

This implication is most obvious when we consider the second group of prerequisites. Every phenomenal world must be ordered categorically, for every act of pointing achieves closure only in virtue of a prior domain specification, which in turn presupposes the existence of a set of most general domain specifications, or categories, to sanction pointing. This condition shouldn't be taken to imply that all phenomenal worlds have the same categorial ordering, for though successful pointing allows us to infer the existence of some categorial ordering in a given phenomenal world, it doesn't determine the particulars of this ordering. When we consider whether there's a single, universal categorial ordering for all phenomenal worlds or rather different possible categorial orderings for different phenomenal worlds, serious questions of justification arise. For in the former case we would have to ask, by recourse to what might the universality of a single categorial ordering be established, while the latter demands an account of how any given categorial ordering might be acquired.

The first prerequisite for learning similarity relations by ostension, ascription, and exclusion required an ability on the part of the pupil to understand ostension *per se.* As far as phenomenal worlds are concerned, this amounts to the requirement that every phenomenal world contain people to ostend, ascribe, and exclude. But it's hard to say what this requirement implies about the categorial ordering of phenomenal worlds.

The third group of prerequisites concerned the similarity classes to be formed. With regard to phenomenal worlds, it requires that every phenomenal world contain not only a global categorial ordering but, in addition, similarities and differences within each category, giving rise to more or less precisely determined classifications of the objects in the phenomenal world.

So the aforementioned prerequisites lead to *endogenous constraints on Kuhn's plurality-of-phenomenal-worlds thesis.* The conditions that must be met in order for a phenomenal world to be constituted by learning immediate similarity relations entail absolute limitations on the possible diversity of phenomenal worlds so constituted. All phenomenal worlds must necessarily have a categorial ordering, classifications within the categories, and evidence of alien intentionality. Kuhn's theory leaves open whether the categories are universal and how they should be understood.

3.4 Social Community

As can be seen from the nonlinguistic nature of the ostension component in the learning of immediate similarity relations, the learning process, far from being an autonomous activity of the pupil, requires an instructor.[75] Indeed, this strict, rigid learning process rests to a large degree on the authority of the instructor, an authority the instructor has by virtue of membership in the appropriate community. No alternatives are even presented in the course of learning, let alone actually evaluated by either instructor or pupil.[76] The instructor transmits those similarity relations already mastered by the members of a given social community. "Social community" might, in accordance with the universal applicability of the learning process in producing phenomenal worlds, mean an entire culture, or smaller groups, all the way down to groups of scientific specialists consisting of fewer than twenty-five people.[77] Such "educational, scientific, or linguistic" communities are "relatively homogeneous," which means just that their members operate with essentially the same similarity relations.[78] These similarity relations are, in a sense, legitimated by their community,[79] which employs them over an extended period,[80] successfully.[81] Learning the established similarity relations of a given group can be understood as part of the socialization process in which group membership is acquired.[82]

Here the scientific community emerges, as stated earlier,[83] as the agent of scientific activity. It is not the scientist, conceived as an individual, who has similarity relations at his disposal; they are rather the property or attribute of particular communities, and the individual has access to them only *qua* member of the community. To belong to a given community means, among other things, to have mastered the same similarity relations as other members of the community.[84]

75. 1974a, *ET* p. 310. —This dependence on an instructor shouldn't be taken to exclude the possibility that learned similarity relations might be productively transformed without supervision, as is done by the vanguard of a scientific revolution. —See Kamlah and Lorenzen 1967, p. 27.

76. 1959a, *ET* pp. 227–230, 232, 237; *SSR,* pp. 5, 136–138, 165–166, 168–169; 1963a, pp. 350–351; 1974a, *ET* p. 312; 1990, p. 314.

77. 1970c, *SSR* pp. 181, 193.

78. 1974a, *ET* p. 309; similarly 1970c, *SSR* p. 193. In those works in which the linguistic aspect of world constitution takes center stage, the community's status as a linguistic community is naturally emphasized; see e.g. 1983a, p. 682; 1989a, pp. 14–15; 1990, p. 302.

79. 1974a, *ET* p. 306; 1970c, *SSR* p. 189; 1974b, p. 509.

80. 1970a, *ET* p. 285; 1970c, *SSR* p. 189.

81. We are not yet ready to explain what "successfully" means here, for we must first relate the function of similarity relations in scientific knowledge; see § 3.7.

82. *SSR,* pp. 10–11; 1970a, *ET* p. 291; 1974a, *ET* p. 313; 1970b, p. 275; 1970c, *SSR* p. 191; 1989a, p. 15; 1990, p. 302.

83. See § 1.1.b and chap. 3, introduction.

84. E.g. 1974a, *ET* p. 305; 1970b, p. 273; 1970c, *SSR* p. 189; 1989a, p. 15; 1990, p. 302.

3.5 Perception

As stated above,[85] the role of perception in world constitution recedes in Kuhn's work after 1969, without vanishing entirely. In this section we will thus primarily consider work completed up to 1969. Afterwards, we will be in a position to consider how the motives for Kuhn's perception theory reappear in the more linguistically oriented conception of the phenomenal world of his later theory of concepts.[86]

In order to understand why and how, according to Kuhn, immediate similarity relations are coconstitutive of perception, we must first examine an important feature of perception. Every completed act of perception is characterized by an as-structure,[87] as manifested by a perceptual act which identifies an object *as* a particular individual or *as* a member of a particular natural family.[88] By the as-structure of perception, a perceptual act can't count as completed if its object hasn't been identified as some particular object. We recall that, for Kuhn, "object" should be taken in a broad sense, not in the narrow sense of "material thing";[89] attributes of material things can also be objects of perception. The appeal of the as-structure thesis should now be clear, for an attempted perceptual act which results in complete indeterminacy, in no identification of its object *as* something, must be taken as a failure, an incomplete perceptual act. To complete a perception thus means to identify an object, by means of the senses, as some object in particular, either as a particular individual object or as a member of some particular kind of objects.

One possible misunderstanding should be preempted here. When we say that perception is characterized by an as-structure, we don't mean, taking visual perception as an example, that perception involves seeing something_1 as something_2; "Scientists do not see something *as*

85. See § 2.3.

86. See § 3.6.d.

87. The term "as-structure" is taken from Heidegger 1927, § 32, p. 149, on grounds of the substantive parallels between Kuhn's theory of perception and Heidegger's explanation of "understanding the world." —For other parallels (and differences) between Kuhn and Heidegger, see Rouse 1981.

88. Kuhn never explicitly formulates the thesis of the as-structure of perception, though he implicitly assumes it. This assumption is most clearly expressed in 1970c, *SSR* pp. 193–194: "One of the fundamental techniques by which the members of a group . . . learn to see the same things when confronted with the same stimuli is by being shown examples of situations that their predecessors in the group have already learned to see *as* like each other and *as* different from other sorts of situations" (emphasis mine).

89. Compare § 2.2.a.

something else; instead, they simply see it."[90] Kuhn is here rejecting a view of perception, suggested by the analogy to perceptual change in gestalt shift experiments, on which visual perception consists in seeing something_1 as something_2. In that case, perception would involve, in a certain sense,[91] the interpretation of some sensually given something_1, with something_2 as its product—just as the actual sensually given in gestalt shift experiments is a set of lines, which can be seen as different figures. But the something_1 that, allegedly, constitutes the "actual sensually given" is, in truth, *not perceptually accessible.* Only objects (in the broad sense), that is, something_2, are perceptually accessible, and they must be our starting point in attempts to find out anything about something_1. Access to something_1 can be gained, if at all, only through scientific theories.[92] In order to avoid this misunderstanding, we might describe the as-structure of perception as a nonrelational as-structure.[93]

Now, according to Kuhn's plurality-of-phenomenal-worlds thesis, object perception in a given external situation isn't fixed by the situation alone. But by the as-structure of perception, a completed perception consists in the unequivocal, sensually mediated identification of an object as some particular object. It follows that the purely object-sided components of perception, which fail to determine unequivocally the product of any perceptual act, must be supplemented by the genetically subject-sided components furnished by learned similarity relations.[94] It is by means of the latter that an act of perception identifies the perceived object as sufficiently similar to some previously known individual object or to the members of some previously known natural family.[95] This process of ascription occurs immediately, according to Kuhn, in the sense that it isn't mediated by reference to any defining characteris-

90. *SSR,* p. 85; similarly 1979c, p. ix.

91. I will return to this certain sense later.

92. See *SSR,* p. 129; 1974a, *ET* p. 308; 1970c, *SSR* p. 196; 1974b, p. 509.

93. See Heidegger 1927, p. 149ff.

94. We should here recall that similarity relations aren't *purely* genetically subject-sided but *also* genetically subject-sided; see § 3.2.

95. In *SSR,* this doctrine is formulated without any explicit reference to similarity relations but instead by reference to paradigms and visual-conceptual experience: "something like a paradigm is prerequisite to perception itself. What a man sees depends both upon what he looks at and also upon what his previous visual-conceptual experience has taught him to see. In the absence of such training there can only be, in William James's phrase, 'a bloomin' buzzin' confusion'" (*SSR,* p. 113). Kuhn makes a similar claim in 1964, *ET* p. 263 n. 33, with reference to Hanson as before. The coconstitution of perception by similarity relations is stipulated, for example, at 1970b, p. 274: "Until we have acquired them [the similarity relations], we do not see a world at all." Kuhn's theory of perception is presented in greatest detail in 1970c, *SSR* pp. 192–196; on p. 196 he asserts that he is dealing with a "hypothesis about vision."

tics of the particular object or group of objects.[96] Learned similarity relations can thus also explain why, despite the plurality of phenomenal worlds, unequivocal perception of objects occurs.

In order to explain how it is that learned similarity relations are, without recourse to the defining characteristics of similarity classes, coconstitutive of perception, Kuhn employs a metaphor borrowed from computer science.[97] The basic idea is that of some programmable and reprogrammable neural apparatus, situated between the organs that receive perceptual stimuli and perception itself, that is responsible for processing perceptual stimuli into actual object perceptions. Learning similarity relations can be understood as a way of programming, or reprogramming, this neural apparatus. The program, which steers the transformation of stimuli into perceptions, groups perceptions belonging to one similarity class closer together in perceptual space relative to their distance from members of other similarity classes. As a result, the members of a similarity class are seen as similar to one another and as dissimilar to members of other similarity classes. The program achieves this differential assignment of relative distances in perceptual space by accentuating certain of the objects' qualities while dampening others;[98] qualities specific to similarity classes are accentuated, qualities specific to differences within similarity classes are dampened. The learning process doesn't tell the program *which* qualities should be accentuated in perceptual space as specific to similarity classes; all that's told the program is that certain exemplary objects should be classed as similar or dissimilar. How, precisely, the program organizes perceptual space so as to reproduce the given similarities and dissimilarities doesn't matter, so long as it reproduces them. The program neither makes use of the defining qualities of similarity classes nor produces them; instead, it makes use of given similarities and dissimilarities and produces an accentuation of certain qualities of similarity classes, without thereby

96. The aforementioned immediacy of ascription to a similarity class doesn't rule out the possibility of discovering defining characteristics for the class *after* it's been formed. The claim is simply that such characteristics aren't constitutive of class formation and hence aren't constitutive of perception either. Successful acts of perception, whose completion rests on immediate similarity relations, are rather a prerequisite for the discovery of such characteristics. In order to discover the defining characteristics of a similarity class, one must first have learned to see its members as similar to one another (1970c, *SSR* pp. 194–195; also *SSR* p. 122; 1970b, p. 274; 1974a, *ET* p. 308; 1974b, p. 511). But see § 3.6.f.

97. For the following, see 1974a, *ET* pp. 308–312; 1970b, p. 276; 1970c, *SSR* pp. 194–197, 201, 204; 1974b, p. 511.

98. See *SSR* p. 125, for an early discussion of the highlighting of qualities in perceptual space.

designating them *defining* qualities and thus distinguishing them from merely empirical qualities.

Successful mastery of perceptually relevant similarity relations thus depends on the program's ability to organize perceptual space in such a way that members of the specified similarity classes cluster together, while keeping their distance from other such clusters.[99] Whether the program is successful in this surely depends in part on the program itself. But Kuhn further insists that the kind of stimuli plays a role:

> In a universe of random stimuli, no processing will create data clusters that endure, for the next stimulus is as likely to fall in empty space as in a cluster.[100]

Still, we can never know what the "nonrandom" distribution of purely object-sided stimuli is actually like. According to Kuhn, from the fact that *perceptual space* contains enduring natural families, we can conclude, and must conclude, that the purely object-sided *is* also differentiated.[101] But as our access to the object-sided as such is blocked, any statements made about it from the scientific standpoint must necessarily import genetically subject-sided moments.[102]

This account of the transformation of perceptual stimuli to perceptions raises a series of questions. First, the notion of a purely material mediation between stimuli and perceptions by the nervous system seems to suppress the urgent philosophical problem of the transition from the purely material sphere of stimuli to the sphere of consciousness.[103] Second, one might ask whether the account proffered accords with findings in the relevant special sciences. Third, one might ask what it means for Kuhn's characterization of the transformation of perceptual stimuli into perceptions by a computer-like, programmable nervous system to contain such a strong metaphorical component, though the account is meant literally.

These questions are intrinsically interesting, but we may safely ignore them. For what matters to Kuhn, above all, is that similarity relations are coconstitutive of perception without any mediating influence from the defining characteristics of their relata. This emphasis emerges from the following, primarily negative description of perception. For Kuhn asserts that we shouldn't attempt "to analyze perception

99. 1974a, *ET* p. 312 n. 20; also 1970a, *ET* p. 286 n. 35; 1974a, *ET* p. 318 n. 21; 1970c, *SSR* p. 197 n. 14.

100. 1974b, p. 509.

101. Compare § 3.2.

102. Compare § 2.2.a.

103. See Dreyfus 1979, especially chap. IV.

as an interpretive process, as an unconscious version of what we do after we have perceived."[104] In his own view, this claim sets Kuhn apart from the received epistemological tradition which began with Descartes.[105] In speaking of a change in one's perception of an object, for example, Kuhn would describe a change in perception *itself*—contingent on a change in similarity relations constitutive of perception—whereas the received tradition would describe a change in the *interpretation* of the sensually given, that which all normal observers observe in an identical manner.[106]

In order to understand Kuhn's goal, we must first ask what "interpretation" means here.[107] The following features of the relation "something$_2$ interprets something$_1$" may be extracted from Kuhn's texts:

1. Something$_1$, that which is to be interpreted, is fixed vis-à-vis the process of interpretation and its product.[108]
2. Something$_1$, that which is to be interpreted, is accessible not only as a whole but also in pieces.[109]
3. The process of interpretation is itself a "deliberative process by which we choose among alternatives" and seek and apply "rules and criteria."[110]
4. Something$_2$, the product of the process of interpretation, is linked "logically or piecemeal" to something$_1$.[111]

All of these features of interpretation fail, according to Kuhn, to apply to perception.[112] For what, for purposes of perception, would be the something$_1$ which remains fixed with regard to different possible interpretations (feature 1)? It certainly couldn't be anything immediately given in perception, if we follow Kuhn's perceptual theory. For the immediate product of acts of perception is object identification, and since this identification depends on learned similarity relations, it isn't fixed in the desired sense. The fruitlessness of all attempts to date to

104. 1970c, *SSR* p. 195; similarly 1979c, p. ix.

105. *SSR*, pp. 121, 126; 1970c, *SSR* p. 195. —Compare § 2.1.b.

106. *SSR*, p. 120.

107. The notion of interpretation to be explicated here doesn't necessarily coincide with the notion of text interpretation, since Kuhn models his use of language on that of scientists, who speak of the interpretation of observations, experimental results, and the like. On this issue, see citations in the footnotes which follow.

108. "Fixed": *SSR*, pp. 120, 122, 126, 128; "stable": *SSR*, pp. 121, 125; "unequivocal": *SSR*, p. 126.

109. "Piecemeal": *SSR*, p. 123; "individual": *SSR*, p. 121.

110. 1970c, *SSR* pp. 194, 195; similarly 1974b, p. 511.

111. *SSR*, p. 123.

112. For the following discussion, see *SSR* pp. 120–129; 1970c, *SSR* pp. 194–195; 1974b, p. 511.

construct pure observation languages, designed to articulate that domain of perception claimed to be entirely free from theory, hence independent of subject-sided variation, is further evidence for this point. Accordingly, different modes of perception can't be characterized as differently linked, as wholes or piecemeal, with elementary perceptual data, either (features 2 and 4). Changes in perceptual mode are rather comparable to gestalt shifts, in which the perceived is transformed as a whole. Finally, different modes of perception can't be adequately understood as resulting from a change in the rules or criteria employed within them (feature 3). Applying the terms "rule" and "criteria" to perception suggests that we have such freedoms as the freedom to break a rule, to misapply a criterion, or to experiment with alternative rules or criteria. But this suggestion doesn't do justice to the experiential quality of perceptual acts, on the grounds that completed perceptual acts usually don't allow their potentially variable subject-sided moments to rise to the conscious level. Instead, the product of a perceptual act appears fully determined by the perceived object. Perception tells us, so to speak, the following: because the object is *one* particular object, so it gives rise to *one* particular perception.[113] In addition, according to Kuhn's theory, perception doesn't work with rules and criteria conceived as defining characteristics of the perceived objects but rather works with similarity classes given by exemplars.

Yet Kuhn by no means wishes to deny that interpretation *of* observations and data plays a central role in science.[114] Certainly such observations and data are, in science, evaluated or examined (for some expected feature, for example) by means of general criteria or rules (feature 3). But this activity factually and temporally presupposes determinate perceptions and is further distinguished from perceptual processes by both means and ends.

Two additional aspects of Kuhn's theory of perception still demand exposition.

1. Immediate similarity relations are, according to Kuhn, coconstitutive of perception; without them one could have no particular perceptions, hence no perceptions at all. This thesis, however, contradicts itself unless properly relativized, for the learning process in which similarity relations are mastered, as Kuhn has it, is itself dependent on the completion of certain perceptions. For example, in order for some individual object to be constructed by the immediate similarity relations holding between individual visual perceptions, these perceptions

113. As suggested by the German word for perception, "*Wahrnehmung,*" which decomposes into "*wahr-*" ("true") and "*-nehmung*" ("taking").

114. *SSR,* p. 122; 1970c, *SSR* pp. 194–195, 197–198.

must already be sufficiently determinate. Similarly, the auditory perception of the names of similarity classes must be sufficiently differentiated, in order for such names to be identified and recognized when heard.[115] It follows that not *every* perception is coconstituted by learned immediate similarity relations. There must rather be a certain domain of perceptions, entirely unlearned, that sanctions the learning of immediate similarity relations in the first place. Learning *different* modes of perception is possible only if this domain of unlearned perceptions is presupposed. How this domain should be characterized is, at least in part, an empirical question beyond the scope of our present inquiry.[116] What is of interest, however, is the fact that the existence of such a domain entails a further endogenous constraint on Kuhn's plurality-of-phenomenal-worlds thesis, though the precise nature of this constraint can't be determined here.[117]

2. We must now ask how Kuhn justifies the central tenet of his theory of perception, which stipulates that those similarity classes which are coconstitutive of perception aren't formed with the help of any defining characteristics of their members. In addition to the "plausibility" Kuhn claims on behalf of his theory of perception,[118] and such indirect support as it gains by avoiding the weaknesses of the interpretation theory of perception, Kuhn employs two further strategies.

For one, computer simulations allow us to model the two methods for forming perceptually relevant similarity classes and then to evaluate their relative similarity to actual processes of perception. The first simulation involves a program for transforming perceptual stimuli into perceptions, in which clustering of perceptual space is accomplished by given transformation rules. In the second case, the program is given, instead of transformation rules, a desired clustering of simulated perceptual space, analogous to the set of similarity classes formed in an actual learning process; the program's task is thus to form its own adequate transformations. Though Kuhn himself attempted computer simulations,[119] he apparently never got beyond "the early stages of

115. See Holenstein 1980, p. 19, and Eimas 1985.

116. On this issue, see Holenstein 1980, in which references to relevant empirical literature are given.

117. Compare § 3.3. —This strategy is the same as that employed in Mandelbaum 1979, especially p. 403. Unfortunately, I only became aware of Mandelbaum's essay shortly before this book went to press.

118. 1974a, *ET* p. 309.

119. The most detailed reports of this experience are found at 1974a, *ET* p. 310, and 1970b, pp. 274–275. The simulation is mentioned in 1969b, p. 944; 1970c, *SSR* pp. 191–192 and 197; 1974b, pp. 508–509 and 511.

such an experiment";[120] at least he never published anything beyond the rough description of his proposed simulation.

Another strategy suggests that Kuhn's is a "hypothesis about vision which should be subject to experimental investigation though probably not to direct check."[121] But, to be sure, Kuhn doesn't place great stock in the prospects for this kind of scientific confirmation, his reservations in this regard perhaps being one factor leading to his shift toward a more linguistically, rather than perceptually, oriented conception of the phenomenal world.[122] For, says Kuhn, "translation," the examination of a particular relationship between languages,

> may . . . provide points of entry for the neural reprogramming that, however, inscrutable at this time, must underlie conversion.[123]

As we shall see, Kuhn's linguistically oriented conception of the phenomenal world better fulfills the intention which motivated his theory of perception and in doing so avoids the latter theory's problems of justification.[124]

3.6 Empirical Concepts

a. Preliminary Remarks

Before explaining Kuhn's theory of the role of immediate similarity relations in learning empirical concepts, I must make three preliminary remarks.

First, Kuhn only begins explicitly to work out the role of immediate similarity relations in learning empirical concepts in his essay "Logic of Discovery or Psychology of Research?" (1970a). What *SSR* merely hinted at is now extracted from its close association with other issues and explicated by means of immediate similarity relations. These efforts also mark the beginning of Kuhn's distancing himself from a traditional concept of meaning. As Kuhn's theory about the learning of empirical concepts is unintelligible without a grasp of his view on the concept of meaning, we must begin by considering the development of this view.

120. 1974a, *ET* p. 310.
121. 1970c, *SSR* p. 196.
122. Compare § 2.3.
123. 1970c, *SSR* p. 204; for the notion of conversion, see § 7.5.e.
124. See § 3.6, especially 3.6.d.

In *SSR,* Kuhn uses "meaning" colloquially, without making an issue of the term. Concepts and words have meanings, meanings which determine both what belongs to a concept's extension and what doesn't, and also what elements of this extension are like.[125] But still—and here we find the germ of Kuhn's departure from the traditional theory of meaning—meanings generally aren't, in scientific practice, fully specified by definitions. For when, in science, an empirical concept is explicated at all, the result generally isn't a complete definition capable of determining how the concept should be used in a given community. This form of explication rather serves an auxiliary function, especially of the pedagogical variety; the meaning of an empirical concept is fully realized only through the concept's relationships with other concepts and practical procedures and through its use in applying theories.[126] The unequivocal use of empirical concepts in the scientific practice of a given community is guaranteed by other means than explicit definition.

A further reason, implicitly given in *SSR,* for the absence in scientific practice of definitions of empirical concepts is developed after 1962.[127] In general, so this line goes, an empirical concept *can't in principle* be given an explicit definition sufficient to determine the concept's application in all conceivable cases.[128] This fact shouldn't be attributed to any vagueness in the meanings of empirical concepts, to some deficit in specificity. The meanings of empirical concepts are rather not the sorts of things susceptible to explication by sets of necessary and sufficient conditions determining their application. Furthermore, the argument proceeds, neither he, Kuhn, nor anyone else, could ever go beyond this negative claim to state positively what exactly should be understood by "meaning."[129] Thus, in his work since 1962, Kuhn has often avoided the word "meaning," replacing it with more or less vague expressions, as in the "conditions of applicability" of terms, "criteria which govern their use," and the like.[130] Such manners of speaking are legitimated by the fact that, in its present state, the received theory of meaning can't offer any tangible distinction between a word's meaning and its mode of use, anyway.[131]

125. This follows indirectly from Kuhn's discussion of meaning change in *SSR;* see especially *SSR,* pp. 101–102, 128–129, 149–150.

126. *SSR,* pp. 46–47, 141–143.

127. *SSR,* chap. 5; on this issue, see § 4.1.

128. See § 3.6.f.

129. 1974b, pp. 506, 516.

130. E.g. 1964, *ET* pp. 259, 260; 1974a, *ET* p. 316; 1970b, p. 266; 1970c, *SSR* pp. 188, 198.

131. 1970b, p. 266 n. 2.

It follows from his approach to meaning that we shouldn't expect Kuhn's exposition of the role played by similarity relations in the learning of empirical concepts to treat them as constitutive of meaning in the traditional sense, meaning which might be explicated in a definition. Nonetheless, in the subsection which follows, 3.6.b, I will begin with Kuhn's theory of the learning of empirical concepts without first canvassing what change is in store for the concept of meaning. This approach accords with the development of Kuhn's thought; the articulation of his views on the learning of empirical concepts began without any reflection on the concept of meaning, and such consequences as these views had for the concept of meaning were drawn only over the course of his learning theory's further development.

Second, Kuhn's theory of the learning of empirical concepts presupposes a distinction that, despite the cogency of its purpose, has proved very difficult to articulate within analytic philosophy of science. This distinction, initially, separated "theoretical" from "observational" concepts. Observational concepts were meant to refer, free of all theory, to the pure sensually given, while theoretical concepts were to acquire their meanings exclusively through correspondence rules linking them with the already meaningful observational terms.[132] Kuhn was always critical of this particular articulation of the desired distinction.[133] Indeed, he saw in the "artificiality [of the distinction between fact and theory] . . . an important clue to several of [*SSR*'s] main theses,"[134] and thus to his philosophy of science. Nonetheless, Kuhn never denied the legitimacy of this articulation's *purpose,* insofar as this was to draw attention to the fact that science addresses itself not only to things accessible to the unaided (if not untrained) eye but also to things not immediately visible, the scientific treatment of which thus requires some sort of contribution by scientific laws or theories.

In *SSR,* reliance on this distinction is usually masked by the fluid transition from the literal use of "to see" to a metaphorical use, which Kuhn explicitly acknowledges in *SSR* only by hinting at an "extended use of 'perception' and 'seeing.'"[135] Since 1969, Kuhn's work has made explicit use of the distinction,[136] and we find him speaking of "basic"

132. On this issue, see e.g. Stegmüller 1970, part C; Suppe 1974, pp. 66–109.

133. *SSR,* pp. 7, 52–53, 66; 1962d, *ET* p. 171; 1970a, *ET* p. 267; 1974a, *ET* pp. 300, 302 n. 11; 1974b, p. 505; 1979b, p. 410.

134. *SSR,* p. 52.

135. *SSR,* p. 117. —In later work, Kuhn increasingly distances himself from this smooth transition between literal and metaphorical uses of 'seeing' (1964, *ET* p. 263 n. 33; 1970c, *SSR* pp. 196–198; 1983a, p. 669), in part because this gloss implicitly equates scientific revolutions with visual gestalt shifts (1983a, p. 669), an equation Kuhn later criticizes; see Kuhn 1989b, p. 50; see § 6.2.

136. See especially 1974a and 1979b.

and "theoretical" concepts, and of "theoretical" or "unobservable" entities.[137] Of course, Kuhn doesn't mean his observational and theoretical concepts to be those implicated in the rejected formulation of the distinction, but rather those corresponding, on the one hand, to "terms that are ordinarily applied by direct inspection," and on the other, to terms where "laws and theories also enter into the establishment of reference."[138] But Kuhn has reservations about this distinction, too, though he doesn't tell us why.[139]

So in what follows I will not make use of any distinction between theoretical and observational concepts, preferring the distinction, made in Kuhn's texts, between different *ways of learning concepts*. Although some empirical concepts are learned while employing laws or theories, such that these laws or theories play a role in learning to use the concepts, the use of other concepts is learned by "direct" perception of the objects to which they apply or fail to apply.[140] It is questionable whether this distinction between different ways of learning concepts corresponds to any (more or less fluid) distinction between two classes of empirical concepts, for presumably some empirical concepts can be used both in applying theory and in direct perception.[141]

Third, our discussion in this section must be sensitive to Kuhn's shift from a more perceptually oriented conception of the phenomenal world to a more linguistically oriented conception.[142] Such care is especially called for in treating those ways of learning empirical concepts in which no use is made of laws or theories. Accordingly, I will discuss the two approaches to these ways of learning concepts, which correspond to the two phases in the development of Kuhn's theory, separately (§§ 3.6.b and 3.6.c) and then consider the relationship between the two conceptions (§ 3.6.d). A discussion of concept learning by application of laws or theories sensitive to the differences between the two conceptions isn't necessary, however, for Kuhn's treatment of this issue doesn't change after 1979 (§ 3.6.e). Next, I will consider the doctrine of the impossibility of defining empirical concepts explicitly (§ 3.6.f), which is of central importance to Kuhn's theory of scientific development and which has consequences for his theory of meaning as it applies to empirical concepts (§ 3.6.g).

137. E.g. 1974a, *ET* pp. 300, 302 n. 11; 1970c, *SSR* pp. 196, 197 n. 14; 1977c, *ET* p. 338.

138. 1979b, p. 412.

139. 1979b, p. 410; q.v. 1981, pp. 21–22 n. 2.

140. Of course, this shouldn't be taken to imply that the latter sort of concept refers to the purely factual.

141. 1976b, pp. 186–187.

142. Compare § 2.3.

b. Concept Learning without Use of Laws or Theories in Kuhn's Work up to 1969

On the perceptually oriented conception of the phenomenal world, concept learning in which no use is made of laws or theories is an immediate consequence of Kuhn's theory of perception. Learning those immediate similarity relations that are coconstitutive of perception leads directly to learning certain empirical concepts.[143] The extension of such a concept is just a set of objects located sufficiently close to one another in perceptual space to be perceived as similar to one another. If someone learns a similarity relation constitutive of perception, together with the requisite dissimilarity relations and an appropriate concept name, that person can then apply or refuse to apply this concept to objects in the same way as other members of his or her community; his or her "correct" application of the concept is learned. The unproblematic use and transmission of such concepts within a community evidently depends, not on learning definitions, but only on the ability to master the basic immediate similarity relations; it must be possible to see the relata of these relations as similar. The meanings (whatever they might be) of such concepts can thus be located, at least preliminarily, in similarity relations, not in definitions.

The dependence of successful concept use in a community on successful object perception seduces Kuhn, in *SSR,* into manners of speaking seemingly at odds with his theory of perception. Thus Kuhn asks, with Wittgenstein,

> What need we know . . . in order that we apply terms like 'chair,' or 'leaf,' or 'game' unequivocally and without provoking argument [from other members of our community]?[144]

This question has both a negative thrust, the rejection of defining characteristics as a necessary condition for unequivocal use of terms within a linguistic community, and a positive thrust, toward Wittgensteinian family resemblances. But Kuhn wishes to go beyond Wittgenstein, and he writes in his footnote to the sentence cited above:

> Wittgenstein, however, says almost nothing about the sort of world necessary to support the naming procedure he outlines.[145]

So what is "the sort of world necessary" for this purpose? According to Kuhn, this would be the kind of world in which

143. 1970a, *ET* pp. 284–286; 1974a, *ET* pp. 309–318; 1970b, pp. 274–275.
144. *SSR,* pp. 44–45.
145. *SSR,* p. 45 n. 2.

> the families we named [never] overlapped and merged gradually into one another—only, that is, if there were . . . *natural* families.[146]

"World" can only refer here to some world accessible to language users, or, as Kuhn explicitly remarks in later passages, a phenomenal world:

> The possibility of immediate recognition of the members of natural families depends upon the existence, after neural processing, of empty *perceptual space* between the families to be discriminated.[147]

In such passages Kuhn speaks as though a particular kind of perceptual world, one in which there are discrete natural families, were a necessary condition for the possibility of a naming procedure which does its job without the help of the defining characteristics of its objects. But this intimation is at odds with other passages, as when Kuhn claims that the

> learned similarity-dissimilarity relationships . . . are parts of a language-conditioned or language-correlated way of seeing the world. Until we have acquired them, we do not see a world at all.[148]

Here the converse dependence is apparently asserted, that the "way of seeing the world" is importantly contingent on language.

But this apparent contradiction is easily resolved, once the scope of the opposing claims is appropriately narrowed.[149] Learned language surely can't be taken as an unqualified necessary condition for all perceptions, because language learning is itself dependent on a certain domain of prelinguistic perceptions. But nothing rules out the possibility that bits of language based on this domain of perceptions might differentiate and transform it, in turn giving rise for further differentiation and transformation of those initial bits of language. The relationship between language and perception thus shouldn't be thought of as either sort of unidirectional dependence but should rather be construed as a dialectical process of differentiation and transformation.

Motivated by the need to discern more precisely the endogenous constraints on Kuhn's plurality-of-phenomenal-worlds thesis, we must

146. *SSR*, p. 45.

147. 1970c, *SSR* p. 197 n. 14, emphasis mine; similarly 1970a, *ET* p. 286 n. 35; 1974a, *ET* p. 312 n. 20 and p. 318 n. 21.

148. 1970b, p. 274; compare § 3.5.

149. Compare this with our qualification of similarity relations' claim to be coconstitutive of perception, in § 3.5.

now ask under what conditions this process can be set in motion.[150] The process surely begins with perception, but we must better determine the precise perceptual capacities which make language acquisition possible. Two kinds of prerequisite perceptual capacity can be distinguished. For one, certain sensory powers of discrimination must be fully developed, especially in the visual and auditory domains.[151] In addition, the capacity to process sense impressions in certain ways must be fully developed, so that these sense impressions don't relate "solely to the subject as the modification of its state"[152] but may rather be taken as a counterpart (however described) of the perceiver.[153] For empirical concepts are to refer to things which act as common, perceivable counterparts for many observers, and subject-sided sense impressions must thus have been processed into object perceptions. What matters here isn't the possibility that different observers might, in accordance with the plurality-of-phenomenal-worlds thesis, see different objects in the same perceptual situation, but rather the fact that they all see some perceptual counterpart, some object.*

These two perceptual capacities, prerequisites for the process of acquiring empirical concepts, result in, among other effects, certain constraints being placed on any phenomenal world constituted by means of such empirical concepts. For one, such a phenomenal world must necessarily exhibit sensually discernable distinctions; a similar constraint emerged from our analyses of ostension and perception.[154] Those constraints entailed by the fact that every phenomenal world is a world of objects, however, are a great deal more difficult to sketch, let alone actually to pin down. Kant's doctrine of transcendentally grounded synthetic a priori propositions is an attempt to explicate these constraints,[155] though for Kant only *one* possible phenomenal world, that of Newtonian physics, was ever at issue. To repeat Kant's analysis under the altered conditions of Kuhn's theory would go beyond the scope of this work. We should keep in mind, however, that the mere fact that observational concepts refer to objects is yet a further source

* Translator's note: This discussion makes use of the etymology of the German word for object, *"Gegenstand,"* which, decomposed, suggests "that which stands opposite": a counterpart.

150. Compare §§ 3.3 and 3.5.

151. See Quine 1960, sections 17 and 18.

152. Kant, *Critique of Pure Reason* A320/B376 [Kemp Smith, p. 314].

153. I am here assuming that a phenomenal world is a world of objects. If the concept of a phenomenal world is relaxed such that this condition no longer holds, the argument which follows ceases to apply.

154. Compare §§ 3.3 and 3.5.

155. On this issue, see Heidegger 1962.

of endogenous constraints on Kuhn's plurality-of-phenomenal-worlds thesis.

c. Concept Learning without Use of Laws or Theories in Kuhn's Work after 1969

By 1969, Kuhn was fully aware of the need to move beyond his merely negative, and hence unsatisfactory, characterization of the concept of meaning for empirical concepts.[156] And so he admits, in the discussion following his presentation of 1974a, that

> We are not going to talk about fitting terms to nature without ultimately having things to say about the problem of meaning, and about that problem nobody knows quite what to say at the moment. I do not myself, but I think that what I am getting at with similarity relations may be one of the tools that will be helpful.[157]

For Kuhn, the need to confront the concept of meaning arises from his conviction that the conceptual changes of special interest to him, those which occur during scientific revolutions, *can't be reduced without remainder to extensional change, conceived as the mere addition or subtraction of elements of an extension.* Kuhn formulates this conviction in different ways at different times. In *SSR,* for example, Kuhn asserts that the transition from Newtonian to Einsteinian mechanics represents a particularly clear example of the kind of shift in the conceptual net typical of scientific revolutions, for no new objects (or concepts) were introduced in the process.[158] The less than completely extensional character of revolutionary conceptual change is asserted in a different way in 1969, at which point Kuhn suspects that scientific revolutions are characterized by the loss of analytic (or "quasi-analytic") status on the part of sentences which formerly had it to some degree.[159] In the context of his discussions of scientific revolutions as linguistic changes resulting in a certain untranslatability, Kuhn asserts that while perfect translations preserve meaning,[160]

> In the present state of the theory of meaning, the distinction between terms that change meaning and those that preserve it is at best difficult to explicate or apply.[161]

156. Compare § 3.6.a.

157. 1974b, pp. 515–516.

158. *SSR,* p. 102.

159. 1974a, *ET* p. 304 n. 14; 1970c, *SSR* pp. 183–184; 1976b, p. 198 n. 9. —On this issue, see § 6.3.a, point 2.

160. 1983a, pp. 670, 671, 672, 680, 682.

161. 1983a, p. 671.

Kuhn thus faces the problem that, while a purely extensional treatment of empirical concepts isn't enough, direct access to that which the concept of meaning is supposed to capture isn't possible. Accordingly, "'meaning change' names a problem rather than an isolable phenomenon."[162] Kuhn's strategy for coping with this problem consists in setting aside the problematic concept of meaning, instead asking after the way in which the *reference determinations* of empirical concepts are learned, the way in which their names "attach to nature."[163] So rather than attempting to determine what meaning actually is, he inquires into the manner in which one of those functions a concept's meaning is supposed to have, the ability to identify referents and nonreferents of the concept, is learned. In continuity with earlier conceptions, he insists on one negative feature of the concept of meaning: the meaning of a concept doesn't consist in a set of necessary and sufficient conditions for its application, and consequently, the traditional theory of meaning is bankrupt.[164] In his 1983 papers, Kuhn attempts to derive hints toward a theory of meaning for empirical concepts from an analysis of the learning of reference determinations. Following Kuhn's procedure, I will first discuss the new conception of the process of concept learning in which laws and theories aren't employed, and I will then go on, in § 3.6.g, to consider the consequences of this new conception for the theory of meaning.

In his works after 1969, Kuhn frames his inquiry into the learning of empirical concepts as follows: how can the immediate similarity relations relevant to the identification of an empirical concept's referents and nonreferents be learned by ostension? By contrast with the earlier conception, this approach doesn't consider whether this learning process gives rise to any change in mode of perception.

The learning of empirical concepts occurs in a "metaphor-like process"[165] in which exemplary elements of the extension of the concept to be learned are juxtaposed, and this "juxtaposition of examples calls forth the similarities upon which . . . the determination of reference depend[s]."[166] The parallel to the workings of metaphors consists in the fact such juxtaposition, too,

> neither presupposes nor supplies a list of the respects in which the subjects juxtaposed by metaphor are similar. On the con-

162. *ET*, p. xxii.

163. 1979b, pp. 409–410; 1981, p. 19; 1983a, pp. 676–677, 680–681, and elsewhere.

164. 1979b, pp. 409, 413; 1983a, p. 685 n. 13; 1983b, p. 713.

165. 1979b, pp. 409, 414, similarly pp. 410, 413; 1981, pp. 20, 21; 1983b, p. 714; 1989a, p. 12; 1990, p. 301.

166. 1979b, p. 413.

> trary . . . it is sometimes (perhaps always) revealing to view metaphor as creating or calling forth the similarities upon which its function depends.[167]

Anyone seeking to learn the empirical concept at issue must thus

> discover the characteristics with respect to which they [the exemplars] are alike, the features that render them similar, and which are therefore relevant to the determination of reference.[168]

It doesn't matter *which* features of the objects authoritatively presented as similar are employed by someone learning the concept in reproducing the similarity. In other words, it doesn't matter what one leans on in identifying the concept's referents and nonreferents, for

> In matching terms with their referents, one may legitimately make use of anything one knows or believes about those referents.[169]

All that matters is that the similarity class presented by exemplars is actually learned as the extension of the target concept.[170] One consequence of this view is that different members of a linguistic community presumably will, in applying the same empirical concepts, employ different characteristics of their concepts' referents to identify these referents and distinguish them from nonreferents.[171]

But what guarantees the unequivocal use of these concepts within the linguistic community, when different speakers employ different features of the same concepts in identifying referents? Two aspects of the learning process play a role here.[172] First, several representatives of the similarity class to be learned must be ostended. In addition, a range of nonmembers of the similarity class must be ostended in order to demarcate the class's boundaries with other classes, especially close neighbors. But whether a process of instruction structured in this way actually teaches the target concepts depends, as noted above, on the nature of the phenomenal world at issue.[173]

Since learning a given concept requires an ostension of members

167. 1979b, p. 409. —On this issue, see e.g. Künne 1983, p. 182.

168. 1979b, p. 413; similarly 1981, p. 20.

169. 1983a, p. 681; similarly 1981, p. 19; 1983a, pp. 685–686 n. 13; 1983b, p. 714.

170. 1983a, pp. 682–683.

171. 1983a, pp. 681, 683; 1989a, pp. 12, 16–17, n. 25; 1989b, p. 51; 1990, pp. 301, 303, 317 n. 22.

172. 1979b, pp. 412–414; 1981, p. 20; 1983a, pp. 680, 682; 1989a, p. 16; 1990, p. 303. —Compare § 3.3.

173. See § 3.3; also 1983a, pp. 681–682.

of neighboring similarity classes, this concept can't be learned in isolation from other empirical concepts. Each concept is rather introduced, either alone or somehow correlated with other empirical concepts, into a preexisting net of mutually correlated concepts.[174] Kuhn calls this intertwining of the empirical vocabulary the "local holism" of language.[175] *Local* holism, for not all other empirical concepts are implicated in a given empirical concept's reference determination. *Holism,* because mutually correlated concepts constitute a whole, such that the individual constituent concepts aren't independently existing parts or elements of this whole but are, rather, mere dependent moments; they are nothing outside the whole. The intertwining of empirical concepts has important consequences, both for the theory of meaning[176] and for the debate over theory choice.[177]

d. The Relationship between Earlier and Later Conceptions of Concept Learning without Use of Laws or Theories

Despite the partially divergent terminology of the two conceptions of concept learning without use of laws or theories that were considered in the last two subsections, there remains an important substantive continuity between them. This continuity follows from the identical role played by ostension, in both conceptions, in the learning of similarity relations.

The only substantive difference between the two conceptions is subtle but not without importance. What, in the earlier conception, was described as a particular organization of *perceptual space,* the production of similarity and dissimilarity by accentuating and dampening features of objects in perceptual space,[178] is now conceived as a discovery of particular features of objects, making possible the community's usual *application of concepts* by identifying their referents and nonreferents.

In the first case, it is perception that grounds the connection with the world; to encounter the world is to see it. The perceptual subject

174. 1979b, pp. 415–416; 1981, pp. 10–12, 18–19; 1983a, pp. 670–671, 676–677, 682; 1983b, pp. 714–716; 1990, p. 314. —See Kamlah and Lorenzen 1967, p. 50.

175. 1983a, p. 682; 1983d, p. 566. —In essence, the local holism of language may already be found in *SSR* (and in the 1969 papers): "scientific concepts . . . gain full significance only when related, within a text or other systematic presentation, to other scientific concepts, to manipulative procedures, and to paradigm applications" (*SSR,* p. 142). Similar passages can be found e.g. on pp. 149, 149–150.

176. See § 3.6.g.

177. See § 7.5.

178. Compare § 3.5.

may be an individual, but his or her perception is molded in a process in which the community of which the individual is (or will be) a member plays a great role. For the learned similarity and dissimilarity relations which belong to or characterize the community are (to an uncertain degree) coconstitutive of perception.[179] This molding of perception by the community is reflected in the particular empirical concepts employed in the community.

In the second case, by contrast, the connection with the world is a product of language; to encounter the world is to capture it linguistically. Insofar as the linguistic community is the proprietary subject of its language, the individual is connected with the world only *qua* member of the community. Perception plays a role in this connection with the world, but rather an ancillary one; it helps the individual to achieve that mode of connection with the world dictated by the language of members of the linguistic community.

With this greater emphasis on language, an important aspect of Kuhn's theory becomes clearer. This has to do with individual differences between scientists belonging to the same scientific community, differences which, under special circumstances, become relevant to scientific practice. The fact that members of a scientific community *use* an empirical concept *in the same way* doesn't imply that they identify referents and nonreferents of the concept *by the same criteria*.[180] The individual differences between members of the community aren't apparent in linguistic practice so long as this practice is successful, in the sense that all members of the community ascribe objects to and exclude objects from the extensions of empirical concepts in the same way.[181] But the difference becomes both apparent and important when criteria for identifying referents and nonreferents that heretofore produced the same identifications begin, in response to certain new phenomena, to produce divergent results.[182] At this point the group may begin to talk at cross purposes, if all speakers no longer match the same words with the same extensions.

179. Compare § 3.4.

180. An analogous claim holds for the older conception. A similar arrangement of relative distances in the perceptual spaces of two different observers doesn't imply that the two also accentuate and dampen the same features of perceptual objects. But on the more linguistically oriented conceptions, the potential for individual difference is both more evident and more easily verified.

181. And, we should add, so long as the members of the linguistic community don't *reflect* on their use of concepts and discuss it with each other.

182. On this issue, see § 7.5, especially 7.5.c. —For two further sources of individual difference between members of a scientific community of epistemic relevance under certain conditions, see §§ 4.3.c and 5.5.a.

e. Concept Learning with the Help of Laws and Theories

For Kuhn, the manner in which concepts may be learned with the help of laws and theories should be explored by inquiring into the way we learn to apply laws and theories. Laws and theories are applied to "problem situations," that is, to the objects or constellations of objects about which these laws or theories make assertions. There are five steps to applying a law or theory:[183]

1. The identification of a problem situation falling within the law or theory's intended sphere of application
2. The selection of one of the many equivalent formulations of the law or theory
3. The specification of this formulation of the law or theory for the particular problem situation
4. Mathematical-logical manipulations of this specification of the law or theory, generating assertions about the given problem situation
5. The assignment of these assertions to the set of experimental or observational statements about the problem situation

For example, treating the free fall of a body in a gravitational field first requires the identification of this problem as a problem of classical mechanics. Next, one of the equivalent formulations of classical mechanics must be chosen, for instance, Newton's second law, "Force is equal to mass times acceleration." This law is then specified for the particular situation, yielding $mg = md^2s/dt^2$. This done, the differential equation must be solved, resulting in $s(t) = gt^2/2 + v_0t + s_0$. Finally, this solution must be evaluated for individual heights or times of fall and compared with observed data.

The meanings of empirical concepts employed in the chosen formulation of the theory or law obviously play an especially great role in the first, third, and fifth steps.[184] A problem situation can only fall within the intended sphere of application of a formulation of a law or theory if the empirical concepts employed in this formulation have referents in the problem situation. Furthermore, the law or theory can

183. For the discussion which follows, see 1976b, where the parallels between Sneed's 1971 and Kuhn's philosophies of science, as worked out by Stegmüller 1973, are discussed. —In his latest work, Kuhn places the emphasis in his account of concept learning during law or theory application somewhat differently; see 1989a and 1990. These efforts also consider an example worked out in detail, the learning of basic concepts from Newtonian mechanics. This shift in emphasis couldn't be considered in the present study.

184. These are not necessarily discrete, temporally sequential steps. The identification of the problem situation and the instantiation of the chosen formulation of the law or theory might, for example, be mutually dependent; see also 1974a, *ET* pp. 301–302.

only be specified for a given situation, and the resulting assertions compared with observations, if the referents of the empirical concepts employed in the law or theory have actually been identified in that situation.[185] Kuhn's question regarding the manner in which empirical concepts may be learned while employing laws or theories can thus be posed as follows: How do we learn to ascribe referents to the empirical concepts employed in a law or theory in the first, third, and fifth steps of the law or theory's application?

In order to get a better idea of just what must be learned in order to identify referents in this way, Kuhn begins by abstracting from the meanings of the empirical concepts employed in any particular formulation of a law or theory. In the formula $F = ma$, for example, F, m, and a are taken as uninterpreted placeholders. Kuhn calls formulations of laws or theories in which the meanings of empirical, though not mathematical or logical, concepts have been abstracted "symbolic generalizations,"[186] or more precisely, "generalization sketches," "schematic forms," "law-sketches," or "law-schemata."[187] These terms hint not so much at the fact that generalizations contain uninterpreted placeholders instead of empirical concepts but rather at the fact that the appropriate symbolic generalization for a given situation must be specified *before* any consequences the law has in that situation can be deduced, and definitely before any comparison with experimental or observational results can be made.[188] Our question concerning the way in which empirical concepts occurring in particular formulations of laws and theories are learned thus becomes: How do we learn to

185. The meanings of implicated empirical concepts can even play a role in the second and fourth steps. In the second step, the formulation of the law or theory might be chosen with an eye toward the easy identification of referents for empirical concepts. In the fourth step, that of logical and mathematical manipulations, the meanings of empirical concepts might play a role, for example, in justifying approximations which, though not legitimate in purely mathematical terms, may be made acceptable by recourse to what, in physicists' jargon, is called the "reasonable behavior" of certain quantities.

186. 1974a, *ET* p. 299; 1970c, *SSR* pp. 182–184. —Talk of "symbolic generalizations" is specific to the 1969 papers, where symbolic generalizations occur as components of the "disciplinary matrix" (see § 4.3). In 1979b, by contrast, the less specific claim that "laws and theories also play a role in establishing reference" is made (p. 412); similarly 1981, p. 8.

187. 1970c, *SSR* p. 188; 1974a, *ET* pp. 299, 300 n. 10; 1970b, p. 272; 1974b, p. 505; 1983a, p. 678. —The terms "law-schema" and "law-sketch" aren't meant to suggest that, once interpreted, these symbolic generalizations become purely synthetic propositions, incapable of fulfilling either expository or stipulative functions. Their status with regard to the analytic/synthetic distinction is rather ambivalent; on this issue, see §§ 3.6.f and 6.3.a.

188. 1974a, *ET* pp. 299–301; 1970b, pp. 272–273; 1970c, *SSR* pp. 188–189; 1974b, p. 517.

assign referents to the placeholders for empirical concepts occurring in law-schemata when applying the schemata to particular problem situations, in such a way as to allow the law-schemata to be specified and, after logical and mathematical manipulation, compared with observation and experiment?

The key to this question lies in the special way in which scientists are trained and in their typical approach to new problems by analogy to problems already solved.[189] In both cases, the *immediate similarity relations between problem situations* play an important role. The student, whose task is to learn the empirical concepts occurring in laws and theories, is first presented with the application of these concepts in solving exemplary problem situations. The student's next assignment is to go beyond the exemplars in solving further problem situations, *with the understanding that they are soluble using the same law or theory*. Just as occurs when, in scientific practice, a familiar theory must be applied to a heretofore unfamiliar problem, so now, too, an analogy between the new problem situation and other problem situations, for which the right specifications of the relevant law-schemata are already known, must be found. As happens in perception, or when concepts are directly applied to objects,[190] analogy accentuates those features of both familiar and unfamiliar problem situations by means of which the new may be seen as similar to the old. These similarities permit the specification of the law-schemata for new situations by analogy with the specifications appropriate to familiar situations. In other words, they make it possible to apply the concepts occurring in the schemata to new problem situations.

Here, as in object perception, an as-structure emerges.[191] The new problem situation must be seen *as* a situation for which some law-schema can be specified in a manner analogous to the specification for some familiar problem situation. This analogy must be immediate in the sense that it has no recourse to characteristics which would define the class of problem situations susceptible to treatment with a given symbolic generalization.[192] The unproblematic use and transmission within a community of the empirical concepts employed in laws and theories thus depends not on the mastery of definitions but only on the possibility of actually mastering the fundamental, immediate similarity

189. *SSR*, pp. 46–47; 1974a, *ET* pp. 305–307; 1970b, pp. 273, 275; 1970c, *SSR* pp. 189–190, 200; 1976b, p. 181; *ET* p. xix.

190. Compare §§ 3.5 and 3.6.b–d.

191. Compare § 3.5.

192. These two parallels to perception, the as-structure and the fulfillment of the attendant requirement that attribution occur without regress to defining features, are likely what makes it possible to use "to see" in both literal and metaphorical, or rather theoretical, senses.

relations. In other words, it must be possible to see the right problem situations as similar to one another. The meanings (whatever they might be) of such empirical concepts should thus, at least for the time being, be localized not to definitions but rather to similarity relations.

The "local holism of language" also holds for concepts whose proper use is learned in applying laws and theories and, indeed, is more complicated for them than for empirical concepts learned without reference to laws and theories. We may distinguish between three sources of holism.

First, we recall that the mutual intertwining of concepts learned without reference to laws or theories emerged from the fact that the appropriate similarity relations can only be learned if members of other similarity classes, especially close neighbors, are ostended. Only in this manner could those features suitable for distinguishing referents from nonreferents be discovered.[193] This form of intertwining has its analogue in the fact that a conceptual system characteristic of law or theory must be learned through its divergence from the conceptual systems of other laws or theories. For the first decision to be made about any problem situation at hand inevitably concerns the theory or law to be used in addressing it.

A second form of intertwining for empirical concepts occurring in laws or theories arises from the fact that they are learned together, and not just in a temporal sense. When a law-schema is applied to a given problem situation, the selection of referents for some constituent concept or another depends in part on the selection of referents for the other concepts occurring in the same law-schema.[194] On grounds of their substantive interdependence, these concepts can only be learned together.

Third, concepts employed in a law or theory are also intertwined with those learned and applied in direct observation of objects. For these latter concepts play a role both in identifying the right law or theory to be used in attacking a given problem situation and in finding referents for the concepts which occur in laws and theories.[195]

f. The Impossibility of Explicitly Defining Empirical Concepts

The doctrine of the impossibility of explicitly defining empirical concepts is of special importance both to Kuhn's theory as a whole and to

193. See § 3.6.c.

194. 1970c, *SSR* pp. 183–184; 1979b, p. 412; 1983a, pp. 675–677; 1983d, p. 566; 1989a, p. 16; 1990, p. 303.

195. Compare footnote 33 to § 3.2 on the interdependence of similarity relations from different domains.

his theory of meaning in particular. This doctrine builds on the weaker thesis which states that, in fact, empirical concepts are generally learned and applied uniformly in the appropriate scientific community without any recourse to explicit definitions.[196] The weaker thesis is evident from three aspects of scientific practice.

First, it seems that

> Very few correspondence rules are to be found in science texts or science teaching.

And furthermore,

> if asked by a philosopher to provide such rules, scientists regularly deny their relevance and thereafter sometimes grow uncommonly inarticulate.[197]

"Correspondence rule" here refers—in a slight departure from the usual use of this expression in the philosophy of science—to everything with pretensions to define an empirical concept explicitly, whether its form is that of an intensional, a mere extensional, or an operational definition.[198]

Second, the function of exercises assigned to students of the natural sciences, and the difficulties students typically experience with them, are best interpreted as evidence less of the application of *previously known* concepts to new problem situations than of students' efforts to master concepts *not yet adequately known* by way of the immediate similarity relations between problem situations.[199]

Third, it frequently happens in the history of science that a new problem is attacked with the same strategy employed for a problem solved earlier, a problem which may concern very different phenomena.[200] The transposition of concepts from previously solved to new problems happens by means of the immediate similarity relations holding between the problem situations, not according to defining criteria.

Formulated in terms of the (Cartesian-) Leibnizian distinction between the *clarity* and *distinctness* of concepts, this doctrine claims that the empirical concepts of science can be clear, unequivocally applicable,

196. E.g. 1970a, *ET* p. 286; 1974a, *ET* pp. 302 n. 11, 312; 1970b, pp. 274, 276; *ET*, pp. xviii–xix; 1979b, pp. 409, 413; 1983a, p. 681; 1989a, pp. 15, 16; 1990, pp. 302, 303. —See Kamlah and Lorenzen 1967, p. 29.

197. 1974a, *ET* p. 305; similarly *ET*, pp. xviii–xix; q.v. *SSR*, pp. 142–43.

198. 1974a, *ET* p. 302 n. 11.

199. *SSR*, pp. 46–47, 80; 1974a, *ET* pp. 305–308; 1970b, pp. 272–274; 1970c, *SSR* pp. 189, 191.

200. Compare § 3.6.e.

without thereby being distinct, without having any known defining criteria.[201]

Up to now we have only set up a thesis about the actual employment of empirical concepts in the sciences. But Kuhn asserts the further claim that empirical concepts *can't in principle be adequately explicated by means of definitions.*[202] Kuhn's doubts on the extent of the analytic/synthetic distinction's applicability, for which he's indebted to Quine, form the background to this claim.[203]

For now we must ask what Kuhn understands by a definition. Kuhn's notion of a definition is *pragmatic;* a definition of a concept does a certain job; it provides necessary and sufficient criteria for the concept's unequivocal application in all conceivable situations.[204] In other words, a definition makes the application of a concept *decidable;* it doesn't matter whether it does this by recourse to intensions, extensions, or operational criteria. Kuhn is well aware that, as far as the claim that empirical concepts can't, in this sense, be defined is concerned, "A negative of that sort scarcely can be proven."[205] The claim will become plausible, however, if we presuppose Kuhn's views on the learning of empirical concepts by means of immediate similarity relations.

Let us consider a concept learned by means of immediate similarity relations, unproblematically used in some scientific community. If we

201. Leibniz claimed, however, that even indistinct concepts could be decomposed into features, though these might be unknown to the concept user. The only exceptions are concepts with the status of a "basic concept, whose characteristic is itself . . . and thus can't be decomposed into characteristics but is grasped only through itself." Yet "colors, smells, tastes, and other particular objects of the senses" aren't among these basic concepts (Leibniz 1684, vol. 4, pp. 422–423, p. 10f. [My translation. —A.T.L.]). By contrast, as we shall soon see, Kuhn denies (with Wittgenstein) that such decompositions could possibly do justice to their goals.

202. *SSR,* chap. 5 (and somewhat more reticently on p. 142); 1964, *ET* p. 259; 1970a, *ET* p. 287, and nn. 36, 37; 1974a, *ET* p. 302 and n. 11, p. 316 and n. 21; 1970c, *SSR* p. 192; 1974b, p. 511; *ET,* pp. xix, xxii; 1984, p. 245; 1989a, p. 21 n. 19; 1990, p. 317 n. 17. —On this issue, see Kant, *Critique of Pure Reason,* A727ff/B755ff, and Kant, *Logic* (ed. Jäsche), § 103, vol. 5, p. 573.

203. 1961a, *ET* p. 186 n. 9; *SSR,* pp. vi, 45; 1964, *ET* p. 258; 1970c, *SSR* pp. 183–184. Kuhn doesn't take these doubts so far as to claim that the analytic/synthetic distinction could, in principle, never be explicated. What he does claim is that this distinction is unsuited to the classification of scientific propositions. Aligning himself with Braithwaite, he asserts that "an inextricable mixture of law and definition . . . must characterize the function of even relatively elementary scientific concepts" (1964, *ET* p. 258). This of course presupposes rather than denies the ideal differentiability of "law" and "definition."

204. *SSR,* p. 45; 1970a, *ET* p. 287 and nn. 36, 37; 1974a, *ET* p. 302 and n. 11, p. 316 and n. 21; 1974b, p. 511.

205. 1974a, *ET* p. 305.

attempt to explicate this concept by means of necessary and sufficient conditions for its application, we will in all probability find different possible sets of such rules, all of which are equivalent with regard to the concept's use in the community *to date*.[206] Which of these alternative sets of rules actually defines the concept in question is undecidable. But their differences are by no means negligible, for they may produce different results as applied to the concept's employment in a *new* situation. The differences between possible explicit definitions show up a certain "openness" on the part of concepts learned by means of immediate similarity relations, an openness no explicit definition can capture. Explicit definitions may serve to narrow empirical concepts, but they can never adequately reflect the way in which they are used in a given community.[207] The learning and use of empirical concepts by means of immediate similarity relations is simply of a fundamentally different kind than learning and use by way of explicit definitions.[208]

One important point of difference between practicing scientists and the philosophers or historians of science whose business it is to reconstruct their practice deserves mention here.[209] Scientists may wish (and, under certain circumstances, be obliged)[210] to define concepts previously employed without explicit definition, thus attempting to bring their linguistic practice under the regulation of a norm; these definitions may reproduce previous linguistic practice, or they may not. But the function such definitions exercise for *scientists* under certain circumstances doesn't permit the philosopher or historian to reconstruct concepts by means of explicit definitions if scientists learned these concepts by means of immediate similarity relations and applied them without definitions. The doctrine of the impossibility of adequate explicit definition of empirical concepts doesn't claim that such definitions are, in general, unsuited to normative purposes. All it claims is that explicit definition isn't the right means by which to reconstruct a linguistic practice in which empirical concepts are employed without recourse to explicit definitions.

But doesn't the lack of explicit definitions for empirical concepts seriously handicap scientific practice? Aren't just such definitions a necessary prerequisite for the unequivocal use of concepts within a scientific community? Any appearance of indispensability on the part of

206. 1974a, *ET* p. 303.
207. 1974a, *ET* pp. 303–304 and n. 13, pp. 313–318.
208. 1974a, *ET* pp. 312–313; 1970c, *SSR* p. 198.
209. 1974a, *ET* pp. 303–304 and n. 13, pp. 313–318; 1974b, pp. 512–513.
210. 1970a, *ET* p. 286; 1974a, *ET* p. 318.

definitions vanishes when we consider that phenomenal worlds are always organized in a particular way; they always contain natural families.[211] In an appropriately organized phenomenal world two things, in particular, are unnecessary in order to apply empirical concepts, both individually and collectively, in an unequivocal way. First, defining characteristics which provide necessary and sufficient conditions for a concept's application in *all conceivable* situations are unnecessary, for rules governing unequivocal concept use in situations ruled out by the very organization of the phenomenal world are superfluous.[212] It is enough for empirical concepts to be unequivocally employed in the phenomenal world for which they were introduced. Expressed negatively, this amounts to the claim that only those misuses of a concept actually possible in this phenomenal world need to be ruled out when the concept is learned. Second, as far as ensuring unequivocal concept use is concerned, there's no need to decide whether the criteria a speaker uses in identifying referents of a given concept are suitable for inclusion in that concept's definition or merely traits accruing to the referents on empirical grounds. Such decision is unnecessary so long as unequivocal concept use is sustained by the particular organization of the phenomenal world.

But not only are explicit definitions of empirical concepts unnecessary when unequivocal concept use in the community can be achieved by learning immediate similarity relations, they would, in many respects, prove a hindrance to scientific practice.[213] For if definitions are taken seriously, they are binding on concept application even in such novel situations as appear in the course of research. For example, unexpected objects might be discovered.[214] Some such discoveries will prove tricky borderline cases for a definition-based criterion for concept application, though they pose no trouble at all for immediate similarity relations. Other cases might call an established classification into question. But, in such cases, we gain nothing from summary decisions by definition. Whether the established classification should be retained or abandoned in light of the discovery of new objects must be decided by empirical examination of the objects themselves, not by definitions given in ignorance of these objects.

211. Compare § 3.3.

212. 1964, *ET* pp. 259–260; 1970a, *ET* pp. 286, 288; 1974a, *ET* p. 312 n. 20 and p. 318 n. 21; 1970c, *SSR* p. 197 n. 14; 1983a, p. 682; 1989a, p. 21; 1990, p. 307.

213. 1970a, *ET* pp. 287–288; 1974a, *ET* pp. 314–316; 1970c, *SSR* pp. 197–198.

214. One of Kuhn's examples is the black swan, which overuse by philosophers of science has leeched of almost all color.

g. Consequences for the Theory of Meaning as Applied to Empirical Concepts

What consequences does the above have for the theory of meaning as it applies to empirical concepts? We are faced with a choice between two alternatives. On the one hand, we might insist that an empirical concept only has meaning in the full sense if, in principle, an adequate definition governing its use can be given. In this case, most of the empirical concepts of science would have no meaning in the full sense. On the other hand, we might reject this concept of meaning as inadequate—at least stated so universally—and assert that those concepts which are used unequivocally in a given linguistic community have meaning in the full sense even if this meaning can't be articulated in a definition. Someone has "learned a concept" or "knows its meaning" just in case he or she can use it in the same way as other members of his or her community. It remains unanswered in this case, however, in what the meaning of an empirical concept consists.

Kuhn chooses the second of these alternatives. His choice makes sense when we consider the way in which he juxtaposes the two options.[215] *Defining* an empirical concept (in the ideal case) sanctions its unequivocal use in *all conceivable* situations. *Learning* an empirical concept by way of immediate similarity relations, by contrast, only sanctions (in the ideal case) its unequivocal use *in all expected situations in a given phenomenal world.*[216] The first alternative's surplus determinacy is obviously no help to unequivocal concept use in a given phenomenal world and can even prove a hindrance. Consequently, an empirical concept learned by way of immediate similarity relations and unequivocally applicable in a given phenomenal world must be deemed well-founded. In other words, it has a determinate meaning.

In these reflections, Kuhn is evidently applying a *pragmatic condition on the adequacy of a concept of meaning:* someone has mastered the meaning of an empirical concept, whatever that might be, if he or she can use the concept correctly, relative to the appropriate linguistic community.[217] Evaluated by this condition, the first alternative—determinate meaning equals definitional explicability—is inadequate and should be discarded.

When we ask, with an eye toward clarifying the concept of meaning, what is involved in the ability to employ empirical concepts correctly relative to a linguistic community, two observations appear

215. See e.g. 1964, *ET* p. 254.

216. Q.v. 1989a, p. 20; 1990, pp. 306–307.

217. See e.g. 1974a, *ET* pp. 305, 312, 316 and n. 21; 1974b, p. 511; 1979b, p. 413; 1989a, p. 12; 1990, p. 301.

worth making. First, the particular content of the criteria employed by an individual speaker in evaluating whether a given concept applies is irrelevant so long as the criteria result in correct concept use.[218] Consequently, the particular criteria employed by individual speakers make no *immediate* contribution to the concept's meaning. Second, in accordance with that feature of empirical conceptual systems called "local holism," the correct use of an empirical concept can't be learned in isolation from the correct use of other concepts.[219] It follows that *a given concept has no meaning when taken as an individual* but rather has it only when regarded as a dependent moment in a conceptual network.[220] Borrowing a term from linguistics, Kuhn calls this network of concepts a "lexicon" or "lexical network,"[221] for, as do the words in the genre of books referred to by that name, the constituents of a system of empirical concepts allude to one another.

Now we are in a position to state what's involved in knowing a concept's meaning.[222] For each individual speaker, the criteria he or she employs for referent determination place concepts in particular relations to one another, relations of extensional exclusion, extensional overlap, genus and species, synonymy, and others. In order for *one* concept to be used correctly, that is, unequivocally within a community, the network of relations *between* concepts must be the same for all speakers in the linguistic community. In other words, this network isn't contingent on any particular choice of those criteria which mediate the interweaving of concepts for the individual speaker. This network of relationships, invariant over the linguistic community, is called the "structure of the lexicon."[223] To know the meaning of a concept, that is, to know how to use the concept correctly, thus means to know the structure of that portion of the lexicon in which the concept occurs, and to know it by means of criteria, particular to the individual speaker, for identifying referents and nonreferents of the concept.

3.7 Knowledge of Nature

We have seen what role immediate similarity relations play in the constitution of a phenomenal world; they are coconstitutive of perception

218. See § 3.6.c.
219. See §§ 3.6.c and 3.6.e.
220. 1983b, pp. 713–714; 1989a, p. 12 and p. 25 n. 25; 1990, p. 301 and p. 317 n. 22.
221. 1983a, pp. 682–683; 1983b, pp. 713–714; 1989a; 1990.
222. 1983a, pp. 682–683; 1983b, pp. 713–714.
223. Ibid.

and of the formation of empirical concepts. Their importance as moments of world constitution emerges from the fact that a phenomenal world is organized perceptually and conceptually.[224] Kuhn variously claims on behalf of immediate similarity relations that, beyond their world-constituting function, they or the concepts they introduce contain "knowledge of nature" or of the "world."[225] Two points of clarification are called for. First, what, precisely, is the content of such knowledge (§ 3.7.a)? Second, in what sense are we here dealing with knowledge—what are the characteristics and capacities of this form of knowledge (§ 3.7.b)?

a. The Content of Such Knowledge

According to Kuhn, the knowledge of nature or the world contained in immediate similarity relations involves two sorts of content: the existence of reasonably well-described similarity classes in a phenomenal world and the regularities which objects constructed by the similarity relations obey.[226] The first component of knowledge I will call *quasi-ontological knowledge,* quasi-ontological because it concerns a phenomenal world with no claim to exclusive reality. The second component I will call *knowledge of regularities.* Both epistemic components are mutually inextricable moments of the knowledge contained in immediate similarity relations.[227] After we have discussed the contents of both components of knowledge, we will return, at the end of this subsection, to consider why this is so.

The immediate similarity relations which lead to the formation of natural families contain, so goes the claim, *quasi-ontological knowledge.* The immediacy of these similarity relations just means that the similarity of the relata is taught and learned not by means of defining characteristics but by repeated ostension of exemplary elements and nonelements of the appropriate similarity classes.[228] This immediacy means that, as far as concept learning without recourse to laws and theories is concerned, it must be possible perceptually to identify the referents

224. Compare Ch. 2.

225. *SSR,* pp. 127, 128; 1964, *ET* pp. 255, 258, 260; 1970a, *ET* p. 285; 1974a, *ET* pp. 312–313; 1970b, pp. 270–272, 274; 1970c, *SSR* pp. 175, 190–191, 196; 1974b, pp. 503–504; 1981, p. 21; 1990, p. 315. —See Kamlah and Lorenzen 1967, p. 169, and Rescher 1982, pp. 18–23.

226. E.g. 1974b, p. 503.

227. Thus, in Kuhn's texts, they are always discussed together. Yet the separate treatment of the two moments of the knowledge contained in similarity relations strikes me as useful for clarity of exposition.

228. Compare §§ 3.2 and 3.3.

of these concepts without serious boundary disputes. There can be no fluid transitions between different natural families of objects, since, if there were, drawing a clear border between natural families would be possible only by convention.[229] This immediacy of similarity relations also means that, as far as concepts learned in applying laws or theories are concerned, there can be no fluid transitions between similarity classes of problem situations, and that the concepts at work in the implicated symbolic generalizations must be applicable without boundary conventions. The quasi-ontological knowledge which similarity relations contain only by virtue of their assumed property of being immediately teachable and learnable consists in the proposition, learned along with the immediate similarity relations themselves, that there are no objects which fall through the cracks between similarity classes.[230] In other words, immediate similarity relations assert a claim to knowledge, on which the phenomenal world—apparently "of its own accord"—is organized in a certain way. This organization seems to be given to perception at every exposure to observable objects. Similarly, problem situations exhibit a high degree of nonarbitrariness vis-à-vis the symbolic generalizations "appropriate" to their resolution and with regard to the determination of referents for the concepts contained in these generalizations. Kuhn's critical agenda, at this point, amounts to the insistence that the knowledge involved here is *quasi*-ontological, in the sense that it concerns only a particular phenomenal world, not all possible phenomenal worlds, and certainly not the world-in-itself.[231]

Quasi-ontological knowledge is gained at the same time that the names of natural families are learned. In fact, these two forms of knowledge, knowledge of language and quasi-ontological knowledge, are distinguishable but not separable; they are "not really two sorts of knowledge at all, but two faces of the single coinage that a language provides."[232] This "coin" is the network of immediate similarity and dissimilarity relations, for these relations are constitutive of the meanings of their respective concepts and thus assert a claim to the appropriate quasi-ontological knowledge.

Concepts learned by way of immediate similarity relations contain, in addition to quasi-ontological knowledge, *knowledge of regularities*, or, to use a term from Kuhn's "A Function for Thought Experiments"

229. Compare § 3.5.

230. E.g. 1970a, *ET* p. 285; 1974a, *ET* p. 312; 1981, p. 21; 1983a, pp. 681–682.

231. The most developed treatment of this point to date is at 1979b, pp. 414, 417–419.

232. 1981, p. 21; similarly *SSR*, pp. 127–128; 1964, *ET* pp. 257–260; 1970a, *ET* pp. 285–286; 1974a, *ET* pp. 302 n. 11, 312–313; 1970b, pp. 270–271, 272, 274, 276; 1970c, *SSR* pp. 190–191; 1974b, pp. 503–504.

(1964), they have "legislative content."[233] What this means is that such concepts contain knowledge about "how the world behaves"[234] or about "the situations that nature does and does not present."[235] The possibility that such knowledge may be contained in empirical concepts arises from the fact that they are used without binding explicit definitions. For when we try to formulate explicit criteria for the use of a given empirical concept[236] on the basis of contemporary conceptual analyses *and* the concept's actual use in concrete cases, we make three discoveries. First, it appears that a plurality of different criteria is required to govern the concept's use in different situations.[237] Second, the conjunction of these criteria has empirical content.[238] Third, these are criteria

> whose coexistence can be understood only by reference to many of the other scientific (and sometimes extrascientific) beliefs which guide the men who use them.[239]

When these three points are taken together, it becomes clear how the lack of any binding explicit definition opens up the possibility of tacitly importing "legislative content" into a concept's use. If the correct application of a concept is always decided by means of a fixed explicit definition, then any actual attribution of a concept to an object only allows us to infer the existence in the present phenomenal world of objects with the features named in the definition. But the fact of attribution alone doesn't tell us anything about the regularities these objects obey. If, however, immediately transmitted similarity relations are used in determining correct concept use instead, the possibility of identifying members of a similarity class on the basis of different, logically independent but empirically co-occurring features arises. This possibility is compatible with correct concept use just in case there is actually an empirical correlation between these features; if the features

233. 1964, *ET* pp. 258, 260.

234. 1970b, p. 274; similarly 1964, *ET* p. 260; 1974a, *ET* p. 312; 1974b, pp. 503–504; 1981, p. 21; 1983a, pp. 681–682.

235. 1970c, *SSR* p. 191.

236. The fact that this explication can't precisely capture the concept isn't really an issue here (see § 3.6.f). What matters here is to show *that* the concept contains knowledge of regularities.

237. 1964, *ET* pp. 258–259; Kuhn's argument here also rests on those of Braithwaite and Carnap. Kuhn explicates two different criteria for the Aristotelian concept of speed in his 1964, especially *ET* pp. 246–247. A similar case is Carnot's concept of caloric; see 1964, *ET* p. 259 n. 30, along with Kuhn's 1955b, alluded to there.

238. 1964, *ET* p. 259. This empirical content amounts, for the Aristotelian concepts of speed, to the claim that all motion is "quasi-uniform"; 1964, *ET* pp. 254–256.

239. 1964, *ET* p. 259.

are subsequently made explicit criteria for concept use, then of course their conjunction will have empirical content. Put in terms of Kuhn's later conception of concept learning,[240] whatever an individual member of a linguistic community uses to identify the referents of a concept is legitimate so long as its result, the similarity class it produces, is correct. Thus all new empirical knowledge about the features of elements of the similarity class can be used for referent identification.[241] It follows that the set of acceptable criteria for referent identification changes with increasing empirical knowledge about the elements of a concept's extension; new criteria may be added and old criteria modified or abandoned without the concept's extension changing. This process of criterion change by no means must proceed in the same way for all members of the linguistic community, and therefore the kind and degree of heterogeneity within the community is subject to change.

The fact that knowledge of regularities resides in empirical concepts has several important consequences. First, the knowledge of regularities residing in one or more empirical concepts gives rise to constraints on the introduction of further concepts. For the knowledge of regularities residing in the conceptual system as a whole must, if contradictions are to be avoided, be consistent. Second, the domain of possible theorizing by means of concepts in which knowledge of regularities resides is constrained, for only those propositions consistent with this indwelling knowledge of regularities can be formulated.[242] Third, certain "compelling" propositions with a peculiar status intermediate between analytic and synthetic now appear possible. These are just those propositions which explicitly articulate the knowledge of regularities implicit in empirical concepts.[243]

A fourth, and closely related, consequence is that we may now understand why the empirical concepts of abandoned scientific theories today appear self-undermining or confused, though they didn't appear so to language users of their period. For those claims to knowledge of regularities sanctioned by concept use may turn out to be only partially justified, or entirely unjustified, with the result that employing different criteria for referent determination in the same situation may lead to contradictory results. The concept itself then appears self-

240. Compare §§ 3.6.c and 3.6.d.

241. On this issue, see 1981, p. 19; 1983a, pp. 681–683.

242. Kuhn illustrates this intertwining of concepts and theory formation by three examples: Aristotelian dynamics, in 1964, *ET* pp. 255–260; 1977b, *ET* p. 20; 1981, pp. 8–12; phlogiston theory, in 1983a, pp. 674–676; and classical mechanics, in 1964, *ET* p. 260; 1983a, pp. 676–677; 1983d, pp. 566–567.

243. *SSR*, p. 78. —I will return to the peculiar status of such propositions in § 6.3.a, point 2.

undermining or confused. But the real problem isn't any *logical* error in concept formation, of the kind which prevents a concept's being applied consistently in all possible worlds. The problem is rather the falsity of certain assumed *empirical* correlations, which only precludes the concept's consistent application in certain special worlds, just those in which the empirical correlations fail to hold.[244]

Kuhn's views on the knowledge of regularities implicit in concept use may be illustrated by an example from contemporary physics. In Csonka 1969 an attempt is made to construct a theory on which the elementary interaction of particles is assumed temporally symmetrical and not, as is generally thought, retarded.[245] This idea, which, since the beginning of the century, had already been proposed and worked out for the special case of electrodynamics,[246] implies that causality, too, must be conceived as temporally symmetrical. It follows that, in addition to the usual temporal ordering of cause and effect, the temporal priority of effect over cause is also permitted.[247] The construction and articulation of such a theory, according to Csonka, runs into the difficulty that "our language is no longer capable of describing a world in which effect may precede its cause," for "the concepts of time and causal sequence have partially merged."[248] Expressed in Kuhn's terms, the problem is as follows. First, we find different criteria for the use and reconstruction of the cause-effect relation, which was introduced without explicit definition; one is the particular "causal generation" which connects cause with effect, and the other is the (merely necessary) criterion which stipulates that effect occurs after cause.[249] Second, the conjunction of the two criteria has the empirical content (though the conjunctive proposition may seem analytic) that causes always precede their effects. Third, the coexistence of the two criteria may be traced to the early phases of modern science, in which final causes were ruled out as scientifically illegitimate.[250] Fourth, one consequence of the triumph of this modern notion of causality is that "some consider

244. 1964, *ET* p. 242. Kuhn's analysis of the apparent confusion or self-undermining of the Aristotelian concept of speed is an exemplary illustration of this point; see *ET*, pp. 253–258.

245. This work makes no reference to Kuhn, though some of its formulations and conclusions show a surprising similarity with the appropriate bits of Kuhn's theory.

246. The classic sources are listed in Csonka 1969 nn. 2–4. The status of the proposal in 1975 is summarized in Pegg 1975.

247. Csonka 1969, p. 1266.

248. Ibid.

249. The problems associated with a precise understanding of this "causal generation" need not concern us here. See Brand 1979.

250. Csonka 1969, pp. 1266–1267.

it inconceivable" that effect might precede cause, and for "self-evident" reasons.[251] And fifth, the modern notion of causality actually places severe constraints on scientifically legitimate concept and theory formation, for it rules out any teleology which fails to reduce to this causality as scientifically illegitimate.[252]

At the beginning of this subsection I claimed that quasi-ontological knowledge and knowledge of regularities were inseparable moments in the knowledge implicit in similarity relations. This claim follows from the immediacy of similarity relations, for empirical concepts can't be precisely explicated,[253] which means that a division of the features of a concept into defining and empirical characteristics isn't possible, nor is the class of such features uniquely determined. In this case, features that allow us to *identify* an object or situation as a particular object or situation can't be distinguished from those behavioral attributes of a previously described class of objects or situations that pertain to the class on merely empirical grounds. But then knowledge of the existence of particular objects or situations becomes inseparable from knowledge of these objects' or situations' behavior, and both knowledge components appear as mere moments in that knowledge implicit in immediate similarity relations.

At last we are in a position to explicate further the interweaving of language (or, more precisely, the system of empirical concepts, or lexicon) with the world, as it is asserted by Kuhn.[254] One of Kuhn's ways of formulating the connection is by asserting that the concepts employed in a given historical community "were not intended for application to any possible world, but only to the world as the scientist saw it."[255] For, to begin with, it is only under certain circumstances that these concepts are learnable at all by means of immediate similarity relations.[256] Furthermore, the use of this conceptual system gives rise to claims to knowledge about nature when empirically correlated fea-

251. Csonka 1969, p. 1266.

252. Compare Spaeman and Löw 1981. —Csonka cites "an example in which our own prejudices aren't involved" to illustrate the "confusion which can result from the unconscious merging of two concepts," "the merging of the concepts of 'north' and 'downstream' in ancient Egypt" (pp. 1266, 1280). This identification was empirically correct for the region around the Nile, and unproblematic for word use, but led to inconsistent descriptions on the discovery of the Euphrates, which flows south. These are preserved on a tablet from the reign of Thutmosis I ((16th century B.C.).

253. Compare § 3.6.f.

254. Compare especially §§ 2.3 and 3.2.

255. 1964, *ET* pp. 259–260, also p. 254; similarly 1970a, *ET* p. 288; 1970b, p. 270; *ET*, pp. xxii–xxiii.

256. Compare §§ 3.3, 3.5, 3.6.b.

tures of elements of a concept's extension are used to identify these elements in applying the concept. Using this conceptual system, laden with knowledge of nature or the world, is fundamentally unproblematic only so long as nature or the world behaves in the manner asserted by this knowledge.

b. The Characteristics of Such Knowledge

We must now clarify in what sense "knowledge" is meant when we talk of the quasi-ontological knowledge and knowledge of regularities contained in immediate similarity relations. Kuhn himself is somewhat reluctant to designate these as "knowledge," for he allows the possibility that "Perhaps 'knowledge' is the wrong word."[257] Let us consider those features and capacities of the thing which prompt us to call it "knowledge."

To begin with, this knowledge is teachable and learnable, and it can also be overlearned.[258] This is a consequence of the property, assumed by Kuhn,[259] of immediate similarity relations being teachable, learnable, and overlearnable. In addition, this knowledge is subject to both confirmation and refutation,[260] for claims to quasi-ontological knowledge or knowledge of regularities can be fulfilled or disappointed. Furthermore, this knowledge is "systematic."[261] Designating it as systematic should ward off any impression that this knowledge, since incapable of being explicated, rests "on unanalyzable individual intuitions."[262] On the contrary, the propositions of such knowledge are the property of particular scientific communities and thus not individually subjective. They are also susceptible to analysis, though such analysis must take into account the fact that these propositions are implicitly contained in immediate similarity relations founded on paradigmatic exemplars. The total organization of scientific knowledge thus has more in common with the systematicity of Anglo-Saxon case law than with that of the codified law of Roman legal tradition[263] or that of an axiomatic system. Accordingly, the analysis of implicit

257. 1970c, *SSR* p. 196.

258. 1970a, *ET* pp. 285–286; 1970c, *SSR* p. 196; 1974b, p. 509.

259. Compare § 3.1.

260. 1970a, *ET* pp. 285–286; 1970c, *SSR* pp. 175, 196; 1974b, p. 509.

261. 1970c, *SSR* pp. 175, 191–192; similarly 1970b, pp. 274–275; 1974b, pp. 510–511.

262. 1970c, *SSR* p. 191.

263. Kuhn mentions the parallels to Anglo-Saxon legal tradition in passing in *SSR*, p. 23, and 1970b, p. 275. These parallels are somewhat further worked out in King 1971, section 4b.

knowledge can't start out from the assumption that it can be adequately reconstructed by means of universal propositions.[264]

Kuhn claims the following capacities on behalf of such knowledge. First, it is capable of generating more or less explicit prognoses and expectations on the future behavior of nature. This applies both to such objects and situations as can be expected and to their attributes and behavior.[265] In addition, such knowledge "provides a basis for rational action."[266] This claim should be uncontroversial, for accurate expectations on the action-relevant behavior of nature can only improve the chances of an instrumentally rational action's success. Finally, such knowledge clearly leaves open the possibility of comparatively evaluating different epistemic claims, choosing one as "more effective."[267]

The listed features and capacities of this form of knowledge grant its designation by "knowledge" every appearance of legitimacy. But one feature generally associated with this designation is lacking, for

> We have no direct access to what it is we know, no rules or generalizations with which to express this knowledge.[268]

This means, first of all, that such knowledge is implicit knowledge or, in terms introduced by Michael Polanyi, "tacit knowledge."[269] Expressed more positively, this knowledge is "built into language" or "intrinsic to language."[270] "Language" here evidently means linguistic *practice*, the learning and use of empirical concepts by means of immediate similarity reactions, that is, without explicit definitions. But in addition, Kuhn also means that such knowledge is, in principle, impossible to *explicate;*

> Its precise scope and content are, of course, impossible to specify, but it is sound knowledge nonetheless.[271]

264. See § 3.6.f. I will shortly return to the question of whether implicit knowledge, like empirical concepts, is in principle not susceptible to adequate explication.

265. E.g. *SSR,* pp. 62–65, along with the other passages cited at the beginning of this section.

266. 1970a, *ET* p. 285; similarly 1974a, *ET* p. 312.

267. 1970c, *SSR* p. 196.

268. 1970c, *SSR* p. 196; similarly 1970a, *ET* pp. 285–286; 1970b, p. 275.

269. *SSR,* p. 44 n. 1; 1963b, pp. 392–393; 1970b, p. 275; 1970c, *SSR* pp. 175, 191, 196. McIntyre claims that Kuhn's philosophy owes a great, but unacknowledged, debt to the works of Polanyi ("a view of natural science which seems largely indebted to the writings of Michael Polanyi [*Kuhn nowhere acknowledges any such debt*]," McIntyre 1977, cited as printed in Gutting 1980, p. 67, emphasis mine.). This less than flattering accusation is refuted in *SSR,* p. 44 n. 1; 1963a, p. 347 n. 1, and 1970c, *SSR* p. 191.

270. 1970b, pp. 270, 271, 272; 1981, p. 21.

271. 1970a, *ET* p. 285; similarly 1963b, pp. 392–393.

Stated more precisely,

> in learning . . . a language, as they must to participate in their community's work, new members acquire a set of cognitive commitments that are not, in principle, fully analyzable within that language itself.[272]

In other words, the knowledge built *into* the language can't be fully explicated *by means* of the particular language. But why should the precise explication of a language's implicit knowledge be impossible within the language? Doubtless because such an explication, as Kuhn elsewhere asserts, would have to articulate this knowledge in the form of "rules or generalizations,"[273] and this kind of articulation would "alter the nature of the knowledge possessed by the community,"[274] for he has "in mind a manner of knowing which is misconstrued if reconstructed in terms of rules."[275]

Naturally, if this knowledge isn't initially stored in the *form* of rules and generalizations, an explication which molds it into such a form will, in a certain sense, change its nature. Furthermore, if being implicit is an essential attribute of the nature of this knowledge, then any explication of such knowledge will, *qua* explication, fail to capture just this kind of knowledge. The philosophy of science shouldn't neglect the possibility, repeatedly emphasized by Kuhn, that the implicitness of the knowledge that is built into language will prove important to an understanding of scientific development.[276]

This having been admitted, however, it still isn't clear why the *content* of the knowledge implicit in a system of empirical concepts shouldn't, in principle, be explicable (or, in Kuhn's terms, "fully analyzable") in *the* language of which these empirical concepts are part. In any case, Kuhn offers no argument for this claim. And in fact, those claims to quasi-ontological knowledge sanctioned by the unproblematic use of empirical concepts learned by means of similarity relations can be explicated. No speaker seems to have any special trouble becoming conscious of the criteria by which he or she successfully distinguishes between the members of neighboring similarity classes. This capacity for distinction, unmediated by definitions, is just the basis for our expectation that there are no entities which fall in the gaps between

272. *ET,* p. xxii.
273. 1970c, *SSR* p. 196.
274. 1974a, *ET* p. 314, similarly p. 318; 1970c, *SSR* p. 175.
275. 1970c, *SSR* p. 192.
276. E.g. 1974a, *ET* pp. 312–313; 1970c, *SSR* pp. 191–198; 1981, p. 21. —See § 7.5.c below.

similarity classes.[277] The same holds for knowledge of regularities; there is no fundamental obstacle to articulating those expectations about the behavior of objects in the world that are woven into the fabric of concept use. Kuhn admits this indirectly. He asserts of someone who has learned to apply "swan" by means of immediate similarity relations that "He can point to a swan and tell you *there must be water nearby*."[278] And this, precisely, is the ability to explicate knowledge of a regularity.

This critique of Kuhn's claim that knowledge implicit in the use of empirical concepts can't be explicated merits three points of qualification. First of all, we have only claimed that Kuhn fails to demonstrate convincingly that such knowledge can't, *in principle,* be explicated. But this is not to deny that such explication may, *in fact,* prove extraordinarily difficult. Second, one shouldn't infer from the in-principle explicability of such epistemic claims that they must therefore be taken as entirely synthetic, that is, completely free of any concept-determining function. Denying this assertion is equivalent to asserting the indefinability of empirical concepts,[279] for the indefinability of empirical concepts rests on the fact that the features serving to identify referents and nonreferents can't be divided into those that are defining and those that are merely empirically correlated. With regard to the analytic/synthetic distinction, the status of those features alleged to represent "knowledge" of given objects is thus equally ambiguous. In other words, the analytic/synthetic distinction isn't meaningfully applicable to the knowledge implicit in empirical concepts.[280] As Kuhn puts it in 1981,

> In much of language learning these two sorts of knowledge—knowledge of words and knowledge of nature—are acquired together, not really two sorts of knowledge at all, but two faces of the single coinage that a language provides.[281]

Third, the claim that the *content* of such knowledge might in principle be explicated doesn't contradict Kuhn's claim that such explication can be very misleading, for such explication would indeed assimilate this form of knowledge to other, explicit forms of knowledge. Such assimilation will be misleading to the extent that the inexplicit form of this knowledge proves important to an understanding of scientific development.

277. See § 3.7.a.

278. 1974a, *ET* p. 312, emphasis mine; similarly 1970a, *ET* pp. 285–286.

279. See § 3.6.f.

280. This may be another reason for Kuhn's hesitancy in subsuming this knowledge under the concept *knowledge.*

281. 1981, p. 21.

3.8 The Nonneutrality of the Analyst's Viewpoint

We must now look back over Kuhn's project of the general analysis of the constitution of phenomenal worlds as a whole, since it was left open from the very beginning what claims to methodological status this analysis might plausibly make. How must Kuhn's general analysis of the constitution of phenomenal worlds be qualified, given the fact that the analyst attempting this project can have no neutral standpoint, divorced from any given phenomenal word? This problem will turn out to be so far-reaching and so complex that our efforts at treating it here must necessarily remain provisional and incomplete.[282]

It became clear in § 2.2.d that the stimulus ontology, in the version on which stimuli were held to be determinate but indescribable by us, led to considerable difficulties. To be sure, positing such stimuli apparently guarantees neutrality toward all possible phenomenal worlds in that it avoids attributing anything to stimuli from the perspective of any given phenomenal world.[283] But the cost of this neutrality is the impossibility of identifying (approximately) equivalent stimulus situations for different observers. This version of the stimulus ontology thus becomes unsuited to the task for which it was introduced in the first place. Stimulus situations can't be judged equivalent or different in any neutral way, independent of phenomenal worlds (at least, if we deny theories of stimuli the ability to capture the one universal reality even approximately as it is in itself).

Similarly, at the beginning of chapter 3, we noted the impossibility of carrying out the general analysis of the constitution of phenomenal worlds from a standpoint outside the analyst's phenomenal world. Initially, this impossibility holds of *the point of departure for such analysis.* Our phenomenal world may contain source texts in the history of science, or assertions made by persons belonging to other cultures, that are partially unintelligible. But only under certain special conditions will this unintelligibility provide us with a motive for attributing a

282. Compare the following discussion with the similar critique leveled by Gerhard Seel, in Seel 1988, against "constructivism in the sociology of science," as maintained by Karin Knorr-Cetina.

283. In truth, even this posit doesn't yield strict neutrality vis-à-vis all possible phenomenal worlds, for it makes the hidden assumption that the genetically subject-sided is *in itself* separate from the genetically object-sided—regardless of how inaccessible the two sides, though separate in themselves, might be to us *qua* separate. This thesis may be neutral relative to a particular class of phenomenal worlds, a class which may indeed include many of the phenomenal worlds assumed in the sciences. But it is questionable whether it even includes all of the phenomenal worlds assumed by the natural sciences, for in certain interpretations of quantum mechanics the observer can't be separated from the observed, even in principle (see e.g. Bohm 1980).

phenomenal world different from our own to the authors of these texts or assertions and for taking the constitution of this alien phenomenal world as susceptible to our analysis.[284]

First, such alien texts or assertions must be examined along the lines suggested by prior experience with texts or assertions from our own culture that, though initially completely or partially unintelligible, eventually turned out to have some palpable sense. The possibility of interpreting certain aspects of the behavior of members of an alien culture as successful communication *among themselves* can contribute to our willingness to ascribe some sense to unintelligible texts or assertions. This interpretation remains hypothetical for so long as we fail to understand what's communicated; but even within our own culture we encounter situations which may reasonably be interpreted as communication, albeit communication in which we have no part. Such situations must be taken as models.

Second, we must in some way, however rudimentary, be able to convince ourselves that the (partial) unintelligibility of the texts or assertions of an alien culture might constitute evidence for a phenomenal world different from our own. The results of perceptual psychology may show me, for example, that some of my perceptions (and those of other members of my culture) are subject to the influence of genetically subject-sided, variable factors.[285] The suggestion that perception *in general* might be influenced by subject-sided factors then becomes plausible, along with the corollary that the subject-sided factors might be systematically different in different cultures, being the product of systematically different learning processes. It follows that the phenomenal worlds of two different cultures can be different, in which case the empirical concepts characteristic of one will find no precise counterparts in the other.

Third, we must convince ourselves that an alien phenomenal world isn't in principle inaccessible to someone not at home in it. We might, for example, be convinced of the accessibility of other phenomenal worlds by our own experience as historians or ethnologists.[286] This conviction may be further supported by ontological and biological arguments.[287] But any conviction that alien phenomenal worlds aren't in principle inaccessible must have certain underpinnings in our own phenomenal world.

But in Kuhn's work even the *execution of a general analysis* of the

284. Some tentative hints at the first of the two points which follow may be found in 1970c, *SSR*, p. 193.

285. Compare § 2.1.b.

286. Compare § 2.1.b.

287. Compare § 2.2.c.

constitution of phenomenal worlds is necessarily and fundamentally molded by the phenomenal world of the analyst. For this analysis demands a host of assumptions plausible *only relative to the analyst's phenomenal world,* because they refer to objects in this phenomenal world. Most of these assumptions are of the anthropological variety. Kuhn assumes a certain learning capacity on the part of humans, allowing them to undergo a process of instruction which significantly influences their conception of reality.[288] In order for this learning process to proceed along the lines proposed by Kuhn, humans must have certain perceptual capacities, including the capacity to understand ostension for what it is,[289] mastery over certain categories that cannot be learned by ostension, and the capacity to understand and use the concepts of similarity and negation.[290]

The apparent self-evidence and near triviality of most of these anthropological assumptions rests on the fact that every person one encounters appears to have the aforementioned capacities. But in the present context, such assumptions aren't nearly as harmless as they might seem. For they are all undoubtedly propositions, gleaned from the natural standpoint and, for the time being, *realistically* interpreted, about objects of a particular phenomenal world, that of the analyst. This implies the further assumption that world-constituting people really exist, independent of the analyst and of his or her theories about them, and have these capacities *in themselves* and not simply in their relation to the analyst. But this is to make assertions with a claim to correctly describing, at least approximately, certain objects of *the one true* reality.

A dilemma similar to that faced by the modified stimulus ontology now appears to confront the general analysis of the constitution of phenomenal worlds.[291] On the one hand, we can accept the aforementioned anthropological assumptions as true in precisely the sense suggested by the natural standpoint, as at least approximately true statements about the one reality itself. In that case, there seems no reason not to accept those statements about other objects of the "mesocosm"[292] that appear just as convincing as our anthropological assumptions, as at least approximately true of reality. But to do so would go against the basic thrust of Kuhn's theory, for Kuhn asserts that, even

288. Compare § 3.1.
289. Compare §§ 3.5 and 3.6.b.
290. Compare § 3.3.
291. Compare § 2.2.e.
292. The term "mesocosm" was coined by Gerhard Vollmer; see e.g. Vollmer 1983, especially § 4.

in the most elementary empirical concepts learned by ostension, genetically object-sided and genetically subject-sided moments interact.[293] So if we are to avoid contradicting central portions of Kuhn's theory, it seems that we can't interpret the anthropological assumptions in which the general analysis of the constitution of phenomenal worlds must be grounded according to Peircean realism.

The alternative which presents itself requires that our anthropological propositions' claims to reality be weakened. It is tempting to do so by analogy to the weakened claims to reality associated with the first kind of stimulus concept in light of the second.[294] More concretely, this alternative wouldn't make the anthropological assertions arbitrary, in the sense of being entirely independent of the proprietary features, and hence of the resistance, of that which is in itself. But the descriptions contained in these anthropological assertions would stand in no describable relation to any features of that which is in itself. Our anthropological assumptions couldn't be taken as having any claim even approximately to capture *the* subject-independent reality in its own determinacy. Their applicability would be confined to the analyst's phenomenal world, and any objects—if there be such—which, in another phenomenal world, corresponded to the humans populating the analyst's own phenomenal world, might be incommensurably different from people as the analyst sees them.[295] In this case no analysis of the constitution of an alien phenomenal world could have any pretensions to capturing it even approximately as it really is. The alien phenomenal world which analysis has apparently made accessible would, in truth, be only a construction from within the analyst's own phenomenal world, to which it would remain ineluctably relative. An analyst living in another phenomenal world could, with equal justification, arrive at completely different results (granted that such an analysis is even meaningful from the perspective of this other world). Social solipsism follows; access to alien phenomenal worlds isn't possible, for our own culture can never be shaken off.

But isn't the position of someone attempting a general analysis of the constitution of phenomenal worlds, thus described, wholly analogous to the epistemic position of the natural scientist, as Kuhn describes

293. Compare §§ 3.2, 3.6.b, and 3.6.c. —The argument for the claim that even the most elementary empirical concepts contain subject-sided moments, and, *therefore,* can't refer to *the one* reality, only applies to Kuhn's theory, and not, for example, to Kant's. For Kuhn apparently holds *all* subject-sided moments of reality to be learned, and thus potentially variable.

294. Compare §§ 2.2.b and 2.2.c.

295. I will return to discuss the notion of incommensurability, here anticipated, in detail in § 6.3.

it, and hence tolerable? The epistemic subject in both cases is concerned with knowledge of some particular object, an alien phenomenal world in the former case, a given natural domain in the latter, and both cases can be attacked in different, incompatible (or even incommensurable) ways. In the natural sciences, different theories can be evaluated comparatively, and though, according to Kuhn, the theory evaluated as better can't make any claim to capture reality to a closer approximation, the natural sciences still qualify as an enterprise characterized by progress.[296] If we assume the possibility of comparatively evaluating different theoretical approaches to world-constitution, then we may judge that this domain, too, might exhibit scientific progress, without thereby making any claim that subsequent attempts to capture the constitution of reconstructed phenomenal worlds succeed in better approximating them as they actually are. Anyone who demands of the general analysis of the constitution of phenomenal worlds that its reconstructions at least approximately capture the originals as they really are demands too much, according to the standard of natural science. Approximating the truth (let alone actually capturing it) is an unattainable goal for humans, even in the natural sciences, and so the analysis of the constitution of phenomenal worlds will also have to be content with less.

But it would be inconsistent with the goals of Kuhn's theory if assertions about alien phenomenal worlds couldn't be interpreted realistically. For the whole purpose of any attempt to understand alien phenomenal worlds is to capture them in all their alterity; it is nothing else but to grasp them as they are in themselves.[297] If we abandon our goal

296. See §§ 5.4 and 7.6 below.

297. One might object that, since the intended goal of epistemic efforts in natural science is, in part, to capture reality as it is in itself, the ineliminable, genetically subject-sided moments in knowledge in the natural sciences contradict this goal. And yet there seem to be two important differences between the natural sciences and cultural sciences. First, in the case of the natural sciences, the presence of genetically subject-sided moments in knowledge can be reconciled with these goals by means of an altered understanding of "empirical reality." Empirical reality, the reality actually accessible to theory and practice, is denied subject-independence *both in everyday encounters and in science* for so long as either Kantian (historically invariant) or Kuhnian (historically variable) genetically subject-sided moments take part in its constitution. In the cultural sciences, by contrast, a similar shift is impossible, for it would consist in the concession that phenomenal worlds are never accessible to us as they are in themselves. But this isn't right, for our own phenomenal world is indeed accessible to us as it is *in itself,* since in the case of this world "in itself" and "for me" collapse (I will shortly return to this issue in the text). Second, regardless of how knowledge is characterized in the modern natural sciences, natural sciences may be ascribed a purpose in the in-principle technical applicability of all their results (see Hoyningen-Huene 1984a, p. 491ff, and Hoyningen-Huene 1989). Such potential technological fruitfulness surely can't be attributed to the general analysis of the constitution of phenomenal worlds.

of understanding alien phenomenal worlds even approximately as they are in themselves, our efforts at understanding seem to lose all purpose. For then the reconstructions we produce are pictures or models that, thoroughly tied to the analyst's phenomenal world, needn't coincide even approximately with phenomenal worlds as they are for members of alien cultures. And this situation, again, would be social solipsism.

And indeed Kuhn seems to interpret the results of his general analysis of the construction of phenomenal worlds realistically. This view resonates with his interpretation of the results of the new internal historiography of science.[298] Nowhere in Kuhn's work do we find any hint that these results might not be interpreted according to Peircean realism.[299] On the contrary, many turns of phrase suggest that the new historiography finally attempts to understand the history of science as it *really* happened. The new historiography's goal was just to "display the historical integrity of . . . science *in its own time*,"[300] or to "analyze an older science *in its own terms*."[301] Kuhn thus has no compunction in claiming that a particular assertion about the history of science "is simply a statement from *historic fact*, based upon examples,"[302] or that some particular kind of historical narrative has "the virtue of great verisimilitude."[303] Relatedly, Kuhn is perfectly willing to class certain results of the older historiography as erroneous:

> [The preceding example] illustrates once more the pattern of *historical mistakes* that misleads both students and laymen about the nature of the scientific enterprise.[304]

The use of "mistake" here is highly significant, because, for one, Kuhn staunchly opposes describing revolutions in (natural) science as learning from mistakes, as Popper would have it.[305] In addition, Kuhn holds the classification of abandoned or obsolete positions as "mistakes" to be inappropriate in epistemology, as well:

298. Compare § 1.2.c.

299. Kuhn's 1980a, alone of all his work, might be taken to suggest the rejection of a Peircean realist interpretation of historiography. For here he alludes to the view that all so-called historical facts contain an interpretive moment, summarized by him in the sentence "History is interpretive throughout" (1980a, p. 184). And yet the point of the pursuit (interpretive and otherwise) of historical fact is to be able to produce "narratives that aim to say what occurred and make it plausible" (1980a, p. 185). The interpretive moment in historiography thus appears to Kuhn to be compatible with its realist reading.

300. *SSR*, p. 3, emphasis mine; similarly 1970e, p. 68; 1977b, *ET* p. 11.

301. *SSR*, p. 167 n. 3, emphasis mine; similarly 1984, p. 250.

302. *SSR*, p. 77, emphasis mine; similarly p. 96.

303. *SSR*, p. 147.

304. *SSR*, p. 142, emphasis mine. Several such mistakes are discussed on this page.

305. 1970a, *ET* pp. 277–280.

> this very usual view of what occurs when scientists change their minds about fundamental matters can be neither all wrong nor *a mere mistake*. Rather it is an essential part of a philosophical paradigm.[306]

All this indicates that Kuhn sees the transition to the new internal historiography of science as a transition to a historiography entitled to realistic interpretation.[307]

Some critics have, understandably, seen Kuhn's realistic reading of historiographic results as the source of an internal contradiction in his theory.[308] What arguments could be adduced in support of a position claiming that the historiography of science and the general analysis of the constitution of phenomenal worlds can be interpreted according to Peircean realism[309] while prohibiting such interpretation for the results of natural science?[310] As far as this question is concerned, is there any relevant difference between the natural sciences and the sciences which describe, analyze, and explain alien world views?

There does appear to be a strategy which would allow us to grant the cultural sciences a status different from the natural sciences. This strategy proceeds from the assumption that cultural scientists always have one case at their disposal in which their knowledge of a particular world view as it is *in itself* can be taken as fact. This is the case of knowledge of our *own* world view, for here and only here "in itself" and "for me" coincide.[311] And so knowledge of alien world views can be represented primarily in terms of their divergence from our own. This representation shows how the components diverging from our

306. *SSR,* p. 121, emphasis mine.

307. Q.v. 1983a, p. 671, where Kuhn claims to see "no limits of principle" on the efforts of historians and ethnologists.

308. E.g. Giedymin 1970, pp. 257–258 n. 1; Holcomb 1987, p. 475; Munz 1985, p. 118; Phillips 1975, pp. 58–60; Radnitzky 1982, p. 71; Scheffler 1967, pp. 21–22, 74, and 1972, pp. 366–367; Shapere 1964, p. 397; q.v. Curd 1984, pp. 3–4. —Also compare the analogous criticism of Feyerabend 1984 in Hentschel 1985, p. 388, with n. 9.

309. Compare § 2.2.e.

310. The idea that the cultural sciences might have *better* epistemic prospects than the natural sciences was, apparently, first expressed by Vico: "Whoever reflects on this cannot but marvel that the philosophers should have bent all their energies to the study of the world or nature, which, since God made it, He alone knows; and that they should have neglected the study of the world of nations, or civil world, which, since men had made it, men could come to know" (Vico 1744, Book 1, part 3, trans. Bergin and Fisch [Ithaca: Cornell, 1968], p. 96). Vico's argument for this thesis makes explicit use, broadly speaking, of certain anthropological assumptions; Kuhn is forced to the same course, as we shall see later.

311. I by no means wish to claim that knowledge about one's own world view is unproblematic. Still, there does seem to be something like "privileged access" to one's own world view.

world view relate to one another in such a way as to produce a total world view whose appeal, like that of our own world view, lies in its virtues as a candidate picture of reality itself.[312] Of course, this kind of reconstruction of an alien world view is subject to error. Still, a more or less accurate reconstruction doesn't seem fundamentally impossible. For, to begin with, we already know one exemplary member of the class of world views that we can attempt to reconstruct just as it is in itself. Our own world view may or may not have an exemplary character with regard to other world views; but it serves above all as a point of departure for any reconstruction of an alien world view. And so the alien world view is presented through its deviations from our own. As to whether these deviations can be captured more or less as they are in themselves, we have evidence from the history of European science for the historical continuity between different world views suggesting that they can. The cultural sciences thus appear actually to have an access to alien world views that allows them to be captured more or less as they are in themselves. By contrast, according to Kuhn, the natural sciences, as far as the real nature of things is concerned, are forever stumbling in the dark, simply because they have no special case that, known for what it is in itself, could serve as a point of departure for further inquiry into nature. Progress in the natural sciences, which Kuhn never denies, can't be interpreted as drawing closer to the truth.[313]

But this argument for separating the epistemic standards placed on the natural and the cultural sciences runs afoul of the problem that it contains certain premises whose justification is less than transparent. For one, the argument makes anthropological assumptions, taking those capacities which play a role in world constitution as more or less strictly universal. In addition, alien world views are accessible only through certain channels, namely sources physically at hand, from which all information on alien world views is derived. These sources must exhibit a degree of physical persistence in order for them to function as bearers of meaning. Both kinds of assumption are normally, that is, from the natural standpoint, relatively unproblematic. They are not unproblematic, however, in the context of our present argument. For under the assumption that physical reality is never accessible to us as it is in itself, it becomes questionable how both anthropological assumptions and the assumed physical persistence of sources can be justified relative to a *plurality* of phenomenal worlds.

If a satisfactory solution to these fundamental problems should

312. See, for example, Kuhn's representation of the Aristotelian view in his 1981, pp. 9–12.

313. See § 7.6.d below.

present itself, the realistic interpretation of both historiography and the general analysis of the constitution of phenomenal worlds would be permissible. One further problem for the general analysis of the constitution of phenomenal worlds remains, however. This analysis makes claims to generality, or applicability to all phenomenal worlds. A proof that the results of such analysis can be interpreted realistically would still not allow us to cash out these claims, because the analysis might stray arbitrarily far from the truth for many (or even all) phenomenal worlds. How might we show that the general analysis of the constitution of phenomenal worlds actually generates results which hold for *all* phenomenal worlds? The most appealing strategy involves confirming the analysis in concrete individual cases. But even if this confirmation is successful for all past phenomenal worlds known to the historian, no real claim to universality can be asserted, for nothing ensures that the discovered factors in the constitution of this class of phenomenal worlds really apply to *all* phenomenal worlds. It might be that phenomenal worlds in the examined class were all constructed in comparable ways because they all belong to a common, more or less narrow tradition. This suspicion is especially troublesome when all test cases have been taken from the history of European science.

An approach that takes the anthropological basis of world constitution into account appears to offer the only prospect for a more rigorous justification of claims to universality. This approach requires that we know which capacities needed for the constitution of a phenomenal world humans come equipped with and which are developed over the course of interaction with the environment. If our knowledge of the anthropological foundations of world constitution is solid enough, we will presumably be able to make universal claims about the processes and structures which must *necessarily* participate in world constitution. It might, for example, turn out that the reason ostension is such a necessary moment in an individual's access to the world is that the individual is, in virtue of his or her capacities and limitations, dependent on an ostension-based learning process in order to gain access to any world in the first place. But this, of course, is mere speculation. In any case, the strategy outlined above faces the problem, discussed earlier, of establishing the universality of its anthropological assumptions in such a way as to remain consistent with other portions of Kuhn's theory, if these portions aren't to be sacrificed.

CHAPTER FOUR

The Paradigm Concept

THE LAST CHAPTER LEFT LARGELY OPEN how a network of similarity and dissimilarity relations, the learning of which is necessary in constituting a phenomenal world, acquires its fixed points. I contented myself with mentioning that, in Kuhn's theory, these fixed points are the objects of ostension and are called paradigms.[1] Now, "paradigm" is the label for what has become the best-known part of Kuhn's theory—best-known, at least, by name. In the most diverse scientific fields, paradigms are discussed or their existence denied;[2] as Margaret Masterman put it so nicely at the end of the 1960s, "in new scientific fields particularly, 'paradigm' and not 'hypothesis' is now the 'OK word.' "[3] "Paradigm" is here being used in a much broader sense than that originally intended by Kuhn. Accordingly, our first task is to ask in precisely what sense and for what purpose Kuhn originally introduced the paradigm concept (§ 4.1).

It's by no means the case that all blame for this broader use of the notion of a paradigm should be laid at the feet of Kuhn's readers. On the contrary, Kuhn himself uses the paradigm concept in broader senses quite shortly after its introduction. The vagueness of what is, after all, a central concept of Kuhn's theory is also a reason this notion so quickly

1. Compare § 3.3.

2. A selection can be found in Gutting 1980's bibliography, pp. 330–339; also see e.g. Barnes 1982; Barnett 1977; Bayertz 1981b; Bluhm 1982; Böhler 1972; Briskman 1972; Bryant 1975; Coleman and Salamon 1988; Crane 1980 and 1980a; Engelhardt 1977; Eysenck 1983; Harrey 1982; Haverkamp 1987; Hodysh 1977; Jauss 1979; Kobi 1977; Merton 1945; de Mey 1982; Meyer-Abich 1986, p. 120; Percival 1976 and 1979; Perry 1977; Postiglione and Scimecca 1983; Roth 1984; Rüsen 1977; Schmidt 1981; Schorsch 1988, pp. 24–31; Seiler 1980; Shrader-Frechette 1977; Spaemann 1983b, pp. 111–122; Strug 1984; Thimm 1975; Toellner 1977; Törnebohm 1978; Trenckmann and Ortmann 1980; Weimer and Palermo 1973; Winston 1976; young 1979, and many others.

3. Masterman 1970, p. 60.

became the focus of criticism.[4] In response, Kuhn attempted to set this criticism to rest by distinguishing the two chief meanings of the paradigm concept. Furthermore, he also sought to shift both meanings slightly away from those given the paradigm concept in *SSR*. In § 4.2 we will trace the development of the paradigm concept from its first and narrowest construal, through the period of its expansion and increasing ambiguity, to its later development in terms of the distinctions undertaken by Kuhn in 1969. This progression can be seen as the second chief line of development in Kuhn's thought.[5]

The 1969 differentiation of the paradigm concept resulted in the notion of a disciplinary matrix. This notion will be discussed in § 4.3. There we will be especially concerned to inquire what relations the so-called components of the disciplinary matrix bear toward one another. In the process, we will come to understand both how the original paradigm concept could become so ambiguous and why Kuhn stops using the term "disciplinary matrix" after 1969.

Finally, in § 4.4, we will discuss the function of paradigms in their narrowest sense, as exemplary problem solutions, in Kuhn's philosophy of science.

4.1 Reasons for Introducing the Original Paradigm Concept

According to Kuhn, the discovery that many historical and contemporary fields, especially fields of modern science, operate in research traditions resting on a relatively firm consensus among the participating specialists[6] serves as the point of departure for the introduction of the paradigm concept.[7] Regardless of whether or to what degree this con-

4. A selection: Austin 1972; Buchdahl 1965; Erpenbeck and Röseberg 1981, pp. 441ff; Hall 1963; Masterman 1970; Röseberg 1984, pp. 25ff; Schramm 1975; Shapere 1964, 1966, pp. 70–71 and 1971; Suppe 1974a; Toulmin 1963; Wisdom 1974. On this issue, cf. Stegmüller 1973, pp. 195–207, who in contrast to many other critics identifies precisely the function Kuhn had in mind for the notion of a paradigm: Kuhn 1976b, p. 182; q.v. Cedarbaum 1983.

5. As Kuhn puts it in 1969, "No aspect of my viewpoint has evolved more since the book was written" (1970b, p. 234; similarly 1970c, *SSR* p. 174). —The other main line of development is the transition from the perceptually oriented conception of the phenomenal world to a more strongly linguistically oriented conception: compare §§ 2.3, 3.6.d.

6. But recall the limiting qualifications placed on the status of apparently competitionless communities, discussed in § 1.1.b.

7. On the history of the paradigm concept before its famous use by Kuhn, see above all Cedarbaum 1983; also Blumenberg 1971; Toulmin 1973, pp. 106–107; Cohen 1985, p. 519. None of these authors mentions the fact that Neurath, Schlick, and Cassirer

sensus will prove truly monolithic on closer scrutiny, phases of scientific development characterized by such consensual traditions can easily be distinguished from those phases in which there is no universal consensus at all among the specialists in a given field.[8] Contemporary examples of fields with no universal consensus can be found in most of the social sciences. But phases in which this universal consensus was lacking can be found even in most[9] of the natural sciences, either by going back to the period before the first consensus or by seeking out phases of scientific development between the collapse of an old consensus and the forging of a new one.[10] In his 1959a, Kuhn calls the phases before consensus and with consensus the "preconsensus phase" and the phase with "firm consensus," respectively.[11] This same distinction is preserved in *SSR* and 1963a, though the two phases are called "preparadigm" and "paradigm" periods;[12] the motivation behind this new terminology will soon become clear.[13]

Two chief questions arise with regard to the existence of such a universal consensus in a given scientific field:

1. What, precisely, is the object of this consensus? This question arises in light of the implausibility of assuming the consensus to apply indiscriminately to all topics of scientific communication.
2. By what means is such consensus produced, both in the realm of

---

also discuss paradigms. In his critique of Popper's *Logic of Scientific Discovery,* Neurath frequently uses "paradigm" in the sense of an "ideal model" (Neurath 1935, pp. 353, 357, 361). Schlick uses it to mean, roughly, an "exemplary case" in his *Allgemeine Erkenntnislehre* and in his Winter 1933–34 lecture (Schlick 1918, second edition 1925, chap. 7, end of first paragraph; Schlick 1933–34, p. 45). Lichtenberg, who according to Toulmin and Cedarbaum first introduced the notion of a paradigm, wasn't unknown to Schlick, as evidenced by Schlick's reference to him in chap. 20 of his *Erkenntnislehre.* But Schlick's use of "paradigm" may also have been mediated by Cassirer's (1910) *Substanzbegriff und Funktionsbegriff;* on p. 243 he speaks of "paradigms" (emphasized in his text) in the sense of exemplary illustrations (of certain principles and theorems of pure mathematics). Schlick cites Cassirer's work in chap. 40 of his *Erkenntnislehre.* —Finally, it should be noted that Kuhn cites a work in *SSR,* p. 63 n. 12, whose title contains the term "paradigm": Bruner and Postman's "On the Perception of Incongruity: A Paradigm."

8. 1959a, *ET* p. 227; *SSR* pp. vii–viii, 43; 1963a, pp. 349, 351; 1963b, pp. 387–388; *ET* pp. xviii–xix.

9. Possible exceptions are those fields formed by the splitting or merging of fields with consensus: 1959a, *ET* p. 231; *SSR,* p. 15; 1963a, p. 353.

10. 1959a, *ET* pp. 230–232; 1961a, *ET* pp. 187 n. 11, 222; *SSR,* pp. viii, 11–15, 21, and elsewhere; 1963a, pp. 353–357.

11. 1959a, *ET* pp. 231–232.

12. 1963a, pp. 353–359; *SSR,* pp. 47, 163, and elsewhere.

13. A further designation for the periods with firm consensus is "normal science." I will return to discuss this notion in detail in § 5.1.

scientific training and within the scientific community at large in concluding a phase of dissent?

These questions are not, of course, independent; as we shall soon see, they can be regarded as two of Kuhn's approaches to the same goal, that of understanding the specifics of universal scientific consensus. The first three publications in which Kuhn discusses the paradigm concept differ in their ordering and weighting of the two approaches, among other points.[14]

The "concrete problem solutions that the profession has come to accept"[15] are basic to the total domain spanned by a scientific community's consensus. Though the scientific community's consensus isn't confined to these concrete problem solutions, they constitute a particularly important part of it, the part of greatest interest to the philosophy of science.[16] As Kuhn puts in his later clarification of the paradigm concept,

> One sense of 'paradigm' is global, embracing all the shared commitments of a scientific group; the other isolates a particularly important sort of commitment and is thus a subset of the first.[17]

The concrete problem solutions include, for one, those solutions encountered by students in the course of their training, in lectures, exercises, laboratory assignments, text books, and so on. They also encompass problem solutions equally accepted by the appropriate community but found only in the technical literature.[18] Consensus on particular problem solutions must evidently consist in at least two different consensual moments: agreement that a particular situation, articulated in a

14. 1959a is almost exclusively concerned with the issue of training, as it was presented at a conference on the identification of scientific talent to an audience in which psychologists predominated. *SSR* begins with reflections on the history of science and only later takes up the issue of training (pp. 46–47). Kuhn's 1963a begins with the issue of training, and then quickly turns to the history of science, as it was presented at a symposium on this field.

15. 1959a, *ET* p. 229, and identically 1963a, p. 351; similarly *SSR,* pp. viii, 10, 11, 42, chap. 5 and elsewhere; 1963a, pp. 351–352; 1963b, pp. 392–393; 1969c, *ET* p. 351; 1974a, *ET* p. 298; 1970b, pp. 235, 272; 1970c, *SSR* pp. 186–187; 1974b, pp. 500–501; *ET,* p. xxii; 1984, p. 245. —In 5.3 we will discuss the genus to which such scientific problems typically belong.

16. 1970b, p. 235; 1974a, *ET* p. 298; 1970c, *SSR* pp. 175, 181, 187.

17. 1974a, *ET* p. 294; similarly 1970c, *SSR* p. 175. —In § 4.3.e we will attempt to resolve the question of whether we can really speak of concrete problem solutions as forming a "subset" of the set of shared commitments.

18. *SSR,* p. 43; 1974a, *ET* pp. 305–307; 1970b, p. 272; 1970c, *SSR* p. 187. —Any division of problem solutions into these two classes won't, of course, remain constant over time.

particular way, constitutes a *scientific problem,* and agreement that a particular way of dealing with the problem constitutes a *scientifically acceptable solution to it.* I will return to these components later.[19]

But first we must ask what it means for a given community to accept concrete problem solutions; what are they accepted *as?* This question demands a specification of the parameters of consensus as it applies to concrete problem solutions. Kuhn's answer is that these concrete solutions to particular problems are accepted *not merely as what they are in themselves but also insofar as they provide guides for research,* as the basis of scientific practice.[20] Two aspects of consensus can now be distinguished. The first is the *locally normative* aspect, according to which a proposed problem solution is accepted as the appropriate solution to a problem. The second is the *globally normative* aspect, which consists in the mandate to generalize the locally normative aspect, projecting it on further research in the relevant field.[21] The globally normative aspect of consensus on concrete problem solutions, their *function as exemplars,* is what led Kuhn to call them "paradigms."[22] In § 4.4 we will discuss how the special way in which paradigms are recognized affects the conduct of research. For now we must ask how Kuhn justifies his claim that concrete problem solutions constitute a particularly important element of the research-governing consensus of a scientific community.

Kuhn's argument proceeds comparatively, by showing that concrete problem solutions take priority over other candidates for the establishment and maintenance of the form of consensus peculiar to a scientific community.[23] The other candidates Kuhn mentions are "concepts,"[24] "conceptual models,"[25] "definitions,"[26] "defining characteristics of . . . quasi-theoretical terms,"[27] "laws,"[28] "theories,"[29] "points

19. See §§ 4.4.b and 4.4.c.

20. E.g. 1959a, *ET* p. 235; *SSR* pp. 10, 100, and many others; 1963a, pp. 351–352. Kuhn repeatedly uses the term "consensus" to indicate the second aspect of consensus.

21. This distinction between the two aspects of consensus only approximate Kuhn's distinction between the "cognitive function" and the "normative function" of paradigms (*SSR,* p. 109). In this passage, "paradigm" is used in its broad sense, as is clear by the end of the paragraph in which this distinction is introduced.

22. The English word "paradigm," however, is inadequate to Kuhn's purposes insofar as it suggests the arbitrary substitutivity of similar exemplars; *SSR,* p. 23; *ET,* p. xix.

23. For what follows, see also Lugg 1987.

24. *SSR,* pp. 11, 46.

25. 1963b, p. 391.

26. *SSR,* p. 47; similarly 1974b, p. 511; *ET,* p. xix; 1984, p. 245.

27. *ET,* p. xviii.

28. *SSR,* pp. 11, 46; 1970c, *SSR* p. 191.

29. *SSR,* pp. 11, 46; 1970c, *SSR* pp. 187–188; 1969c, *ET* p. 351.

of view,"[30] "(explicit) rules,"[31] "(fundamental) assumptions,"[32] "principles,"[33] "(explicit) generalizations,"[34] "rationalizations (of paradigms),"[35] "logically atomic components (of paradigms),"[36] and "abstract characteristics (of paradigms)."[37]

What does this rather heterogeneous list address? Kuhn aims to establish that the globally normative aspects of consensus involve more than consensus over *explicit and unequivocal guides to action* (whether absolute or conditional). For paradigms serve "*implicitly* to define the legitimate problems and methods of a research field for generations of practitioners."[38] It follows that consensus isn't primarily, let alone exclusively, established by the following sorts of elements:

—by explicit definitions of concepts, that is, by explicit necessary and sufficient criteria for concept application; these would provide explicit guides to action in concept use;

—by laws or theories conceived in abstraction from their consummated application to concrete individual cases, where such theories are considered unequivocally applicable to all possible cases;

—by any kind of explicit, unequivocal methodological precepts, such as recipes for problem choice, the evaluation of problem solutions, crisis identification, theory improvement, theory evaluation, theory comparison, theory rejection, and so forth.[39]

Kuhn calls all such explicit guides to action "rules," in slight departure from common usage. Concrete problem solutions consensually accepted by the community, by contrast, play their globally normative role in an implicit way, serving as exemplary models for scientific practice, as a source of analogies.[40]

How might one demonstrate that concrete problem solutions, not rules in the abovementioned sense, are dominant in the establishment

30. *SSR*, pp. 11, 42.

31. *SSR*, pp. 42–49, 88; 1974a, *ET* pp. 302–307, *ET* pp. 318–319; 1970b, pp. 272–273; 1970c, *SSR* pp. 175, 187, 191; 1974b, p. 511; 1984, p. 245.

32. *SSR*, pp. 42, 44, 45, 46, 49, 88; 1963b, p. 391.

33. *SSR*, p. 43.

34. *SSR*, p. 43; 1970a, *ET* pp. 284–288; 1974a, *ET* pp. 302 n. 11; 1970b, pp. 274–275.

35. *SSR*, pp. 44, 49.

36. *SSR*, p. 11.

37. *SSR*, pp. 44, 46.

38. *SSR*, p. 10, emphasis mine; similarly pp. 16–17. Compare references to Polyani's "tacit knowledge" in n. 269 to § 3.7.

39. See § 4.3.c below.

40. 1984, p. 245. —See § 4.4 below.

and maintenance of scientific consensus? Kuhn's demonstration consists in the following four arguments.

1. When a historian (or a contemporary) examines a research tradition with an eye toward its consensual moments, finding a core of common, concrete problem solutions is, in general, unproblematic. The coherence of the research tradition associated with this core is a consequence of the fact that scientists find and process their research problems by forming analogies to these exemplary problem solutions.[41] By contrast, the search for a set of *rules which exhaustively determine a community's research practice* generally isn't sussessful; individual such rules may exist, but they aren't enough to explain the research tradition's coherence. Even the attempt to reconstruct a set of rules assumed to underlie research practice *implicitly* generally misses its mark.[42]

The factual possibility of concrete problem solutions' taking precedence over rules (in the above sense) is grounded in the truth of the following proposition.[43] A consensus over the concrete problem solutions that govern a common research practice doesn't imply that there's also agreement about the characteristics of this research practice, characteristics that, because of their inarticulacy, can only be gained by reflecting on the practice itself. In short, sharing a common practice doesn't imply sharing a common theory about that practice. For research guided by paradigms, this means that explicit definitions of concepts and methodological precepts of the kind that can only be extracted by reflection on the research tradition aren't necessarily part of the consensus. They are thus ruled out as the primary vehicles of a research tradition's coherence.

2. The fact cited in (1) also has its expression in scientific training. The contents of science—concepts, laws, theories—are always learned together with exemplary problem solutions, concrete problem solutions which function as more than mere illustrations. They serve, above all, to fix the meanings of the concepts they employ. An important moment of scientific knowledge thus resides in them.[44]

41. Compare § 3.6.3. —The fact that we are indeed concerned with basic science becomes important here, for the identification problems in the applied sciences, especially, may be determined by external factors; compare § 1.1.a.

42. *SSR*, pp. 42–46; 1970a, *ET* pp. 284–288; 1974a, *ET* pp. 302–319; 1970b, pp. 273–275; 1970c, *SSR* pp. 187–198; 1979b, pp. 412–415; 1981, pp. 19–21; 1984, p. 245. —Also see § 3.6.f, as well as Dreyfus 1979, chap. 8, on the whole complex of problems surrounding rules.

43. Compare *SSR*, p. 44.

44. *SSR*, pp. 46–47, 80; 1970a, *ET* pp. 284–288; 1977b, *ET* p. 17; 1974a, *ET* pp.

3. The primacy of concrete problem solutions over rules as vehicles for the globally normative aspect of research consensus is also indirectly confirmed by the fact that debates over rules arise primarily in times when consensus over concrete problem situations is lacking or oscillates. The need for explicit rules comes about only when the actual bearers of consensus fail to fulfill their function.[45]

This historical fact about science initially only allows us to infer that an existing research consensus isn't primarily borne by *explicit* rules. But we must further accept that the aforementioned debates generally don't lead to any consensus about explicit rules. This failure is easily explained under the assumption that there are no rules which serve, even *implicitly,* as sufficient guides to action in research.

4. Finally, the great abundance of sciences, and especially the abundance of specialties within larger fields, can be better explained by reference to problem solutions than to rules. For different groups of specialists who agree on general research methodology and on which theories to apply differ precisely in their particular applications of theory, in the concrete problem solutions on which the work of each group is based.[46]

But this argument for the primacy of problem solutions over rules in establishing scientific consensus isn't convincing. For different problem solutions could fulfill their social function, the microstructuring of scientific communities, through the operation of unconscious rules. In his 1969 papers, Kuhn modifies the argument accordingly.[47] He now asserts that, by comparison with certain rules also subject to consensus,[48] concrete problem solutions are most suited to fixing the microstructure of the scientific community. We may want to agree, but this claim is irrelevant to the question currently under discussion.

The first of the four arguments is the most important and the one to which Kuhn dedicates by far the most space. This argument attempts to demonstrate directly the priority, in the consensual practice of research, of concrete problem solutions over rules, where "rules" is understood to refer primarily to explicit definitions of concepts and to general laws and theories conceived in abstraction from any application to individual cases. We may indeed find laws and theories as components of a consensus, but taken in themselves, apart from any concrete

306–307; 1970b, pp. 272–275; 1970c, *SSR* pp. 187–191; 1974b, p. 501. —Compare §§ 3.6.e and 3.7.

45. *SSR,* pp. 47–49.

46. *SSR,* pp. 50–51.

47. 1974a, *ET* p. 307 n. 17; 1970c, *SSR* p. 187.

48. The "symbolic generalizations" are meant here; see § 4.3.a below.

application, they are too little specified or have too little content to determine consensually either how they should be applied or in which problem situations they should be applied with sufficient precision.[49] But in what way is this missing content, the content over which a scientific community with a common research practice must have mastery, supplemented? Certainly not by explicit definitions of the appropriate concepts, for these play a minimal role in scientific practice. Nor can this content be *adequately reconstructed* by means of explicit definitions.[50] It consists rather in the immediate similarity relations learned by means of concrete problem situations.[51] This same relationship between rules and concrete problem solutions is also expressed in the second argument, but from the different perspective of scientific training.

Have we thus cashed out Kuhn's claim to have established the priority of concrete problem solutions over rules? As it stands, this question has no simple affirmative or negative answer. We must first distinguish between different types of rules.

1. Concrete problem solutions take no priority over rules in the sense of *laws or theories,* where these laws and theories are used in determining the referents of their component empirical concepts; indeed, this is precisely the role Kuhn wishes to reserve for them.[52] In this case, concrete problem solutions and laws or theories form an indivisible unit: on the one hand, a law or theory is unintelligible *without* an adequate number of applications, for its empirical concepts are underdetermined; on the other, individual problem solutions are unconnected with one another and only partially intelligible *without* the law or theory whose applications they embody to lend them unity.[53]

2. Concrete problem situations do, however, have priority over rules in the sense of *explicit definitions,* for more or less adequate explicit definitions of employed empirical concepts can only be gleaned by abstraction from concrete problem solutions.[54] But still, concrete problem solutions can't be taken as independent of the laws and theories which participated in the determination of the referents of the empirical

49. 1970a, *ET* p. 284; 1970c, *SSR* p. 188. —Compare § 3.6.e.

50. Compare § 3.6.f.

51. Compare §§ 3.6 and 4.4.a below.

52. Compare § 3.6.e.

53. In Hegelian terms: Theory or law on the one hand and applications on the other, as a *concrete*-universal, form a dialectical unity, whose moments are inseparable from one another; each moment contains the other within it. I will return to this relationship in § 4.3.e.

54. Compare § 3.6.f.

concepts occurring in them; these laws and theories are rather a moment in the problem solutions.

In conclusion, we note that Kuhn sometimes refers to "the priority of paradigms"[55] in another sense. In this sense, the expression concerns paradigms in the broad sense, namely, the whole of a scientific consensus. Paradigms in this sense enjoy priority over all three kinds of rules: explicit definitions, laws and theories, and methodological precepts. For such "rules" can be read off the totality of scientific consensus only by abstraction,[56] and to this extent they are indeed secondary.

4.2 The Development of the Paradigm Concept

Kuhn's writings didn't stick to the original meaning of the paradigm concept. Paradigms in their original sense are, according to their first appearance on the scene, "concrete problem solutions that the profession has come to accept."[57] But Kuhn soon used the paradigm concept in other senses, in works composed around the same time as *SSR* and even in *SSR* itself, without being fully aware that he was doing so. Finally, in his 1969 papers, he consciously changed the paradigm concept in response to criticism provoked by its ambiguity. These conscious changes can be seen as a second chief line of development in Kuhn's thought to date,[58] though this development consists far more in the correction of earlier formulations and the explication of earlier allusions than in any substantial retraction. "On fundamentals my viewpoint is very nearly unchanged,"[59] says Kuhn ten years after first introducing the paradigm concept; this concept, as before, remains "the central element of what [he takes] to be the most novel and least understood aspect of this book [SSR]"[60] and thus of his philosophy of science as well. In this section, I will sketch this second line of development in Kuhn's thought. Later we will be in a position to try to make sense of certain aspects of the development.[61]

55. Such is the title of chap. 5 of *SSR;* see also p. 11.

56. *SSR,* pp. 11, 42, 43. In § 4.3.e it will become clear that this process is indeed one of abstraction, not simply the separate inspection of parts, components, or elements of the whole.

57. Compare § 4.1.

58. The first chief line of development was the transition from the more perceptually oriented to the more linguistically oriented conception of the phenomenal world and its construction; compare especially §§ 2.3 and 3.6.d.

59. 1970c, *SSR* p. 174.

60. 1970c, *SSR* p. 187; similarly 1974a, *ET* p. 294; 1970b, pp. 234–235.

61. See § 4.3.e.

Before Kuhn introduced the concept of a paradigm in his 1959a, those passages of his work in the history of science where one would later have expected to find talk of paradigms make quite traditional use of theories, theory applications, belief, views, and, as a gloss on "theories," "conceptual schemata."[62] This emphasis on the conceptual moment of theories is largely continuous with the predominance of the conceptual aspect in the later notion of a paradigm,[63] and accordingly, the use of such expressions as "conceptual transformation" with regard to scientific revolutions remains prominent as well.[64]

In examining the development of Kuhn's paradigm concept after 1959, we should distinguish two parallel streams. One involves the bifurcation of the paradigm concept into "disciplinary matrix" and "exemplar" (§ 4.2.a), while the other leads to a weakening of the paradigm concept (§ 4.2.b).

a. From "Paradigm" to "Disciplinary Matrix"

One development in Kuhn's thought, initially unnoticed by him, took place between the first use of the paradigm concept in his 1959a and its famous appearance in *SSR*. This development was, in his own judgment, responsible for much of the confusion surrounding the paradigm concept.[65] It is especially in evidence in his 1963a, which was composed in 1961, immediately after the first version of *SSR*.[66] The point of departure for this development is the paradigm conceived as a generally accepted exemplary problem solution, of the sort found in scientific textbooks.[67] Next Kuhn uses "paradigm" to refer not only to such problem solutions but also to the scientific classics serving as models for scientific practice.[68] "Paradigm" is then applied to the theories contained in these classics.[69] Finally, "paradigm" encompasses everything: a generally accepted theory including exemplary problem solutions,

62. E.g. 1952a, pp. 27, 36; 1957a, pp. 36–41, 75–77, and elsewhere (see index, p. 293).

63. See § 4.4.a below.

64. E.g. 1952a, p. 14; 1955a, p. 95; 1957a, pp. vii, 1, 183, and elsewhere; see chap. 6 below.

65. 1974a, *ET* pp. 318–319; 1970b, p. 234; 1970c, *SSR* pp. 181–182; *ET*, pp. xviii–xx, and elsewhere. —Here and in what follows, I will cite 1974a, 1970b, 1970c, and 1974b in the order of their composition, not their publication; see 1974b, p. 500 n. 2 and *ET*, p. xx and n. 8. on this order.

66. 1963a, p. 347 n. 1; 1970b, p. 249; *ET* p. xviii n. 6.

67. 1959a, *ET* pp. 229–238 passim; 1963a, p. 351.

68. The transition to this new meaning can be found in 1963a, pp. 351–352.

69. 1963a, pp. 356 and n. 2, 367.

governing research, with implications for what there is in the world, how it behaves, what questions we may ask about it, what methods may be used in pursuit of these questions, and what answers we may expect.[70] "Paradigm" now refers, as Kuhn later admits, to everything subject to professional consensus in a given scientific community.[71] In *SSR,* too, "paradigm" is used in narrow and broad senses, in both of which paradigms have a variety of functions.[72] Thus Margaret Masterman was led, in a remark often quoted (sometimes with delight), to claim that Kuhn uses "paradigm" in at least twenty-one or twenty-two meanings in *SSR.*[73]

In his 1970a, first presented in 1965, Kuhn begins to reverse the expansion of the paradigm concept that went unnoticed for so many years,[74] hoping, on his part, to reverse the associated confusion as well.[75] In his 1969 papers, with the aim of recapturing "the original sense of the term,"[76] he introduces the distinction between "disciplinary matrix" and "exemplars." "Disciplinary matrix" replaces "paradigm" in its broadest sense, while "exemplars" replaces "paradigms" in its narrowest sense.[77] The disciplinary matrix, as the totality of all objects of scientific consensus, contains, in addition to other elements, the exemplars to which alone the term "paradigm" was, philologically speaking, properly applicable. Having lost control of that term, he is forced to abandon it, but that to which it refers, the exemplary objects of ostension, ascription, and exclusion, remains indispensable up to the present.[78]

Where Kuhn uses "paradigm" at all in his work since 1969, it usually is in what he takes to be the more basic sense of "exemplar,"[79] though the broad sense still makes an occasional appearance.[80] But he has now resumed talking of "theories" and "theory choice," where in

70. 1963a, pp. 358, 359, 362.

71. 1974a, *ET* p. 294; *ET* p. xix.

72. See § 4.4 below.

73. Masterman 1970, p. 61. —The widely held but false view that Masterman identified at least twenty-*two* (as opposed to twenty-one) meanings of the paradigm concept can be traced to Kuhn himself; 1974a, *ET* p. 294, and 1970c, *SSR* p. 181.

74. *ET,* p. xviii n. 6.

75. 1970a, *ET* pp. 267 n. 3, 284.

76. *ET,* p. xx.

77. 1974a, *ET* pp. 293–294, 297–298; 1970b, pp. 271–272; 1970c, *SSR* pp. 175, 182–187.

78. 1974a, *ET* p. 307 n. 16, and p. 319; 1970b, pp. 235, 272. —Paradigms are described as "(concrete) examples" (in the sense of objects of ostension) in 1989a, p. 15 n. 11, and 1990, p. 316, n. 9, pp. 302, 314.

79. 1983b, p. 715; 1984, p. 245.

80. 1971a, pp. 144–145; 1977c, *ET* p. 320.

SSR he usually, though not always, substituted "paradigm."[81] "Disciplinary matrix" isn't used at all after 1969.

b. *The Retraction of the Property of Universal Acceptance*

Parallel to the aforementioned development of the paradigm concept, though doctrinally largely independent, is a second dimension of change in the paradigm concept of Kuhn's 1969 papers. In *SSR,* as in 1959a and 1963a, the notion of a paradigm is closely related to that of normal science, where normal science describes a research practice resting on a consensus among essentially all participating specialists.[82] As we saw in § 4.1, Kuhn claims that this universal consensus has a core, by reference to which the special research practice of normal science may be explained. This consensual core is the paradigm or set of paradigms. With normal science in view, Kuhn describes the paradigm concept as a core of *universal* consensus, consensus carried by all affected specialists; in their original sense, paradigms are "concrete problem solutions that the profession has come to accept."[83]

But Kuhn later realized that scientific schools in competition with each other may have an internal consensus with just the same elements as consensus in competitionless communities.[84] However, neither the school's consensual core nor the totality of its consensual elements may be called a paradigm, since both lack the universal acceptance throughout the profession provided for in the definition of a paradigm. As a consequence, the emergence of a competitionless community out of a school had to be described as Kuhn described it in *SSR,* as the "acquisition" or "emergence" of a paradigm.[85] But such formulations have made "a paradigm seem a quasi-mystical entity or property which, like charisma, transforms those infected by it."[86] In order to avoid this and other misunderstandings, Kuhn in 1969 abandons the requirement of *universal* acceptance as a feature of the paradigm concept, in both narrow and broad senses:

> The members of all scientific communities, including the schools of the "pre-paradigm" period, share the sorts of elements which I have collectively labelled 'a paradigm.'[87]

81. E.g. 1977c; 1979b, pp. 415–417.
82. See chap. 5 below.
83. 1959a, *ET* p. 229, and identically 1963a, p. 351; compare § 4.1.
84. 1974a, *ET* p. 295 n. 4; 1970b, p. 272 n. 1; 1970c, *SSR* pp. 178–179.
85. E.g. *SSR,* pp. 11, 13, 15, 18, 19.
86. 1974a, *ET* p. 295 n. 4; 1970b, p. 272 n. 1.
87. 1970c, *SSR* p. 179.

This weakening of the paradigm concept is already implicitly contained in *SSR* and 1963a and may be elicited from two features of these texts. First of all, Kuhn doesn't strictly hold to his initial definition of the paradigm concept. We find him talking, for example, of "the first nearly *universally accepted* paradigm for physical optics," of "first *firm* paradigms," or of the "first *universally* received paradigms."[88] But if the original definition of a paradigm is actually presupposed, such formulations are redundant. In these formulations, Kuhn apparently already has in mind the weakened paradigm concept, which indeed requires the listed qualifications in order to describe a universal consensus. Furthermore, in his preface to *SSR,* written at the end of the book's composition,[89] Kuhn begins to distance himself from the view that the preconsensus phase can be distinguished by means of the paradigm concept from the phase which follows. This distinction, he claims, has been made

> much too schematic. Each of the schools whose competition characterizes the earlier period is guided by something much like a paradigm. . . . Mere possession of a paradigm is not quite a sufficient criterion

for marking the transition from preconsensus to consensus phase.[90] And so the way is prepared for the weakening of the paradigm concept undertaken explicitly in the 1969 papers.

But even in 1969, Kuhn isn't of the view that the difference between universal and school-specific paradigms is only social, completely external to the paradigms themselves. Paradigms characteristic of the consensus phase are rather of another "sort" of "nature."[91] Only this kind of paradigm permits the kind of scientific practice which, characteristic of the consensus phase, distinguishes it from the preconsensus phase.[92]

88. *SSR,* pp. 13, 15, 15, 61; 1963a, pp. 351–352, 365, all emphasis mine. Also see e.g. *SSR,* p. 123, where Kuhn claims that new paradigms are born in flashes of intuition. Here, too, it should be clear that universal acceptance isn't part of the paradigm concept.

89. The preface is dated February, 1962; p. xii.

90. *SSR,* p. ix; similarly 1963a, p. 353.

91. 1970c, *SSR* p. 179. —In this passage we also find the following, highly misleading sentence: "What changes with the transition to maturity is not the presence of a paradigm but rather its nature." But of course those exemplary achievements of science that, initially accepted only within a school, become accepted by the entire community of scientists don't change their "nature" in the course of this transition, not in the sense of changing with regard to some essential attributes.

92. See chap. 5 below, especially §§ 5.2 and 5.5.b.

4.3. The Disciplinary Matrix

The uncontrollable expansion of the paradigm concept and its resulting ambiguity[93] are what motivated the distinction of "disciplinary matrix" from "exemplars" in the relevant papers composed in 1969. "Disciplinary matrix" is meant to replace "paradigm" in the broad sense by encompassing the totality of all objects of scientific consensus. Still, "paradigm" in the broad sense isn't entirely congruent with "disciplinary matrix," for as Kuhn notices, *consciously* addressing himself to the totality of all objects of scientific consensus, there are moments in this consensus which were never taken into account in the *uncontrolled* expansion of the paradigm concept. The function ascribed to the disciplinary matrix of a given scientific community is to account

> for the relative fullness of their professional communication and the relative unanimity of their professional judgments[94]

Even Kuhn's choice of the word "matrix" is meant to suggest that the disciplinary matrix consists of different "elements," "constituents," "components," or "subsets."[95] Kuhn's list, with no claims to completeness,[96] names four sorts of components: symbolic generalizations, models, values, and exemplary problem solutions. These four components will be discussed, in order, in the four subsections which follow. Afterward, we will have to ask what determines the relationships the four components of the disciplinary matrix bear to one another (§ 4.3.e).

a. Symbolic Generalizations

By "symbolic generalizations," Kuhn understands the formalized or easily formalizable *universal* propositions regarded by a scientific community as natural laws or the fundamental equations of theories.[97] Conceived as part of the disciplinary matrix, these universal propositions

93. Compare § 4.2.a.

94. 1970c, *SSR* p. 182; similarly 1974a, *ET* p. 297; 1970b, p. 271.

95. 1974a, *ET* pp. 294, 297–298; 1970b, pp. 271–272; 1970c, *SSR* pp. 175, 182. —Although I find Kuhn's descriptions of what, as we shall see, are *moments* of the disciplinary matrix as "elements," "components," etc., inadequate, I will continue to follow Kuhn in calling them "components" of the disciplinary matrix until I demonstrate the inadequacy of this label in § 4.3.e.

96. 1974a, *ET* p. 297; 1970c, *SSR* p. 182.

97. 1974a, *ET* pp. 297–302; 1970b, pp. 271–273; 1970c, *SSR* pp. 182–184, 188–189; 1974b, pp. 516–517. —In the cited passages, Kuhn doesn't distinguish between fundamental equations and natural laws, and he makes explicit reference only to natural laws (1970c, *SSR* p. 183). But fundamental equations, generally not described as natural laws,

must, in contrast to their use in scientific practice, be read as purely formal, divorced from the meanings of all constituent nonlogical, nonmathematical symbols by an abstraction from all empirical concepts that results in a string of uninterpreted signs.[98] Symbolic generalizations are thus *artifacts constructed by the philosopher of science,* for reasons we will discuss shortly. As to their function, we should not assume that, once interpreted, they operate only as empirical laws; on the contrary, symbolic generalizations also have concept-determining functions, inextricably linked with their character as laws.[99]

In setting laws and fundamental equations as a component of the disciplinary matrix, why does Kuhn abstract from the meanings of their empirical concepts? His reason is that a scientific community's consensus over laws or fundamental equations contains two distinguishable consensual moments with a certain degree of mutual independence. One moment of this consensus involves agreement over the *logical form* of laws or equations, what Kuhn calls symbolic generalizations. The other concerns the *logical interpretation* of these formulae, the ascription of meaning to them. Different scientific communities may agree on symbolic generalizations but differ on the meanings of the empirical concepts to be inserted in them.[100] It thus makes sense to distinguish the two consensual moments contained in the disciplinary matrix, since Kuhn's interest lies especially in the problem of the meaning of empirical concepts; the bearers of meaning for empirical concepts, according to Kuhn, should be sought above all in another component of the disciplinary matrix, that of exemplary problem solutions.[101]

b. Models

"Models," for Kuhn, is a seemingly heterogeneous designation. On the one hand, it encompasses *heuristic* models and analogies in accordance with which phenomena from a given class may be treated *as if* they were something else entirely. On the other, *ontological* (or, equiva-

---

are also at issue; consider his use of the Schrödinger equation (1974a, *ET* pp. 298, 307 n. 17) or, similarly, Maxwell's equations (*SSR,* p. 40) as examples. —In Kuhn's own judgment, *SSR* seriously neglects the role of "the formal elements of a science," such as laws (1974b, pp. 516–517). The serious consideration they deserve is inserted with the treatment of symbolic generalizations as components of the disciplinary matrix in the 1969 papers.

98. 1974a, *ET* p. 299. —That there is no abstraction from the meanings of mathematical and logical symbols is evident from the fact that the acceptance of symbolic generalizations implies the acceptance of the results of (correct) mathematical and logical manipulations carried out on them (ibid; 1970c, *SSR* p. 183).

99. 1970c, *SSR* pp. 183–184; also compare § 3.6.e.

100. 1974a, *ET* pp. 299, 307 n. 17; 1970c, *SSR* p. 188; 1974b, pp. 516–517.

101. See § 4.4.a below.

lently, metaphysical) convictions, convictions regarding what there is and what its fundamental characteristics are, are also subsumed under "models."[102] While *SSR* deals with the ontological convictions derived from paradigms,[103] it hardly mentions heuristic models at all.

Kuhn's reason for treating both of these heterogeneous moments of consensus under the heading of models is that they perform similar functions for the scientific community; they play similar roles in the identification of unsolved problems and in gauging the acceptability of proposed solutions to them.[104] Models, in Kuhn's sense, perform these functions in virtue of being a source of similarity relations. In the case of heuristic models, they serve as a source of "external" similarity relations, of similarities between essentially different objects and situations. As for ontological commitments, these are a source of "internal" similarity relations, of similarities between objects or situations of the same ontological type.[105] Such similarity relations sanction an analogous application of concepts to the objects or situations they deem similar and are thus coconstitutive of the meanings of these concepts.[106]

For Kuhn, models, so defined, are different from the other components of the disciplinary matrix in one important respect. Such models frequently belong to the consensus within a scientific community but by no means always do so. Thus in certain stages of scientific development we may find coherent research traditions in which there is no agreement on basic ontological questions.[107]

c. Values

Kuhn often alludes to the fact that scientific knowledge is the product of highly specialized groups and therefore that certain of its characteris-

102. 1974a, *ET* pp. 297–298; 1970b, pp. 271–272; 1970c, *SSR* p. 184. —This use of the words "ontological" and "metaphysical" is, of course, most frequent in the realm of analytic philosophy.

103. E.g. *SSR* pp. 4–5, 7, 40, 41, 103, 109.

104. On this issue, see §§ 4.4.b and 4.4.c below.

105. 1981, p. 20.

106. 1979b, p. 415; 1981, pp. 20–21; also 1974b, p. 506.

107. 1970b, p. 255; 1970c, *SSR* pp. 180, 184. This appears to be a softening, in the 1969 papers, of the position expressed in *SSR* and 1963a, where Kuhn still claimed that "Effective research scarcely begins before a scientific community thinks it has acquired firm answers to questions like the following: What are the fundamental entities of which the universe is composed? How do these interact with each other and with the senses? . . . At least in the mature sciences, these are firmly embedded in the educational initiation that prepares and licenses the student for professional practice "(*SSR,* pp. 4–5; similarly 1963a, p. 359). Likewise, he then maintained "The commitments that govern normal science specify not only what sorts of entities the universe does contain, but also, by implication, those it does not" (*SSR,* p. 7). Q.v. *SSR,* p. 109.

tics may be understood only by recourse to the features of these communities.[108] Each group is special primarily in virtue of its special, socially binding values. Many features of scientific development result from the fact that the underlying decisions in favor of certain alternatives were made on the basis of particular scientific values.

Values form that component of the disciplinary matrix[109] least subject to variation, both from scientific community to scientific community and over time.[110] Since the value systems of all scientific communities share a common core, they may be given an abstract-universal description. This is also the reason all (natural) scientists form, in a certain sense, a *single* community.[111]

Scientific values operate on two levels.[112] Scientific practice contains both evaluations of individual applications of theory and evaluations of whole theories. Evaluations of individual applications of theory (as more or less successful) take place all the time, while the evaluation of whole theories is reserved for special phases in scientific development.[113] In such phases, evaluations of individual applications of theory serve an instrumental function in the global evaluation of theories. Given Kuhn's overarching interest in scientific revolutions, scientific values are important above all as a basis for the global evaluation of theories.

What scientific values can the disciplinary matrix be said to contain when described in abstract-universal terms? As far as the *contents* of scientific values are concerned, Kuhn's answer isn't, in his judgment, any different from the traditional answer in the philosophy of science.[114] But his conception of the manner in which these values operate

108. *SSR,* pp. 153, 167–169; 1971a, p. 146; *ET,* p. xx; 1983c, p. 28.

109. 1970b, p. 272; 1970c, *SSR* pp. 184–186. —Kuhn also calls the total set of values a "value system" or "ideology" (e.g., 1970a, *ET* p. 290; 1970b, p. 238).

110. 1970b, p. 238; 1970c, *SSR* p. 184; *ET* pp. xxi–xxii; 1977c, *ET* pp. 335–336.

111. 1970c, *SSR* p. 184.

112. 1970a, *ET* pp. 288–291; 1970b, pp. 237–238, 248, 272; 1970c, *SSR* pp. 184–185, 186, 205; *ET,* pp. xx–xxi; 1977c. —In some of these passages, Kuhn uses the word "prediction" in its broad sense, common among natural scientists. "Predictions" in this sense are assertions about events or states derived *from* theories, regardless of the temporal relationship between these events or states and the time of assertion.

113. In anticipation of part 3: The governing theory isn't evaluated in the course of normal science, but rather enjoys the self-evident acceptance of its guiding role. A crisis consists in the evaluation of the theory which previously governed research as needy of revision, usually because certain anomalies have been evaluated as important anomalies. In a theory choice situation, different theories are comparatively evaluated with a view toward their relative performance.

114. 1970b, pp. 261–262; 1970c, *SSR* p. 199; 1971a, p. 146; *ET* pp. xxi–xxii; 1977c, *ET* pp. 322, 333.

is very different from that of this tradition, as we will demonstrate later. Making no claims to completeness, Kuhn lists the following values in his 1977c, in which he considers scientific values in greatest detail:[115]

—*Accuracy:*[116] Applications of theory, assertions about factual situations derived from theory, should be both qualitatively and quantitatively accurate. This value carries great, though not overwhelming, weight; the history of the sciences is characterized by an increase in the importance of quantitative accuracy.[117]
—*Consistency:*[118] A theory should be free of internal contradiction and compatible with other accepted theories.
—*Scope:*[119] A theory should have a broad domain of possible applications.
—*Simplicity:*[120] A theory should provide unifying perspectives for the ordering of apparently unrelated groups of phenomena and have the simplest possible conceptual and technical apparatus and procedures for application.
—*Fruitfulness:*[121] A theory should encompass new phenomena or new relations between previously known phenomena.

In addition to these values, Kuhn occasionally cites others, such as the unity of science,[122] explanatory power,[123] naturalness,[124] plausibility,[125] and above all a theory's capacity to define and solve as many theoretical and experimental problems as possible, especially of the

115. *ET* pp. 321–322. —At the corresponding places in *SSR,* Kuhn's talk is predominantly of "arguments" rather than "values" (pp. 153–159), in accordance with the argumentative function values have in theory choice debates (1970b, pp. 261–262);' see § 7.4. —See McMullin 1983, pp. 15–16.

116. Also 1961a, *ET* pp. 212–213; *SSR* pp. 138, 147, 153–154; 1970a, *ET* p. 289; 1970b, p. 272 and elsewhere; 1970c, *SSR* pp. 185, 206; 1971a, p. 146; 1977c, *ET* pp. 322–323; 1983d, p. 564.

117. 1977c, *ET* p. 335; q.v. Boos and Krickeberg 1977, and Hoyningen-Huene 1983.

118. Also 1970a, *ET* p. 291; 1970c, *SSR* pp. 185, 206. —In 1970c, *SSR* p. 185, Kuhn remarks that he gave this value too little attention in *SSR.*

119. Also 1970b, pp. 241, 261; 1970c, *SSR* p. 206; 1983d, p. 564.

120. Also *SSR,* pp. 155–156; 1970a, *ET* p. 291; 1970b, pp. 241, 261; 1970c, *SSR* pp. 185, 206; 1971a, p. 146; 1977c, *ET* p. 324.

121. Also *SSR,* pp. 154–155, 159; 1970b, p. 261; 1971a, p. 146; 1980a, pp. 189–190.

122. 1970a, *ET* p. 289.

123. 1970a, *ET* p. 289.

124. 1971a, p. 146.

125. 1970c, *SSR* p. 185.

quantitative variety.[126] These values, especially the last, are at least partially implied by those listed above, just as the values in the list itself evidently aren't entirely independent of one another. But I won't pursue questions of the design, reciprocal entailments, weighting, and completeness of scientific values any further, since an abstract-universal description of scientific values, independent of any given historical situation or scientific community, wouldn't be especially rich, anyway.[127]

But according to Kuhn, we run into indeterminacy even in the value system of a particular scientific community made concrete in some particular way in a particular historical situation. This indeterminacy is inherent in such value systems and fulfills important functions in scientific development. On this point, Kuhn's departure from the received tradition in philosophy of science is clear. For the indeterminacy reveals itself in the fact that a given such value system might, in its concrete application, generate different evaluations *depending on the individual evaluator.*[128]

There are two reasons for the indeterminacy of concrete scientific value systems. First, the individual values of a given scientific community may legitimately be interpreted in different ways by different community members; what, precisely, is meant by "simplicity," or which aspects of a given theory should be given most attention with regard to this value, isn't uniquely determined by the mere acceptance of the value by the community.[129] Second, different values may contradict each other in a given concrete application, and so a relative weighting is needed. But this weighting isn't uniquely given by community agreement over the value system.[130] The indeterminacy of a scientific value system thus amounts to the incomplete determination of the concrete evaluations it sanctions.

In making a concrete judgment, the individual scientist remedies this underdetermination under the influence of factors which may vary from scientist to scientist.[131] Prominent among these is the individual scientist's particular professional experience, determined, for example, by which fields he or she has worked in, how successfully, and for

126. *SSR,* pp. 155, 169–170; 1970a, *ET* pp. 289, 290; 1970c, *SSR* pp. 185, 205; 1983d, pp. 563–564.

127. Compare § 1.2.c.

128. *SSR,* pp. 155, 159; 1970a, *ET* pp. 288, 291; 1970b, pp. 238, 241, 248, 262; 1970c, *SSR* pp. 185–186, 199–200; 1971a, p. 146; *ET,* p. xxi; 1977c, *ET* pp. 322–331. See Salmon 1990 for an attempt to reconcile Kuhn's position with the traditional view.

129. 1970b, p. 262; 1970c, *SSR* pp. 185, 205; *ET* p. xxii; 1977c, *ET* pp. 322–324. —See Sintonen 1986.

130. 1970a, *ET* pp. 290–291, and previous footnote.

131. *SSR,* pp. 3–4, 152–153; 1970b, p. 241; 1970c, *SSR* p. 185; 1977c, *ET* p. 325.

how long. In addition, extrascientific convictions of the religious, philosophical, or even national variety play a role. Finally, certain personality traits, such as adventurousness or timidity, and specific personal preferences help determine choice. We may thus distinguish two evaluative moments in a value judgment made by an *individual* scientist. One evaluative moment is shared with other members of the same community, as explained by the similarity of their personal histories, while the other isn't shared and is explicable only by reference to idiosyncratic features of the scientist's history.[132]

Kuhn frequently makes use of two not entirely unproblematic formulations in discussing the interplay of collectively shared and individually variable evaluative moments. On the one hand, he speaks of the fact that "values may be shared by men who differ in their application."[133] This claim borders on the paradoxical, for one might ask what it means for people to agree on values when as soon as these values become operational they behave differently, making different value judgments. On the other hand, Kuhn also asserts that the criteria of theory choice "function not as rules, which determine choice, but as values, which influence it."[134] By "rule," Kuhn here means an unequivocally executable, logicomethodological guide to action—an algorithm.[135] Decisions based on rules of this sort come out the same for everyone who applies the same rules; in this sense, they are intersubjective. As far as this intersubjectivity is concerned, it doesn't matter whether the rules are deterministic or merely probabilistic. But decisions reached on the basis of shared values need not be intersubjective, and this, for Kuhn, is what it means for scientists applying shared values to "differ in their application."

The underdetermination of choice by scientific value systems isn't, for Kuhn equivalent to some form of incompleteness, demanding repair; on the contrary, it serves important positive functions in scientific development.[136] For however they turn out, decisions in which scientific values must play a role are often fraught with risk. For example, scientists in theory choice situations might often have to decide in favor of or against a new theory at a time when it's not yet fully worked out

132. 1977c, *ET* pp. 324–325.

133. 1970c, *SSR* p. 185; similarly 1970b, p. 262; 1971a, p. 146; 1977c, *ET* pp. 324–325, 331, 337.

134. 1977c, *ET* p. 331; similarly p. 333; 1970b, pp. 238, 262; 1970c, *SSR* pp. 186, 199–200; 1971a, p. 146. Also see § 4.1.

135. 1970a, *ET* pp. 280–283, 288; 1970b, pp. 234, 241; 1977c, *ET* pp. 326, 328–329, 331, 332–333. See Finocchiaro 1986; McMullin 1983; Stegmüller 1986, p. 350ff; Tianji 1985.

136. 1970b, pp. 241, 248–249, 262; 1970c, *SSR* p. 186; 1977c, *ET* pp. 330–332, 334.

and empirical evidence fails to confirm (or refute) it unequivocally. In this situation, no one can be certain to have chosen the theory that ultimately will convince all members of the community.[137] Under such circumstances, it is necessary that some scientists choose the old theory and some the new, for *both* must have a chance to develop their potential, as they do only when both are explored. Thus scientific development "*requires* a decision process which permits rational men to disagree."[138] Otherwise science would be in danger either of stagnating in *one* tradition or of forever prancing from theory to theory, never stopping to exhaust the potential of any of them.[139] It follows that many disagreements in the history of science which can hardly be understood when unequivocal rules are presupposed, since (at least) one party to the dispute appears to be behaving "unscientifically" (or "irrationally," "unreasonably," "dogmatically," and the like), can now be intepreted as cases of legitimate scientific difference of opinion.

So the value system accepted by a given scientific community leaves room for individuals to make different value judgments;[140] it constrains the space of possible decisions without coercing individual choice. In addition to such choices by *individual* scientists, Kuhn also speaks of the relationship between a value system and the *behavior of a scientific community*, as, for example, when he claims "it is the community of specialists rather than its individual members that makes the effective decision."[141] Such claims merit discussion, for they may contribute to the clarification of one of Kuhn's fundamental theses. This thesis emerges at different points in Kuhn's theory, especially where he speaks of "the sociological base of my position."[142] Just what do these locutions mean?

137. On this issue, see also Lakatos 1971, § 1 (d).

138. 1977c, *ET* p. 332.

139. 1970b, pp. 246–247; 1970c, *SSR* p. 186; 1977c, *ET* p. 332.

140. Another dimension of individual difference between members of a scientific community, epistemically relevant under certain conditions, was discussed in § 3.6.d; members of a scientific community may differ with respect to the criteria they employ in determining the referents of empirical concepts.

141. 1970c, *SSR* p. 200.

142. 1970b, pp. 241, 253; similarly 1959a, *ET* pp. 227–228 n. 2; *SSR*, pp. vii, 153, 162; 1963b, pp. 392, 394–395; 1969c, *ET* p. 351; 1970a, *ET* pp. 290, 291; 1970b, pp. 237–238, 240; 1970c, *SSR* pp. 178, 200, 209–210; 1971a, p. 146; *ET*, pp. xx–xxii; 1979c, pp. x–xi; 1983c, pp. 28, 30; see also § 7.4.c below. —In the title of 1970a, "Logic of Discovery or Psychology of Research?" Kuhn appears to oppose Popper's "logic of discovery" with a "psychology of research" (*The Logic of Scientific Discovery* being, of course, the English title of Popper's opus). Similarly, in the text of 1970a, Kuhn asserts that "explanation must, in the final analysis, be *psychological or* sociological" (*ET* p. 290, emphasis mine). Since such statements led some, for example Lakatos, to misunderstand

As we have seen, the decisions of individual community members made under the influence of their community's shared value system may vary. But still, Kuhn claims, "*group behavior* will be decisively affected by the shared commitments."[143] This statement agrees with Kuhn's characterization—as opposed to Lakatos' criticism—of his manner of inquiry:

> The type of question I ask has . . . been: how will a particular constellation of beliefs, values, and imperatives affect *group behavior?*[144]

Our own inquiry into this question (and, of course, into the answers to it) must make an issue of "group behavior," as used here.

Talk of group behavior obviously makes sense when there is some sort of analogy between the group and an individual. And this is indeed the case for the situations in which Kuhn is primarily interested, transitions from a phase of scientific development in which a community is governed by one consensus to a phase in which it's governed by another.[145] Such transitions are analogous to individual changes of opinion, and, as in the individual case, we might ask, what caused the group to change its opinion? But the answer to this question regarding a change in group opinion must be given *at the level of the individual,* for even given consensus, the group isn't a subject conceived apart from its members. And members of the community differ with regard to the individually variable evaluative moments that participate in theory choice. So how should Kuhn's claim that "shared values, though impotent to dictate an individual's decision, may nevertheless determine the choice of the group which shares them"[146] be understood by us?

All semblance of paradox in such claims vanishes when we consider that Kuhn here has a historical process in view, a process which begins with dissent over theory choice and ends with the community's newfound consensus over the better of the candidate theories, regardless of whether the better theory is the old one or some alternative.[147]

---

him as wanting to explain scientific development by means of *individual* psychological features, Kuhn restricts himself, after 1970b, to using "sociology" in this context (on this issue, see 1970b, pp. 240–241). —See Jones 1986 and his citations regarding other misunderstandings arising from this manner of speaking.

143. 1970b, p. 241, emphasis mine; similarly p. 239; 1970c, *SSR* pp. 186, 199; *ET*, p. xxii.

144. 1970b, p. 240, emphasis mine.

145. Compare § 1.3.

146. *ET* , p. xxi.

147. On the following, see also § 7.4.b.

During the period of dissent, scientists work with the various alternative theories. *Over time,*[148] this work generates enough empirical and theoretical arguments to produce a new consensus on which theory should be chosen. But this new consensus must be qualified in two respects. First, the community's new consensus doesn't have to be truly unanimous from the beginning. Instead, those who diverge from the majority opinion may be excluded from the community,[149] or the community may split.[150] Second, the fact that the individual scientists who make up a (newly formed or previously existing) community all exhibit the same choice behavior doesn't imply that their individual choices were all made for *precisely* the same reasons or motives. On the contrary, each individualized version of the collective values which serve as grounds for theory choice is retained.[151] What matters is only that consensus come about because of the collective value system;

> To understand why science develops as it does, one need not unravel the details of biography and personality that lead each individual to a particular choice, though that topic has vast fascination. What one must understand, however, is the manner in which a particular set of shared values interacts with the particular experiences shared by a community of specialists to ensure that most members of the group will ultimately find one set of arguments rather than another decisive.[152]

d. Exemplary Problem Solutions

Exemplary problem solutions, or paradigms in the narrowest sense are, in the 1969 papers, a component of the disciplinary matrix, or the paradigm in the broadest sense.[153] The importance of exemplary problem solutions to Kuhn's philosophy of science was considered in detail in § 4.1, and so we may safely waive further discussion here. In § 4.4 we will analyze what, precisely, it means for paradigms to be accepted as governing research. But first we must clarify the relationship between concrete problem solutions and other "components" of the disciplinary matrix.

148. This period may last up to a generation; *SSR,* p. 152.

149. *SSR,* pp. 19, 150–152; 1970a, *ET* p. 291.

150. 1959a, *ET* p. 231; *SSR,* p. 15; 1970a, *ET* p. 291; 1977c, p. 329.

151. Kuhn sees a subtle source for the plausibility of his critics' views in the fact that similar decisions mask the dissimilarity of the reasons behind them; 1977c, *ET* pp. 328–329, 333.

152. 1970c, *SSR* p. 200.

153. 1974a, *ET* pp. 301–319; 1970b, p. 272; 1970c, *SSR* pp. 186–191; 1974b, pp. 500–501.

e. The Relationship between "Components" of the Disciplinary Matrix

Kuhn's use of such terms as "elements," "constituents," "components," and "subsets,"[154] not to mention "disciplinary matrix" itself, suggests that the disciplinary matrix contains a number of "parts," where the relation of parts to the whole is conceived along the lines of the element-set or subset-set relationship.[155] This implies that the parts *may be conceived as separate from one another and thus subjected to separate, individual examination*. To be sure, according to Kuhn, the constituents of the disciplinary matrix

> form a whole and function together. They are, however, no longer to be discussed as though they were all of a piece.[156]

Put differently, and now with reference to "theories" instead of the "disciplinary matrix,"

> Theories . . . cannot be decomposed into constituent elements for purposes of direct comparison either with nature or with each other. That is not to say that they cannot be analytically decomposed at all, but rather that the lawlike parts produced by analysis cannot . . . function individually in such comparisons.[157]

Can the relationship between symbolic generalizations, models, values, and exemplary problem solutions really be conceived as a relationship between separable, hence independent, elements that impinge on each other only in exercising their functions?[158] The answer to this question will prove crucial to our understanding of the development of the paradigm concept.

In the context of my discussion of the priority of problem solutions over rules, I considered the relationship between laws and theories on the one hand and their particular applications on the other, revealing it to consist in the inseparable unity of distinct moments.[159] How can laws or theories and exemplary problem solutions become separate "elements" of the disciplinary matrix in Kuhn's 1969 papers? Kuhn

154. E.g. 1974a, *ET* pp. 294, 297–298; 1970b, pp. 271–272; 1970c, *SSR* pp. 175, 182.

155. Shapere 1971, p. 707, draws special attention to the problems associated with this relationship, especially the problem of the components' unity.

156. 1970c, *SSR* p. 182.

157. 1977b, *ET* pp. 19–20.

158. Kuhn at one point in *SSR* uses a formulation which very closely approximates the view I am now arguing: "In learning a paradigm the scientist acquires theory, methods, and standards together, usually in an *inextricable mixture*" (p. 109, emphasis mine). Also see *SSR*, pp. 10, 11.

159. Compare § 4.1.

attempts to separate the two moments by finesse, removing all *empirical* meaning from the concepts employed in a law or theory and ascribing it all to problem solutions. What's left is a symbolic generalization, something "uninterpreted . . . empty of empirical meaning or application."[160] Only the meanings of the mathematical and logical symbols are preserved.[161]

But this step doesn't really permit the total removal of all empirical meaning from symbolic generalizations. Kuhn is emphatic in insisting that the (interpreted) generalizations—laws and theories—have a function in fixing the meanings of the empirical concepts they contain. But if this is the case, the identity relation in a symbolic generalization (for instance, that which holds between the left and right sides of an equation) would be one vehicle for expressing the meanings of empirical concepts contained in the generalizations. And so the *separation* of symbolic generalizations from their applications in problem solutions reveals itself to be impossible. To be sure, the line Kuhn draws allows symbolic generalizations to be conceived as independent elements, but the problem solutions, whose concepts now lack an essential moment in their meanings, may not be so conceived. If *all* meaning, including that of the identity relation in symbolic generalizations, is consigned to the side of problem solutions, then, to be sure, problem solutions may be conceived as independent elements. But then all that remains opposite them are utterly empty strings of signs, necessarily deprived of any epistemological function. If instead, in direct opposition to Kuhn's own intentions, all meaning were consigned to the side of symbolic generalizations, then concrete problem solutions, in turn, would necessarily be void of meaning, deprived of all epistemological function. We conclude that symbolic generalizations and concrete problem solutions *can be distinguished only as linked moments of a single unity;*[162] calling them "elements" or "components" of a disciplinary matrix misses this point.

160. 1974a, *ET* p. 299.

161. Compare § 4.3.a.

162. To avoid misunderstandings: Kuhn's arguments in support of the claim that the distinction between laws or theories and their applications is just the distinction between symbolic generalizations and exemplars (1974b, pp. 516–517) are unaffected by this criticism. All that are affected are formulations which attempt to make this *distinction* into a *separation*. This is precisely the point at which Putnam and Achinstein are right in their criticisms of Kuhn (though their formulations slightly miss the mark), where they argue that, by introducing symbolic generalizations, Kuhn makes, against his intentions, too great a concession to positivism (in Kuhn 1974b, pp. 513, 516). For the *separation* of symbolic generalizations from exemplars is possible if and only if laws and theories exercise no function in fixing the meanings of concepts or, in other words, just in case sentences are *either* analytic *or* synthetic. Kuhn rejects precisely this dichotomy; see § 3.6.f.

The situation is just the same with respect to the relationship between exemplary problem solutions and models. The analogies employed in heuristic models can only be truly understood when applied to concrete problem situations, and conversely, problem solutions attained by means of such analogies can only be understood by reference to them. Similarly, the ontological convictions of a scientific community can be understood, both in content and in degree of conviction, only by reference to the problem solutions governing the community.

The relationship between values and exemplary problem solutions exhibits similar features. The best evidence for the character of a given community's scientific values is found, by and large, in the theories the community evaluates favorably, which in turn reveal community values primarily through the concrete problem solutions they helped achieve.[163] Only the degree of quantitative accuracy to be expected realistically might conceivably be precisely articulated apart from any application; but in fact, in scientific practice, this degree of accuracy is demonstrated by reference to problem solutions,[164] for it is only relative to them that it can be dubbed a realistic expectation.

What follows from all this is that we can't regard the relationship between symbolic generalizations, models, values, and concrete problem situations on the one hand and the disciplinary matrix on the other as that of elements or subsets to a set. The relationship between the former items is that of *linked moments of a single unity;* though individual moments can be distinguished, they can't even be conceived as separate. Concrete problem solutions *can* be seen as the central moment, especially given their world- and science-constituting function.[165] But symbolic generalizations might equally be taken as the central moment, especially when we consider their role in establishing unity among the various moments of the disciplinary matrix. However, there is no need to puruse the rather vague question of the central moment of the disciplinary matrix any further.

The claim that the disciplinary matrix is a single unit composed of linked moments and not a collection of otherwise separate elements or subsets isn't an item of mere nit-picking; on the contrary, it's essential, if we want to understand the development of the paradigm concept.[166] The point of departure for this development is the observation that

163. "[All scientists] must be supplied with examples of what their theories can, with sufficient care and skill, be expected to do" (1970b, p. 248), "It is vitally important that scientists be taught to value these characteristics [scientific values] and that they be provided with examples that illustrate them in practice" (1970b, p. 261; also 1971a, p. 146).

164. 1961a, *ET* pp. 183–86.

165. See § 4.4.

166. Compare § 4.2.a.

paradigms, in the sense of concrete problem solutions, make up the foundation of the consensus which governs research, while explicitly defined concepts or methodological rules, laws, or theories, conceived apart from their concrete applications, do not. But such paradigms include laws, theories, models, and values *as moments*. When we consider the functions concrete problem solutions exercise in scientific practice, their various implicit moments become explicit. And thus paradigms, in the sense of concrete problem solutions, transform themselves from a small but important *sub-domain* of scientific consensus into the *entire* consensus; this transformation is the expansion of the paradigm concept that initially went unnoticed by Kuhn himself. When we consider this result apart from the process which engendered it, it appears unavoidable that there are two meanings of the paradigm concept that, because importantly different, seem worth keeping apart. The original mistake lay in a certain one-sidedness, in the failure to distinguish the *moments contained* within problem solutions from the problem solutions *themselves*. This one-sidedness is a natural reaction to the opposed one-sidedness of the received tradition in the philosophy of science, according to which problem solutions themselves weren't an issue at all and the moments contained within them, where they were even considered to begin with, were kept strictly separate.

So, in his 1969 papers, Kuhn turns the various moments contained in problem solutions into independent components of a disciplinary matrix, hoping thereby to correct the apparent ambiguity of the paradigm concept. But then the merging of the so-called narrow and broad meanings of "paradigm" becomes just as unintelligible in retrospect as the tremendous resonance with which the original introduction of the paradigm concept was received. It then seems puzzling how any author could presume to throw such obviously different things as problem solutions, theories, values, and models together without a substantial portion of his readership—including the professional scientists in this group—raising much in the way of objection.[167] For in fact the attempt at correction Kuhn undertakes in 1969 is just a new form of one-sidedness, in which distinct but inseparable moments are made into separate components of a disciplinary matrix. This new one-sidedness, too, is understandable, for the notion of a unity of distinct but inseparable moments, which this sort of one-sidedness could avoid, was entirely alien to Kuhn, at least in 1969. This shouldn't surprise us in an American historian and philosopher of science of the latter half of the

167. Of course this might be explicable, but only with the help of auxiliary hypotheses really drawn from mob psychology (Cf. Kuhn's debate with Lakatos over mob psychology: 1970b, pp. 262–263.).

twentieth century.[168] But Kuhn seems to have noticed the inadequacy of the notion of a disciplinary matrix, which he never uses in his work after 1969.

In the following discussion of the function of paradigms in the sense of problem solutions, it should become clear that the sense of "paradigm" Kuhn had in mind is just that sense in which laws or theories, models, and values are incorporated as moments.

4.4 The Functions of Paradigms in the Sense of Exemplary Problem Solutions

Paradigms, in the sense of concrete problem solutions accepted by a community, play a globally normative role in that community. This means that paradigms are accepted not just as particular solutions to particular problems. Future research must rather refer itself to these problem solutions, proceeding by analogy from them.[169] An analysis of what, precisely, it means for research to proceed by analogy from exemplary problem solutions yields three globally normative functions. First, future research must employ the conceptual system specific to the exemplary problem solutions (§ 4.4.a). In addition, exemplary problem solutions are used to identify as yet unexplored research problems (§ 4.4.b). Finally, exemplary problem solutions play a role in evaluating the acceptability of proposed solutions to these problems (§ 4.4.c).

a. The Lexicon of Empirical Concepts

The globally normative function of those concrete problem solutions governing a scientific community's research—a function that will prove most important in our later characterization of scientific revolutions—is derived from the fact that a particular system of empirical concepts is employed in articulating both problems and their solutions.[170] In works composed since 1982, Kuhn calls this system a "lexicon" (or "lexical net-work").[171] This term, borrowed from linguistics, addresses the relation of mutual dependence which empirical concepts

168. Kuhn remarks in his preface to *ET* that he and his (North American) philosophical colleagues had little access to European post-Kantian philosophy, on grounds of its opacity (ET p. xv). —On this issue, see Gutting 1979.

169. Compare § 4.1.

170. 1984, p. 245.

171. 1983a, pp. 682–683; 1983b, pp. 713–714; 1989a; 1990. Compare § 3.6.g.

bear toward one another, which Kuhn also calls the "local holism of language."[172]

Now, exemplary problem solutions don't merely apply, or illustrate, some given, previously fixed lexicon of empirical concepts. On the contrary, some of the concepts in the lexicon acquire their particular meanings only with the articulation of the exemplary problem solutions.[173] The functions of the lexicon are thus, indirectly, also functions of the exemplary problem solutions (of which symbolic generalizations must be conceived as moments).[174] We may distinguish between the following three functions of the lexicon.

First, the lexicon of empirical concepts is constitutive of the phenomenal world.[175] In order for it to exercise this function, at least two exemplary problem solutions belonging to each symbolic generalization are required. By means of the conceptual system already available, together with ostension, an instructor can draw attention to the implicated problem situations and to the subsituations and objects occurring within them, thus evoking the immediate similarity relations which link the two situations both as wholes and through their moments. If the pupil is successful in learning these similarity relations, he or she has mastered the concepts belonging to them. The concepts, in turn, structure a particular region of the phenomenal world, making it accessible to the pupil. All research problems subsequently explored must refer back to this region of the phenomenal world.

Second, the concepts in the lexicon contain *implicit* knowledge of nature, knowledge of the present phenomenal world implicit in these concepts.[176] This implicit knowledge expresses itself in the demand that future candidate problem solutions not contradict it.

Third, concepts in the lexicon must be used in articulating *explicit* knowledge about the present phenomenal world, as for example in the formulation of those quantitative regularities which hold between the referents of these concepts under certain conditions.[177]

b. The Identification of Research Problems

Concrete exemplary problem solutions exercise a further important function in research practice by sanctioning the identification of new

172. Compare §§ 3.6.c and 3.6.e.

173. Though this shouldn't be taken to imply that empirical concepts acquire this meaning *individually;* compare § 3.6.g.

174. § 4.3.e.

175. Compare chap. 3.

176. Compare § 3.7.

177. 1961a, *ET* p. 198; 1977b, *ET* p. 19; 1981, pp. 18–19.

research problems for exploration.[178] Here, too, symbolic generalizations,[179] models (in Kuhn's sense),[180] and values[181] play a role, but only by way of the exemplary problem solutions whose moments they are. The identification of a research problem in this way involves two things. First, it requires a conceptual system both licensed by the appropriate scientific community to articulate research problems in general and suited to the articulation of a particular research problem. Second, in order to be identified as a problem, a research problem needs a contextual background from which its problematic character emerges.[182] Regardless to what species a research problem may belong—instrumental, experimental, mathematical, conceptual, or explanatory—its problematic character emerges only given a background of previously accepted instrumentation, theorems, theories, ontological and methodological convictions, and the like, irrespective of the extent to which the community takes these for granted. In scientific communities operating without competition,[183] the further requirement that the problem have some prospects for solution also presents itself.[184]

Exemplary problem solutions may participate in all such moments constitutive of a given research problem. We dealt with the conceptual system licensed for the articulation of exemplary problem solutions in the last subsection. Just as the conceptual system in which the paradigms are articulated may be extended to further situations by way of similarity relations for the articulation of further problems, so the particular problematic character of an exemplary problem may be used, by analogy, in the construction of further research problems. In § 5.2.b, the expectation of a problem's solubility will be seen to emerge from this same analogy.

A scientific community must further be able to evaluate the relative importance of those problems deemed soluble, albeit neither unequivocally nor even always compellingly. Certain trends in the general assessment of the relative importance of research problems are necessary, lest the community flounder in its work. In such assessments, however, exemplary problem solutions play only a subordinate role.[185]

178. *SSR,* pp. 6, 18, 37, 103–110.
179. *SSR,* p. 40.
180. 1970c, *SSR* p. 184.
181. 1970a, *ET* p. 290.
182. Q.v. 1971b, *ET* p. 29.
183. Compare § 1.1.b.
184. I won't discuss the expected solubility requirement and the role exemplary problem solutions play in it any further here. This requirement is characteristic only of the normal science phase and of the special exemplary problem solutions employed in it, and an explication of it must thus wait until chap. 5 (§ 5.2.b).
185. See *SSR,* pp. 25–34.

c. The Acceptability of Solutions to Research Problems

Just as new research problems can be identified by means of previously solved exemplary problems, so these same exemplary solutions offer implicit standards for the acceptability of candidate problem solutions, whether experimental or theoretical.[186] To be sure, assigning this role to the exemplary problem solutions makes sense only if here, too, they are conceived as encompassing the other moments of the disciplinary matrix. For a judgment regarding the acceptability of a proposed problem solution is an evaluation of this proposal undertaken relative to accepted symbolic generalizations, ontological commitments, the heuristic models current in the community, and accepted standards of accuracy, simplicity, and the like.

186. *SSR,* pp. 6, 38–42, together with chap. 5, pp. 103–110; 1970c, *SSR* p. 184; 1971b, *ET* p. 29.

SUMMARY OF PART II

It is of central importance to Kuhn's theory that the world, *qua* object of scientific knowledge, is a phenomenal world, for only a phenomenal world (of a certain type) may be meaningfully said to change in a scientific revolution. Such change, as we shall see in part 3, is among the most important characteristics of scientific revolutions. The claim that the real world of a given community is a phenomenal world can be made plausible by exploring the process whereby members of the community gain access to it. Immediate similarity relations play a central role in this process; mastery over them assists in the constitution of objects in the appropriate world, of perception, and of the empirical concepts by means of which this world is described; at the same time, they contain implicit knowledge of this world. In order to learn the relevant similarity relations, reference must be made to objects and problem situations representative of the phenomenal world in question. These reside in the paradigms, and thus the key role of paradigms in Kuhn's theory resides in their ability to fix the network of similarity and dissimilarity relations. Paradigmatic problem solutions also serve as models for the research tradition built on their foundation.

Kuhn's analysis of the constitution of phenomenal worlds runs afoul of the methodological problem regarding which assumptions may be made for purposes of such analysis. While these assumptions must be powerful enough to make the analysis possible, they must also be weak enough to preserve neutrality toward the plurality of different possible phenomenal worlds. Whether there are such assumptions and, if so, what they are, has to be left open.

PART THREE

The Dynamic of Scientific Knowledge

AS RELATED IN CHAPTER 1, Kuhn's theory of scientific development prepares the way for further discussion by targeting the structure of scientific development in the sense of a universal phase model of that process. But its central issue is the structure of scientific revolutions, their essential features and progressive phases. Revolutions are the breaks that occur between two successive phases of normal science. Thus many of their characteristics may be derived from the peculiar relationship between two such phases. So we must first determine what normal science is, as we will attempt to do in chapter 5. We will then be in a position to pin down the concept of a scientific revolution specific to Kuhn's theory in chapter 6. In conclusion, we will discuss the course revolutions take and the important forces that drive them in chapter 7.

CHAPTER FIVE

Normal Science

IN THIS CHAPTER, we will begin with a few provisional characterizations of normal science (§ 5.1). Next we will develop the fundamental "puzzle-solving" metaphor Kuhn uses to describe the specifics of normal science (§ 5.2) and classify the research problems of normal science (§ 5.3). We will then be in a position to understand why cumulative epistemic progress occurs in normal science (§ 5.4) and to discuss the conditions under which normal science is possible (§ 5.5). In conclusion, we will inquire into the function of those quasi-dogmatic features typical of normal science (§ 5.6).

5.1 Normal Science: Provisional Characterization

Normal science is a phase in scientific development which received less-than-satisfactory attention before Kuhn's *SSR*.[1] And yet, on Kuhn's view, without an understanding of normal science, significant features both of scientific training and of scientific development must remain unintelligible.[2] One central characteristic of normal science is—to a purely negative first approximation—that, though it somehow ignores the essential fallibility of human knowledge, it still isn't an inferior form of science. Kuhn provocatively draws attention to this

1. Compare this with, for example, Popper's admission in his 1970, p. 52, despite his misunderstandings concerning what, precisely, Kuhn means by normal science (on this issue, q.v. Kuhn 1970a, *ET* p. 272). —In essence, the juxtaposition of normal and revolutionary scientific development can already be found in Conant: "Usually our conceptual schemes grow by an evolutionary process, by the gradual incorporation of a series of amendments, so to speak. In this case [that of Torricelli] a completely new idea came along, and rendered obsolete the older one" (Conant 1947, p. 36). Scheibe demonstrated in his 1988, pp. 142–143, that this juxtaposition can be found as early as Boltzmann's 1895 eulogy to Joseph Stefan (Boltzmann 1905, p. 94ff).

2. 1963a, p. 369; *SSR* p. 24; 1970a, *ET* p. 272.

feature in the title of his essay, "The Function of Dogma in Scientific Research" (1963a).[3] Less provocatively, he speaks of research for which "the time for steady criticism and theory proliferation has passed,"[4] "tradition-bound" practice,[5] research conducted "within a framework,"[6] or "the result of a consensus among the members of a scientific community."[7] The suggestion is that certain elements of scientific knowledge aren't open to dispute in normal science, since their validity is a matter of consensus within the scientific community.

Even this largely negative characterization of normal science serves to explain why the two dominant strands of pre-Kuhnian philosophy of science failed, or mostly failed, to recognize it. For both logical positivism and critical rationalism saw science as an enterprise molded by the *persistent* awareness of the fallibility of human epistemic claims. This view of science is evidenced by the weight these approaches placed on the problems associated with protocol sentences or basic statements, and on testing or confirmation theory, respectively.[8]

The existence of a dogmatic form of scientific practice—dogmatic in a sense we have yet to clarify—became apparent to Kuhn in comparing communication in communities of natural scientists and social scientists.[9] Constant debate over foundational questions, such as which fundamental concepts are applicable, which questions appropriate, and which research methods legitimate, is typical of the latter sort of com-

3. To be sure, Kuhn partially retracts this provocative formulation when, in the text of 1963a, he confines himself to talk of "quasi-dogmatic commitments"; e.g. p. 347 n. 1, p. 349. In the discussion following his presentation of 1963a, Kuhn eventually retracts his use of "dogma" entirely, in consideration of its inappropriate connotations; 1963b, pp. 390, 392.

4. 1970b, p. 246.

5. 1959a, *ET* pp. 227, 232, 234; *SSR,* p. 5; 1970c, *SSR* p. 208; *ET* pp. xvii–xviii; or similarly in the title of 1959a; *SSR,* pp. 11, 65, 90, 144; 1962b, p. 383; 1963a, p. 351; 1970a, *ET* pp. 267–268 n. 4, 268; 1971a, p. 138.

6. 1970b, p. 242, similarly p. 243.

7. E.g. *ET* p. xviii; similarly 1961a, *ET* pp. 221–222, 227, 231–232; *SSR* pp. 4–5, 10, 11; 1963a, p. 349.

8. On this issue, consider the illustrative problems Kuhn faced even in explaining what he meant by normal science, especially to the critical rationalists (e.g. 1970b, pp. 233, 236, 242, with reference to Watkins 1970 and Popper 1970; q.v. 1976b, p. 197 n. 1). For the critical rationalists, of course, scientific activity which fails to test human knowledge constantly under the lasting assumption of its fallibility can only be inferior science, or strictly speaking, not science at all: "If Sir Karl [Popper] and I disagree at all about normal science, it is over this point. He and his group argue that the scientist should try at all times to be a critic and a proliferator of alternative theories. I urge the desirability of an alternate strategy which reserves such behaviour for special occasions" (1970b, p. 243). Also cf. Stegmüller 1986, pp. 118ff.

9. 1961a, *ET* p. 222; *SSR,* pp. vii–viii; *ET,* pp. xviii–xix; also *SSR,* p. 164.

munity. Such controversy is highly correlated with the fragmentation of the scientific community into competing schools. In the contemporary natural sciences, by contrast, such foundational questions recede from scientific controversy for long periods of time, though eventually the underlying consensus shatters, belying claims to finality.[10] Kuhn calls sciences that have reached this stage the "mature" sciences.[11] "Normal science" refers only to that stage of mature science in which scientific practice is sustained by a broad-based community consensus on foundational issues.[12]

Normal science is thus simultaneously distinguished from two other stages of science, or forms of scientific practice. On the one hand, it's distinct from the form of scientific practice typical of fields in which the conduct of research has never yet been sustained by any universal consensus. This way of conducting science I will later call "prenormal" science.[13] Normal science is also distinct from phases of fundamental dissent within a science in its mature stage, dissent resulting from the collapse of a previous universal consensus. Kuhn calls this form of scientific practice "extraordinary science" or "science in crisis."[14]

What sort of positive characterization is appropriate to normal science? In *SSR* and other works written around the same time, Kuhn designates paradigms as the foundation of the universal consensus mentioned above, the consensus which makes possible the particular scientific practice characteristic of normal science.[15] At the same time, his exposition of both broad and narrow paradigm concepts includes the provision that paradigms be subject to universal consensus among the relevant specialists.[16] This exposition of the paradigm concept in the early 1960s leaves the terms "paradigm" and "normal science" so closely associated that "normal" and "paradigm-governed" science become synonymous,[17] for the *universal* acceptance of those scientific achievements called paradigms is both necessary and sufficient for the

10. *SSR*, pp. vii–viii. —Compare § 4.1, along with chap. 6 and chap. 7.

11. E.g. 1959a, p. 235; *SSR*, pp. 11, 12, 15, 69; 1964, *ET* p. 261; 1970b, pp. 244–245; 1970c, *SSR* p. 179. —In other passages, Kuhn refers to the "postparadigm" stage; e.g. *SSR*, pp. ix, 32, 87; 1963a, p. 354; 1970b, p. 272 n. 1; 1970c, *SSR* p. 178; 1974a, *ET* p. 295 n. 4.

12. But recall the limiting qualifications placed on the status of apparently competitionless communities in § 1.1.b.

13. See § 5.5.b.

14. See § 7.3.b.

15. *SSR*, chap. 3.

16. Compare §§ 4.1 and 4.2.b.

17. See e.g. *SSR*, pp. 11, 18, 23, 25, 100; 1963a, pp. 359, 361–362, 364. The close association of the notion of a paradigm with that of normal research is explicitly indicated in the relevant introductions of the paradigm concept; *SSR*, pp. viii, 10; 1963a, p. 358.

possibility of normal science. In his 1969 papers, however, Kuhn dissolves this close association by removing the provision that the research-governing achievements must be *universally* accepted.[18] And accordingly, after 1969, it isn't enough to characterize normal science as "paradigm-governed science," for this attribute may also pertain to prenormal science.

But the intention which originally led Kuhn to characterize normal science as paradigm governed is left untouched by the weakening of the paradigm concept. For the purpose of characterizing normal science as paradigm governed was to select the most salient feature in virtue of which this practice could be clearly distinguished not only from prenormal and extraordinary science but also from other creative fields. The task of normal science, as of all other phases of scientific development, is to strive toward the most general yet detailed account of nature, that can be achieved aiming in the process at the highest possible degree of both internal coherence and correspondence with the world.[19] The characteristic feature of normal science now emerges as the fact that it performs this task in a manner fundamentally similar to puzzle-solving.[20] In the section which follows (§ 5.2), I will develop the analogies between puzzle-solving and normal science. These analogies, however, fail to tell us anything about the typical contents of research problems faced by normal science. In order to further characterize normal science, the contents of its research problems must be pinned down; this is the task of § 5.3.

5.2 Analogies to Puzzle-solving

At several points in Kuhn's work, the "puzzle-solving" metaphor for normal science is assigned a central role.[21] Kuhn has crossword puzzles, jigsaw puzzles, or chess problems in view as the basis of the analogy.[22]

18. Compare § 4.2.b.

19. Compare § 1.2.c.

20. 1970c, *SSR* p. 179.

21. E.g. 1959a, *ET* pp. 234, 235, 237; 1961a, *ET* pp. 192, 221; *SSR,* chap. 4, pp. 52, 69, 79, 80, 82, 100, 135, 140, 144–145, 152, 166; 1963a, pp. 349, 362, 363–364, 368; 1969c, *ET* pp. 343, 346, 347; 1970a, *ET* pp. 270–271, 273, 274, 276, 277, 278; 1970b, pp. 246–247, 248, 254, 256; 1970c, *SSR* pp. 178, 179, 205, 209; 1974b, p. 455; *ET,* p. xvii; 1983c, p. 30. —The title of chap. 4 of *SSR* best expresses the great weight placed on the puzzle metaphor: "Normal Science as Puzzle-Solving." —Also compare this metaphor with the analogy between research and game-playing in Staudinger 1984.

22. *SSR,* p. 36, 38; 1970a, *ET* p. 271 n. 6. —The term "puzzle" (and doubtless also Kuhn's choice of such examples as crossword puzzles) has led many authors to the view that the research problems of normal science weren't, on Kuhn's view, "very serious or

For Kuhn, this analogy cuts deep enough to serve as a demarcation criterion separating the sciences from other creative enterprises such as the arts, broadly conceived, or philosophy.[23]

The analogies between puzzle-solving and normal science can thus serve as a guide in the concrete determination of the type of research activity peculiar to normal science. But we are well advised to observe the limits of these analogies. The analogy between normal science and puzzle-solving involves the existence of regulations (§ 5.2.a), expectations of solubility (§ 5.2.b), lack of any receptivity to fundamental innovation (§ 5.2.c), the inadequacy of describing these activities as tests or confirmations (§ 5.2.d), and the peculiar motivation of anyone who either solves puzzles or engages in normal science (§ 5.2.e).

a. The Existence of Regulations

Research in normal science, like puzzle-solving, consists in working out a certain kind of problem. The first analogy addresses, in a very general way, the type of problem-solving involved in both cases. The analogy consists in the fact that problem solving in both normal science and puzzle-solving is subject to certain *regulations;* that is, it operates under normative constraints. Regulations in the activity of normal science have the same function as the "rules of play" stipulated in puzzle-solving; in both cases, the permissibility of approaches to solution and of the solutions themselves is regulated. In other words, neither the approach to a problem nor the acceptance of a proposed solution is entirely left to the whim of the player or scientist.[24]

Of what sort are the regulations which must be observed in research practice of normal science? In *SSR,* Kuhn distinguishes four different kinds of regulation operating in this domain.[25] First, and most obviously, the scientist must adhere to those general propositions which explicitly express theories, laws, or definitions of concepts. Sec-

very deep" (Popper 1970, p. 53 n. 1; similarly Putnam 1974, p. 261 and Röseberg 1984, p. 27). But this only applies if we assign seriousness and depth exclusively to scientific activities which aim toward conceptual innovation.

23. 1970a, *ET* pp. 270, 272–277 (or, indirectly, *SSR,* p. 22). To be sure, according to Kuhn, one couldn't expect this (or any other) criterion to serve as an unequivocal demarcation for all situations.

24. 1959a, *ET* p. 237; *SSR,* pp. 5, 38–42; 1962b, p. 383; 1963a, pp. 349, 362–363; 1970a, *ET* p. 273.

25. *SSR,* pp. 40–42. —In this passage, Kuhn speaks of "rules," a manner of speaking he admits, at the beginning of the passage, is "a considerably broadened use of the term 'rule' " (*SSR,* pp. 38–39) and retracts entirely at the passage's end (p. 42, and chap. 5). By the end of this section it should be clear why Kuhn ultimately abandons even the term "rule," and why, at the same time, I choose instead to talk of "regulations."

ond, the scientist is bound to "a multitude of commitments to preferred types of instrumentation and to the ways in which accepted instruments may legitimately be employed."[26] Such commitments presuppose, for one, the validity of those theories which justify putting instrumentation to a particular use. In addition, the usual uses of equipment also involve a large number of implicit expectations regarding which situations may and, especially, which situations *may not* occur.[27]

Kuhn calls the third kind of regulation the realm of "quasi-metaphysical" commitments, convictions regarding the nature of that which physically exists, and, by implication, that which does not. Such ontological commitments also have a far-reaching methodological or scientifically normative meaning, for they dictate the nature of fundamental laws, prescribe which sorts of explanation are legitimate, and decree which sort of research problem should be attacked. Fourth, there are highly abstract, timeless regulations which arise under the assumption that the enterprise in question is meant to be science at all.[28]

Throughout the phase of normal science, consensus over these regulations reigns over the relevant community. Since paradigms form the core of any scientific consensus, Kuhn claims in *SSR* that such regulations are "commitments that scientists derive from their paradigms."[29] Taken together, these regulations are thus nothing other than that which Kuhn attempted to capture, in his 1969 papers, under the notion of a disciplinary matrix: the sum of "all the shared commitments of a scientific group."[30]

We ought still to consider the proper limits on the analogy between puzzle-solving and normal science as regards the existence of regulations. The analogy breaks down when it comes to the category to which the governing regulations belong. In the paradigm cases of puzzle-solving, regulations are placed on the relevant class of puzzle in the form of *explicit* rules of play; in a chess problem, for example, they stipulate which moves are allowed and how many may be made before checkmate. But one of the cornerstones of Kuhn's philosophy of science is the thesis that research governed by paradigms isn't a practice determined entirely by explicit or explicable rules.[31] Many of the regulations constraining the researcher's whims are given *implicitly*, in the

26. *SSR*, p. 40.
27. *SSR*, pp. 59–60.
28. On these timeless regulations, compare § 1.2.c.
29. *SSR*, p. 40.
30. 1974a, *ET* p. 294; compare §§ 4.2.a and 4.3.
31. *SSR*, pp. 42, 88. —Compare § 4.1.

form of exemplary problem solutions with all their various moments, and can't, in principle, be adequately restated as explicit guides to action, whether conditional or absolute. For this reason, I have chosen to call them "regulations" rather than "rules," so as to preempt our mistaking them for explicit guides to action.

b. Expectations of Solubility

The expectation that there exists one (and perhaps only one) solution to a chosen problem which accords with regulations is constitutive of the conduct of both puzzle-solving and normal science.[32] This analogy breaks down, however, when we inquire into the source of such expectations. In the case of puzzle-solving, the human constructor (or inventor) of the puzzle, together with the institutions which make the puzzle accessible to the puzzle-solver, serve as guarantors of the puzzle's solubility. The constructor of such puzzles often proceeds with an assumed solution in mind, and, taking the regulations the puzzle-solver will have to follow into account, invents an appropriate puzzle.

In the case of the research problems of normal science, however, expectations of solubility are grounded in something else. Prospectively soluble research problems of science aren't "invented" or "constructed." Neither do they lie so close to hand as to be simply "given," though what are given in this sense are problems for which there is *no* justified expectation of their solubility. The research problems of normal science are rather "discovered," "found," or "selected," namely as guided by paradigms in the sense of exemplary problem solutions.[33] Once a given phenomenal world has been constituted (which process also involves a moment of construction),[34] the identification of soluble research problems is a matter not of constructing but rather of finding them. The expectation that a given research problem is soluble arises in part from its analogy to previously solved paradigmatic problems. This analogy initially only allows us to suppose that a solution to the problem is possible *in principle*. The solution must, in addition, be *practically feasible*, given available instrumentation, personnel, funding, and time. Here, too, analogies with previously solved problems provide some points of reference.

It is a historical fact that, in the course of normal science, expecta-

32. 1959a, *ET* pp. 234, 235; *SSR* pp. 28, 37–38, 42, 68, 96, 151–152, 164; 1963a, p. 362; 1964, *ET* p. 262; 1969c, *ET* p. 346; 1970a, *ET* pp. 270–271; 1970c, *SSR* p. 179; 1974b, p. 455.

33. Compare § 4.4.b.

34. Compare chap. 3, especially § 3.2.

tions regarding the solubility of chosen research problems grounded in this way tend, over long periods of time, not to be disappointed. How this fact ought to be understood is another question, one which remains open within the scope of Kuhn's theory.[35]

c. No Intention of Fundamental Innovation

Like puzzle-solving, the solution of research problems in normal science is conducted without any intention of introducing fundamental innovations. "Fundamental innovations" here refers to innovations affecting the governing regulations. So within normal science, neither the conceptual system employed in exemplary problem solutions and symbolic generalizations, nor the generalizations themselves, nor the means of identifying and solving research problems is under review. Regulations are constitutive of both puzzle-solving and normal science and thus aren't open to dispute over the course of these activities. Given the absence in normal science of any receptivity to fundamental innovation, it's understandable that many features of the problem solutions of normal science will be anticipated.[36]

The commitment to regulations doesn't, however, exclude the possible discovery, *within* the bounds set by them, of new methods of solution. This possibility obviously also applies to puzzles. Neither is normal science always an uncreative, merely repetitive routine.[37] On the contrary, overcoming often extraordinarily difficult problems of the instrumental, conceptual, and mathematical varieties[38] sometimes requires a superlative degree of innovative fantasy. New phenomena and entities are also discovered in normal science. But these aren't "unexpected discoveries" "which could not be predicted from accepted theory in advance and which therefore caught the assembled profession by surprise."[39] It is rather anticipated phenomena and entities that are discovered in the course of normal science, and accordingly, such discoveries are "an occasion only for congratulations, not for surprise."[40]

35. *SSR,* p. 173.

36. 1959a, *ET* pp. 225, 227, 232, 233–234, 235; 1961a, *ET* pp. 187, 192, 197; *SSR,* pp. 24, 29, 35, 36, 52–53, 59, 61, 64, 65, 66–67, 76, 96, 122, 163–164; 1962d, *ET* pp. 166–167; 1963a, pp. 348, 362, 363, 364, 365, 368.

37. 1959a, *ET* p. 235; 1961a, *ET* pp. 192–198, 208–209; *SSR,* pp. 36, 38, 135; 1962d, *ET* p. 167; 1970b, p. 246; 1976b, p. 197 n. 1.

38. See § 5.3 for a taxonomy of the research problems of normal science.

39. 1962d, *ET* p. 166; similarly *SSR,* p. 52; 1963a, p. 365. —See § 7.2 below on unexpected discoveries.

40. *SSR,* p. 58; similarly 1962d, *ET* p. 166–167; 1963a, p. 365. Here Kuhn also cites his standard example for the sort of discovery which occurs in normal science: the discovery of new chemical elements whose existence had been predicted by the periodic table.

This analogy with puzzle-solving has the following limitation. Not only is normal science not oriented toward fundamental innovation, it also exhibits a certain tendency to repress such problems and phenomena as might call basic regulations into question.[41] Similarly, during the phase of normal science, scientists reject suggested changes in the governing regulations. Here the dogmatic tendency of normal science is most pronounced.[42] Puzzle-solving apparently offers no analogy to this phenomenon. Yet a further disanalogy consists in the fact that, in normal science, critical phenomena and problems can't be repressed forever. For according to Kuhn, normal science presses beyond itself toward scientific revolution,[43] a tendency which puzzle-solving apparently lacks.

Normal science, then, doesn't seek basic innovations but instead works within certain set bounds. This feature grounds three further metaphors employed by Kuhn in characterizing normal science. For one, Kuhn talks of normal science's being "an attempt to force nature into the preformed and relatively inflexible box that the paradigm supplies."[44] Kuhn also describes normal science as "mop-up work" or "mopping-up operations" in which the expectations of success engendered by previous solutions are actually fulfilled.[45] And finally, he compares the activity of normal science with an attempt "to elucidate topographical detail on a map whose main outlines are available in advance."[46]

d. Neither Test nor Confirmation

Closely associated with the above analogy between puzzle-solving and normal science is the further parallel that neither puzzle-solving nor normal science can properly be seen as a "test" or "confirmation" of its governing rules.[47] This claim may be understood in at least three ways.

1. For one, the claim might concern the *individual motivation* of the puzzle-solver or normal scientist. In the case of the puzzle-solver, it's obviously true; for the puzzle-solver (at least in the normal case) has

41. *SSR,* p. 5–6, 24, 62, 64; 1963a, pp. 348, 360, 364; 1964, *ET* p. 262.

42. Compare § 5.1. —This dogmatic tendency is also manifest in normal science's approach to training; see § 5.5.a below.

43. On this issue, see § 7.1 below.

44. *SSR,* p. 24, similarly pp. 5, 151–152; 1970a, *ET* pp. 270–271; 1970b, pp. 260, 263; 1981, p. 9; also 1961a, *ET* pp. 193–198.

45. 1961a, *ET* p. 188; *SSR,* p. 24.

46. 1959a, *ET* p. 235; similarly *SSR,* p. 109.

47. 1961a, *ET* pp. 187, 192, 197; *SSR* pp. 80, 144–145, 146; 1970a, *ET* pp. 270–272.

no intention of confirming or testing the rules of play (whatever that might mean). Neither, according to Kuhn, is the desire to confirm or test the governing regulations the dominant component of the normal scientist's individual motivation. In § 5.2.e, we will return to discuss the analogy between the individual motivations of those who solve puzzles and those who engage in normal science.

2. The claim that neither puzzle-solving nor normal science consists in testing or confirming the governing regulations might also be taken as referring to the *institutional goals* of puzzle-solving and normal science.[48] Once again, the claim is obviously true of puzzle-solving. Kuhn understands normal science as having a different institutional goal, that of bringing accepted theory into improved and expanded correspondence with nature.[49] Of course, "nature" here refers to a phenomenal world, the phenomenal world to which the members of the appropriate community have been socialized by their culture and special training.[50] The regulations binding on the conduct of normal science make just this institutional goal explicit;

> this srong network of commitments—conceptual, theoretical, instrumental, and methodological . . . provides rules that tell the practitioner of a mature specialty what both the world and his science are like.[51]

This commitment to a set of regulations means that describing their role as that of something to be tested or confirmed would be a category mistake; for the testing or confirmation of some thing is possible only when that thing is open to dispute. The regulations which govern normal science, however, are constitutive of this form of scientific practice and are thus not open to dispute within it.[52]

This institutional goal of normal science has three consequences.

48. What I call "institutional goals" might also be dubbed "functional purpose." Kuhn often speaks of "functions." Examples include the titles of 1961a, "The Function of Measurement in Modern Physical Science," 1963a, "The Function of Dogma in Scientific Research," and 1964, "A Function for Thought Experiments."

49. 1959a, *ET* p. 233; 1961a, *ET* p. 192; *SSR,* pp. 36, 52, 64–65, 80, 122, 135, 152; 1963a, pp. 353, 360–361, 363, 367; 1964, *ET* p. 262; 1968a, *ET* p. 119; 1970b, pp. 237, 246, 250; compare § 5.1. —*Exactly* what "improved correspondence" might involve for a given scientific community is given by the community's characteristic version of scientific values (Compare § 4.3.c).

50. Compare Ch. 3, especially § 3.4, along with § 5.5.a below.

51. *SSR,* p. 42.

52. Normal science has another, indirect institutional goal of great importance to Kuhn's theory, that of meeting the preconditions for the renewal of its governing regulations or preparing the way for a scientific revolution; on this issue, see § 7.1 below.

First of all, the commitment to regulations allows a highly directed and thus also highly detailed, efficient, and cooperative form of research to attack both problem selection and problem solving, a form of research which seemingly can't be realized without such commitments.[53]

Second, this same commitment explains the often esoteric character of work in normal science.[54] For such work is intelligible only to one who understands the conceptual and instrumental details of the exemplary problem solutions and, along with them, the governing regulations; and this understanding is precisely what most outsiders lack.

Third, this commitment explains why it's legitimate, in normal science, to set aside certain unsatisfactorily solved or, for the moment, insoluble problems, attacking instead problems which appear soluble.[55] For

> no theory ever solves all the puzzles with which it is confronted at a given time; nor are the solutions already achieved often perfect.[56]

Such behavior would stand in opposition to the institutional goal of normal science, if this goal were to test regulations. If it were, *every* empirical deviation from theory would have to be pursued until a definite decision was reached on whether or not the problem is soluble within the bounds set by governing regulations. But in practice, the large number of possible reasons for such empirical deviation often makes these definite decisions impossible.[57] As far as the goal of improving the corespondence of theory with nature is concerned, however, what counts are the successful applications of theory; problems which, for the moment, can't be solved at all, or can be solved only unsatisfactorily, may be set aside.

3. Finally, the claim that neither puzzle-solving nor normal science serves as a test or confirmation of the governing regulations can be understood in yet a third sense. To be sure, puzzle-solving and normal science aren't *manifestly* directed, either individually or institutionally,

53. 1959a, *ET* pp. 235–236, 237; *SSR,* pp. 18, 22, 24–25, 42, 64–65, 76, 96, 152, 163–164; 1963a, pp. 356–357, 363; 1964, *ET* pp. 261–262; 1970b, pp. 243, 247; 1970c, *SSR* p. 178. —See § 5.6 below.

54. 1959a, *ET* pp. 235, 236; *SSR,* pp. 11, 18, 20, 23, 24, 42, 64, 163–164; 1963a, pp. 357, 363; 1970b, pp. 247, 254; 1970c, *SSR* p. 178.

55. 1959a, *ET* pp. 234, 236; 1961a, *ET* pp. 202, 203–204; *SSR,* pp. 18, 79–80, 81–82, 96, 110, 146–147; 1963a, p. 365; 1963b, p. 392; 1964, *ET* p. 262.

56. *SSR,* p. 146; similarly pp. 79, 81, 110, 147.

57. 1961a, *ET* p. 202. —On this issue, also see Lakatos 1970, § 2a.

toward any kind of test or confirmation. But a test or confirmation of the governing regulations might still be an unintended by-product of puzzle-solving or normal science. Testing or confirmation of regulations might prove a *latent* function of normal science.[58] Puzzle-solving and normal science might then be *conceived* as testing or confirmation, in the sense that, though this interpretation of such activities doesn't reflect the manifest intentions of either individuals or institutions, it still captures unintended aspects of them.

Let us consider first an interpretation of normal science that attempts to view it as an unintended *test* of regulations. The fact that the failure of individual attempts at problem-solving is taken not as a sign of flawed regulations but rather as a sign of incompetence on the part of the implicated scientist speaks against this interpretation.[59] Though this claim seems rather odd at first blush, it may be unpacked with the help of a fact cited under point 2 above, that normal science always contains arbitrarily many problems deemed soluble "in principle" whose actual solution remains, for whatever reason, impossible. The scientist must thus be in a position to recognize the practically soluble problems for what they are; the more surprising the solution, the more original its exploitation of the limits set by regulations, the greater the scientific achievement. Accordingly, a miscarried attempt at solution can initially only be taken as evidence either that the problem isn't soluble at present or that the implicated scientist is too poorly endowed to solve it. In either case the failure reflects back on the scientist, in the former case because of his or her inability to recognize the problem's practical insolubility early enough and in the latter because of his or her inability actually to provide the solution. Of course it's possible, in principle, that the accepted regulations really are inadequate and should be replaced, just as it's possible that an error in construction or printing is responsible for a crossword puzzle's intransigence. But this parallel to puzzle-solving sheds light on what's unique about research in normal science: initially, this possibility, though "in principle conceivable," appears a mere excuse.

Under these circumstances, the activity of normal science thus can't be conceived as the unintended testing of regulations (at most, what's tested is the scientist!). But neither, under these circumstances,

58. See Merton 1968, pp. 73–138, on the notions of "manifest" and "latent" functions; for a discussion of latent functions as a subset of unintended consequences of action, see ibid., p. 105, n. ⋆.

59. 1959a, *ET* p. 234; 1961a, *ET* p. 192; *SSR,* pp. 35, 35–36, 37, 80; 1963a, pp. 362, 365; 1970a, *ET* p. 271, with n. 6, p. 273; 1974b, p. 455. —As to how matters stand with regard to the significant anomalies which often lead to a departure from normal science itself, we will have to wait until later; see § 7.1.

may we speak of normal science as a *de facto*, though unintentional, *confirmation* of the governing regulations. For confirmation, in the strict sense, can only occur against a background of possible refutation. Just this possibility is ruled out within normal science. Similarly, the rules of play for established kinds of puzzles aren't "tested" or "confirmed" by the success, failure, or abandonment from boredom of individual attempts at solution.

And so the claim that neither puzzle-solving nor normal science can be conceived as a test or confirmation of the governing regulations is valid under all surveyed interpretations. In one respect, though, the analogy breaks down, for just what does it mean to claim that a puzzle's rules of play aren't tested or confirmed? Justified as this question is, it doesn't really touch the heart of the analogy. This consists in the fact that the *calling into question* implicit in our concepts of testing and confirmation is appropriate in neither puzzle-solving nor normal science. What both activities have in common is that they're conducted under the assumption that such questioning of regulations has been ruled out.

e. Individual Motivation

The last analogy between normal science and puzzle-solving concerns the individual motivation of someone who solves puzzles or engages in normal science.[60] Neither puzzle-solving nor normal science is motivated primarily by some intended product of the activity *outside* the activity itself. Again, this is obviously true of puzzle-solving, for a solved puzzle generally doesn't have an especially high intrinsic value. The puzzle's consructor knows its solution to begin with, anyway, and if the solution were all that mattered, there would be many simpler ways of acquiring it. The problems of normal science behave similarly; since they serve neither testing nor confirmation,[61] and since many features of their solutions are known in advance,[62] the substantive worth of solved problems of normal science isn't great enough to explain the care and devotion scientists exhibit in attacking them.[63] What

60. 1959a, *ET* pp. 234, 235, 237; *SSR*, pp. 24, 36, 37–38, 42, 135; 1963a, pp. 362–363; 1970a, *ET* p. 290; 1970b, p. 246.

61. Compare § 5.2.d.

62. Compare § 5.2.c.

63. This should by no means be taken to mean that the solutions to problems of normal science are of no scientific value. For they are naturally seen in light of the institutional goal of normal science, whence they derive their purpose (I will come to the classification of the problems of normal science, and their scientific worth, in § 5.3). Neither is it being claimed that the prospects for solving problems of normal science constitute the chief attraction of the scientific profession; such often-cited motives as intellectual curiosity and the like play a role here (*SSR*, pp. 37–38). And yet such drives,

is decisively motivational in both cases, our analogy tells us, is the fact of a solution's *success* (and perhaps some peculiarities of the strategy which produced it). Since the existence of a solution to the problem has been presuppposed,[64] whether and, if so, how the goal is reached depends entirely on the puzzle-solver's or researcher's abilities. The fascination and personal challenge typically asserted by both puzzles and normal science—on a certain kind of person, anyway—is thus accounted for. The chief motivation toward the activity of normal science consists in the opportunity to prove one's own ability under conditions known to and accepted by the community.[65] These conditions set the stage for a transparent competition for reputation within the community.[66]

5.3 The Research Problems of Normal Science

To begin with, the research problems of normal science may be characterized negatively: such problems aren't selected for social, economic, or other reasons external to science.[67] For one feature of scientific communities in the phase of normal science is that they tend to be extremely isolated from the rest of society, as compared with other communities.[68] Concretely, this isolation means, for a given community, that

—The vocabulary and techniques employed in the community are largely inaccessible to outsiders.
—Scientific communication by members of the community is exclusively directed at other members of the community.
—Only members of the community are accepted as competent to evaluate the quality of scientific work.
—Engagement with the problems of normal science is prescribed by goals *internal* to science.

It is obvious enough that the institutional goal of normal science is internal to science, for it consists simply in expanding and improving the correspondence of theory with nature.[69]

---

if they survive the training period, are hardly satisfied by individual participation in normal science.

64. Compare § 5.2.b.

65. Compare § 5.2.a.

66. *SSR*, p. 36; 1963a, p. 363; 1970a, *ET* p. 290; also 1983c, p. 30; 1984, pp. 250–251.

67. 1959a, *ET* pp. 237–238; *SSR*, pp. 36–37, 96, 164; 1962b, p. 383; 1964, *ET* pp. 261–262; 1968a, *ET* p. 119.

68. *SSR*, pp. 164, 168; 1963a, pp. 359–360; 1968a, *ET* pp. 118–119; 1970b, p. 254; 1971c, *ET* p. 160; 1976a, *ET* p. 36; *ET*, p. xv.

69. Compare § 5.1 and § 5.2.d, point 2.

The primary way of attaining this institutional goal is to take the success contained in exemplary problem solutions as a grail for the scientific endeavor.[70] Why should such an approach be regarded as realistic in pursuit of this goal? Its acceptance is grounded in the scientific community's conviction that the successful solution of an exemplary problem is neither coincidental nor singular. On the contrary, this success is the result of a correct focus on central aspects of the world; "Normal science . . . is predicated on the assumption that the scientific community knows what the world is like."[71] Such previously attained knowledge of the world is expressed in the exemplary problem solutions.[72] The work of normal science thus consists in exploiting the knowledge potential implicit in (and not, in principle, entirely separable from)[73] exemplary problem solutions, by employing the immediate similarity relations to identify soluble research problems.[74]

It is the dependence of even the *problematic character* of the problems of normal science on exemplary problems and their solutions[75] that accounts in large part for the historicity of science. For, with a change in exemplary problem solutions and the consequent change in phenomenal world, the questions which may meaningfully be posed and addressed generally also change.[76]

According to Kuhn, normal science exploits the potential of exemplary problem solutions in three overlapping domains. Within each of these domains, the more observational or experimental problems can be distinguished from the more theoretical problems.[77]

The first domain is that of the theoretical or experimental determi-

70. In § 4.1, I called this goal the "globally normative aspect of consensus over concrete problem solutions"; see also § 4.4.

71. *SSR* p. 5; similarly 1963a, 149.

72. According to the current self-image of the sciences, previously procured valid propositions are contained in the theory or model used in problem solving (where the assumed correctness of the theory or model may have qualitative or quantitative limits). Kuhn's correction of this self-image targets two issues. First, the self-image misses the fact that the concepts employed in the theory or model are constitutive of the world. Second, it fails to notice that, for example in physics, such previously acquired propositions aren't just the fundamental equations of the current theory, for the meanings of empirical concepts contained in these equations can't be extracted from prior successful applications of the theory; see §§ 3.6 and 4.3.e.

73. Compare § 4.3.e.

74. 1970b, p. 275; 1970c, *SSR* p. 200. —Compare §§ 3.6.e, 4.4, and 5.2.

75. 1959a, *ET* p. 235; *SSR*, pp. 5, 6, 109, 140–141, 166; 1963a, pp. 349, 359, 361–362; 1963b, p. 389; 1964, *ET* pp. 261–262.

76. I will discuss the difficulties associated with such changes in detail in chap. 6 and chap. 7.

77. 1959a, *ET* p. 233; 1961a, *ET* pp. 187–201; *SSR*, pp. 24, 25–34, 68, 97, 122; 1963a, pp. 360–361; 1968a, *ET* p. 119; 1970a, *ET* p. 276; 1970b, p. 246; 1977b, *ET* p. 19; 1981, p. 8.

nation of those facts, such as the parameters of relevant materials, the knowledge of which is a necessary condition for any concrete application of theory. The relevance of such facts can be inferred from the appropriate exemplary problem solutions.

The second domain concerns the improvement of correspondence between theory and observation or experiment. This involves both the quantitative improvement of existing comparisons between theory and empirical results and the procurement of new opportunities for comparison. In both cases, theoretical as well as experimental or observational work is called for.

Third, normal science seeks to articulate further the theory it employs in problem-solving. On the part of experiment and observation, this task involves the more precise determination of physical constants, measurements in the service of a quantitative formulation of laws previously only formulated qualitatively, and experimental efforts toward applying theory to classes of phenomena for which the possibility of such application has been posited but not yet accomplished. More theoretical investigations include attempts to clarify existing theory by producing new, equivalent formulations.

5.4 Progress in Normal Science

Before I consider whether and, if so, in what sense normal science exhibits progress, I will attempt a general exposition of the notion of progress. This exposition is needed because the truth of claims about the presence or absence of progress must naturally depend on which notion of progress they employ.

The assertion that a given process, or a given thing (individual, institution, group of people, project) engaged in a process, makes progress has three implications.[78] First, it implies that later phases of the process exhibit gain in one or another respect, as compared with earlier phases. Progress always involves gain in some particular respect and is distinguished from its complement, "mere change," by the latter's lack of any direction toward gain. Second, that respect in which gain is manifest is no mere contingent aspect of the process. On the contrary, it must in some sense be crucial to the thing involved in the process, the "reference point"[79] of the process. Third, applying the term "progress" mostly involves a positive value judgement; gain in that respect important to the reference point of the process is seen as positive relative to

78. See Hoyningen-Huene 1984b.

79. See Lübbe 1977, p. 75ff, on the notion of a "reference point" [*Referenzsubjekt*].

some particular value scale. This positive value judgement is sometimes lacking, as when we speak of the progress of a disease. But for our purposes we may neglect any such ambiguity in the notion of progress.

Ignoring for the moment one qualification (to be discussed later), normal science is an enterprise which, for Kuhn, unquestionably exhibits progress. Such progress is "both obvious and assured."[80] The latter attribute is meant to suggest that progress is an "inevitable" product of the methodological endowment of normal science.[81] For Kuhn, this progress is of a paticular variety, that of "linear," "additive," or "cumulative" epistemic progress.[82]

In order to understand these features of the epistemic progress of normal science, we must consider the characteristics of normal science together. We recall, first of all, that normal science is a research practice in which fundamental innovation is neither intended nor even permitted.[83] The newly gained insights of normal science are thus compatible with previously available knowledge, and they replace not earlier (apparent) knowledge but only ignorance—at least to the extent to which they weren't anticipated already.[84] Furthermore, research problems are selected in accordance with the institutional goals of normal science; each successfully solved problem improves or expands the correspondence of theory with nature.[85] Finally, only those problems whose solution is actually possible given the available means are addressed. Problems whose solution is impossible on practical grounds are neglected, along with those that are later discovered insoluble in principle with the means available in this form of normal science.[86]

When we consider these characteristics of normal science in light of our earlier general exposition of the notion of progress, Kuhn's

80. *SSR,* p. 163; 1970b, p. 245.

81. *SSR,* p. 166.

82. *SSR,* p. 52, 53, 96, 139; 1962d, *ET* p. 175; 1970b, p. 250; 1977b, *ET* p. 19; 1981, pp. 7–8, 18–19.

83. Compare § 5.2.c.

84. *SSR,* p. 95; 1962d, *ET* pp. 166–167, 175; 1963a, p. 365; 1977b, *ET* p. 19. —This presentation of normal science as a practice in which no knowledge is corrected or replaced but rather ignorance makes room for new knowledge is somewhat idealized. In fact, previously available knowledge may be corrected even in normal science, as Kuhn explicitly admits in later work (1981, p. 19). But these corrections occur in such a way as to leave the present phenomenal world untouched. What's meant by "untouched" can relatively easily be made clear by reference to examples. These corrections are limited, for example, to minimal quantitative corrections of physical constants, with no far-reaching consequences. Still, the problem of a general distinction between "normal" and "revolutionary" scientific development remains, and I will return to it in chap. 6.

85. Compare §§ 5.1, 5.2.d, point 2, and 5.3.

86. Compare § 5.2.d, point 2.

conception of progress in normal science becomes entirely transparent. Given the mode of problem selection operative in normal science and the compatibility of new knowledge with old, increase in knowledge of the (current phenomenal) world must, for so long as it is possible at all, be *cumulative* and *inexorable.* Since this increase in knowledge is guided by the institutional goal of normal science, it must, trivially, be both crucial to and positively valued within normal science. The three general features of progress discussed above are thus all present. It follows that normal science necessarily exhibits cumulative epistemic progress. Since the cumulative nature of this epistemic progress is so closely associated with the typical characteristics of normal science, normal science might even be *defined* as that form of scientific practice that proceeds cumulatively.[87]

But with its variety of different, and at first blush contradictory, assertions about progress in normal science, Kuhn's *SSR* raises certain problems. On the one hand Kuhn claims, just as we've so far been led to expect, that

> In its normal state, then, a scientific community is an immensely efficient instrument for solving the problems or puzzles that its paradigms define. Furthermore, the result of solving those problems must inevitably be progress. There is no problem here.[88]

But on the other hand, *SSR* also contains sentences such as these:

> Viewed from within any single community . . . the result of successful creative work *is* progress. How could it possibly be anything else? . . .
>
> . . . During those periods [of normal science] . . . the scientific community could view the fruits of its work in no other way.
>
> With respect to normal science, then, part of the answer to the problem of progress lies simply in the eye of the beholder.[89]

Here Kuhn appears to be speaking as if the appearance of progress in normal science were the product only of a certain perspective, and what's more, a perspective guilty of massive prejudice.

The apparent tension between these assertions dissolves when we recall our original analysis of the notion of progress. In a certain sense, the progress exhibited by normal science is indeed *objective.* Relative to the phenomenal world in which normal science is conducted, to

87. Kuhn at least approximates such definition in his 1981, pp. 7–8, 19; also see *SSR,* p. 92, and 1970b, p. 250.

88. *SSR,* p. 166.

89. *SSR,* pp. 162–163, original emphasis.

previously available knowledge of this phenomenal world, and to the institutional goal of normal science, a positively valued increase in knowledge must necessarily occur. In this sense, there really is no problem here. The question remains whether the premises under which such progress is established are actually accepted. Of course they are accepted by the community engaged in normal science in the appropriate phenomenal world, for these are the only premises under which this community's work can be conducted. To this extent, "the scientific community could view the fruits of its work in no other way." An external observer, however, who saw prior knowledge of this phenomenal world as riddled with error, and thus as unworthy of expansion, would reject these premises. Though even this observer couldn't deny that progress occurs by the standards of the community, he or she would discard these standards.

This kind of relativity presumably obtains wherever claims of progress are made. For such claims always rest on certain premises, the acceptance of which isn't mandatory. In view of this situation, progress in normal science has two special features. For one, within normal science, the requisite consensus over the premises which sanction claims of progress actually exists. Furthermore, normal science is an enterprise in which knowledge-producing communities are largely isolated from other communities.[90] And so all opportunities for convincingly applying external standards to normal science are quite limited. Thus normal science appears *in fact* to be largely exempt from doubts regarding its progress, though these remain possible *in principle*. For this reason, our current concept of science even includes progressiveness as one component in its meaning.[91]

5.5 What Makes Normal Science Possible?

The question, What makes normal science possible? aims at the conditions which must be met in order for normal science to occur. I will here consider conditions of two kinds. First, one might ask after the details of the form of training which enables the individual scientist to practice normal science successfully (§ 5.5.a). One might also ask, from a historical perspective, under what conditions normal science can arise to begin with (§ 5.5.b). The conditions under which a shattered scientific consensus may be reconstituted, that is, the conditions under

90. Compare § 5.3.
91. *SSR*, p. 160–162.

which a resumption of normal science is possible, will have to wait until chapter 7.

a. Training Preparatory for Practicing Normal Science

Let us first consider the medium of instruction typical of training for normal science.[92] This medium is the scientific textbook written especially for students. The dominance of textbooks in training for normal science leads, first of all, to the almost complete insulation of students from the primary literature, from those publications in which scientists originally communicate, or communicated, their results.[93] The student continues to read almost no original work in the research tradition for which he or she is being trained, right up to shortly before he or she begins productively engaging in scientific work. Those original reports he or she finally reads relay, as a rule, the latest work in the field and not the work which founded the field to begin with. The student generally has absolutely no contact with the primary literature thought to be scientifically obsolete, for such literature treats the field with different basic concepts, lines of inquiry, theories, and standards for solutions.

Instead, textbooks present, first and foremost, those scientific achievements subject to consensus within the appropriate scientific community. In particular, these include certain problems along with their solutions, where the community's current theories, models, assumptions, and the like are expressed in these solutions. These problem solutions are presented (in good textbooks, at least) in their most modern form and not in that most faithful to the historical originals. Students are then assigned problems similar to those presented to solve by themselves, using theory or experiment. Different roughly contemporary textbooks don't exhibit any important differences in their selections of concrete problems and solutions, let alone in the theories they present, but rather differ mostly in their didactic details. Textbooks neither systematically discuss the range of currently unsolved problems and appropriate methods of attacking them nor treat those scientific concepts, questions, theories, and standards deemed obsolete in any detail, let alone in a historically adequate manner.[94]

92. On the following, see 1959a, *ET* pp. 227–230, 232, 237; 1961a, *ET* pp. 186–187; *SSR* pp. 5, 43, 46–47, 80–81, 111–112, 136–138, 165–166, 168–169; 1963a, pp. 350–351; 1963b, pp. 391, 393; 1970b, pp. 250, 272–274; 1970c, *SSR* pp. 188–191; 1974a, *ET* pp. 305–308; 1976b, p. 181; 1981, pp. 20–21; 1985, p. 25; 1990, p. 314.

93. With regard to the degree of insulation, there are differences from discipline to discipline, though these aren't especially far-reaching; 1963a, p. 350 n. 1.

94. Compare § 1.2.a.

Now let us ask what *isn't* learned in the course of such training. Three gaps are particularly evident. First of all, the student receives no adequate view of the past of his or her field. On the contrary, this form of training often leaves scientists with a sometimes even grotesquely distorted impression of their past. Second, during the course of this training, the student receives no general overview of the present status of research in his or her field, or at most, receives one at the very end of training. In general, though, this overview is acquired only over the course of the student's own productive scientific efforts. Third, this form of training promotes highly "convergent" modes of thought. Such convergence discourages any comparative evaluation of different possible ways of doing science in the given field. To attempt such an evaluation, the student would have to be as thoroughly familiar with historical and contemporary alternatives to the dominant research practice as he or she is with accepted tradition. He or she would also have to be in a position to estimate the actual potential for solving previously unsolved problems afforded by the tradition into which he or she has been inculcated. Finally, the goals and techniques of comparative evaluation would have to be transmitted and practiced. The form of training currently under discussion offers none of these; instead, it merely earnestly initiates the student into a single tradition.

But what is the positive yield of this kind of training process? Two items are especially noteworthy. First, the student gains access to the (region of the) phenomenal world relevant to the work of his or her community.[95] For, especially by means of the demonstrated exemplary problem solutions and the student's own exercises, the student appropriates the similarity relations constitutive of this region of the phenomenal world. At the same time, he or she acquires the available explicit and implicit knowledge about this same region. Second, the student gains an extraordinarily efficient mastery over the techniques employed in finding and solving the problems of normal science in the relevant region of the phenomenal world, techniques which employ the analogies, furnished by the immediate similarity relations, between previously solved and unsolved problems.[96] This technique permits the employment of knowledge of the relevant region of the phenomenal world implicit in previously solved problems for identifying and solving new problems.

The kind of education which enables someone to practice normal science contains, in virtue of its orientation toward convergent modes of thought, a highly authoritative moment. Accordingly, Kuhn also

95. Compare chaps. 3 and 4.
96. Compare §§ 3.6.e, 4.4, 5.2, and 5.3.

calls it "a narrow and rigid education, probably more so than any other except perhaps in orthodox theology."[97] For what the student receives is

> a relatively dogmatic initiation into a previously established problem-solving tradition that the student is neither invited nor equipped to evaluate.[98]

Scientists who have undergone the training which enables them to do normal science are thus socialized toward carrying on a particular tradition of problem-solving in a particular phenomenal world without gaining any access to alternatives. This training, however, obviously leaves room for individual differences in degree of identification with the given tradition of normal science. The internal distance with which an individual practices normal science and the degree to which he or she retains (private) conceptions of alternative forms of science, or alternative worlds, are irrelevant so long as he or she continues to adhere to the regulations of normal science.[99] During the training process, only those individuals who are either unwilling to undertake or incapable of engaging in the appropriate puzzle-solving activity are weeded out.

This powerful moment of authority in training for normal science, preserved throughout the conduct of normal science itself, is what leads to the description of normal science as a form of quasi-dogmatic practice.[100] How this quasi-dogmatic element of normal science ought to be assessed remains open here; I will return to this question in § 5.6.

b. The Emergence of Normal Science out of Prenormal Science

First, however, we must discuss the notion of prenormal science and the transition from prenormal to normal science.

1. When normal science first comes about in a given field, it doesn't emerge from a state whose candidacy as science may be rejected out of hand. But for Kuhn, the form of science which precedes normal

97. *SSR*, p. 166, similarly p. 136.

98. 1963a, p. 351; similarly 1959a, *ET* pp. 229, 237; *SSR*, pp. 5, 80; 1963b, pp. 386–387.

99. It's even conceivable that a scientist might deliberately practice a given form of normal science solely for the purpose of reaching its limits. Delbrück's early work in molecular biology might constitute an example of this approach; see Fischer 1985 and Kay 1985, p. 245. —Here is yet a third realm of individual differences between members of a scientific community which might, under certain conditions, prove epistemically relevant; see §§ 3.6.d and 4.3.c.

100. Compare §§ 5.1 and 5.2.c.

science is a practice "something less than science."[101] In this phase of scientific development, the consensus so characteristic of normal science doesn't yet exist. Accordingly, in his 1959a, Kuhn calls this phase of science the "preconsensus phase," distinct from phases with "firm consensus."[102] Since paradigms, in the sense of universally accepted exemplary problem solutions with all their various moments, are the heart of consensus in normal science, Kuhn, in 1962, also calls the preconsensus phase the "pre-paradigm" phase.[103] In 1969, however, Kuhn weakens the paradigm concept to the point where not only communities of the consensus phase but also the schools of the preconsensus phase are paradigm governed.[104] As Kuhn himself notes, the preconsensus phase may now no longer meaningfully be called the "pre-paradigm" phase.[105] So as to remain neutral with regard to the different versions of the paradigm concept, I will call the phase of scientific development prior to the first emergence of a universal consensus *pre-normal science.*

We also find a further characterization of prenormal science in Kuhn's work, one somewhat narrower than the notion of a preconsensus phase. Kuhn describes prenormal science as a form of scientific practice in which there's *competition between schools,*[106] schools addressing roughly the same object domain[107] from different, mutually incompatible standpoints. This description of prenormal science obviously goes beyond that given above, which only demanded the absence of consensus. Margaret Masterman has criticized this characterization, suggesting that the missing consensus of the prenormal phase required further differentiation. Her proposal would involve distinguishing "non-paradigm science" (which lacks exemplary problem solutions) from "multiple-paradigm science."[108] Though this distinction indeed makes sense, it's without consequence for Kuhn's theory. For Kuhn is primarily concerned with that stage of prenormal science which *immediately* precedes the onset of normal science and with the transition from

101. *SSR,* p. 13, similarly pp.101, 163; 1963a, p. 355.

102. 1959a, *ET* pp. 231–232.

103. E.g. *SSR,* pp. ix, 20, 47, 61, 76, 163; 19063a, pp. 354, 358, 363.

104. Compare § 4.2.b.

105. 1974a, *ET* p. 295 n. 4; 1970b, p. 272 n. 1; 1970c, *SSR* pp. 178–179. —In his 1970b, Kuhn calls sciences in the prenormal state "proto-sciences" (1970b, pp. 244–246).

106. *SSR,* pp. ix, 4, 12–13, 16, 17, 47–48, 61–62, 96, 163; 1963a, pp. 355–356; 1974a, *ET* p. 295 n. 4; 1970b, p. 272 n. 1; 1970c, *SSR* pp. 178–179.

107. What it means for them to operate in "roughly the same object domain" can only be made clear later, since the various schools may, in part, presuppose mutually incommensurable points of view (*SSR,* p. 4), and the clarification of the notion of incommensurability is reserved for § 6.3.

108. Masterman 1970, pp. 73–75.

this stage to that of normal science.[109] According to his case studies, the clash of the schools, as a rule, immediately precedes a phase of normal science.[110] Nothing here denies the possibility of more diffuse conditions obtaining before the emergence of the schools.

With the admission that schools of the prenormal phase also enjoy (local) consensus and (locally) accepted exemplary problem solutions, the question arises whether and, if so, how the scientific work of such schools distinguishes itself from the work of normal science, performed in competitionless communities.[111] According to Kuhn, one consequence of missing consensus in the prenormal phase is that each school is constantly obliged to explicate and legitimate the foundations of its own approach. In addition, schools are relatively free in their selection of relevant observations and experiments, since there's no universal consensus over precisely which phenomena belong to the appropriate object domain and which don't. Schools are thus inclined to take only fairly easily accessible facts into account, whether they're the product of easily conducted observations or experiments or facts gleaned with greater effort in other areas of scientific practice. Furthermore, it's not clear which aspects of the assembled facts will prove most important for the further development of a given scientific field. In this phase new, often speculative and more or less precisely articulated theories and hypotheses are developed, ones that may lead to the discovery of previously unknown phenomena. Relatedly, publications must contend with a broad range of competing approaches and not just with their actual issue, the chosen object domain. The isolation of prenormal scientific communities from the rest of society, given the relative weakness of direction by factors internal to their field, remains substantially less severe than that typical of normal science.[112] Both the kind of problem explored and the conceptual system employed may be strongly influenced by factors external to science. Finally, the lack of scientific consensus is evidenced by the fact that the typical medium of scientific communication is the book, not the relatively short article in a technical journal.[113]

All in all, the picture of prenormal science which emerges lacks

109. Kuhn says as much in the title of chap. 2 of *SSR,* which discusses prenormal science: "The Route to Normal Science." Furthermore, Kuhn only retraces the case studies of individual scientific fields adduced in this chapter back to the phase *immediately* preceding the first period of normal science.

110. With the exception of those cases in which normal science results from the splitting or merging of previously existing fields of normal science (*SSR,* p. 15).

111. On the following, see *SSR,* pp. 13–18, 47–48, 61, 76, 163.

112. 1968a, *ET* p. 118.

113. 1961a, *ET* p. 187 n. 11; *SSR,* pp. 20, 21.

the analogies to puzzle-solving characteristic of normal science.[114] Attractive research problems can't be unequivocally identified, nor are previously solved problems any sure guide in the solution of future problems, nor do problems selected for sensible reasons offer any justified expectation of solution. And so it becomes clear why prenormal science, with its clash of the schools, appears "something less than science"; the scientific progress normal science so obviously produces[115] isn't attained at all, or is attained only very indirectly and covertly, by prenormal science.[116]

2. How does it happen that the clash of the schools characteristic of normal science vanishes in favor of the consensual scientific practice of normal science? In general, the victory of one of the competing schools over the others heralds the advent of normal science.[117] This school must have produced an achievement so convincing that members of other schools begin to defect; the school must, above all, be in a position to attract the next generation of scientists. In this way, the earlier competing schools gradually disappear, even though not all of their members need choose to join the victor. The new consensus may also come about, to a certain degree, by the majority's inattention to those who fail to share it and by the inability of dissidents to recruit an adequate succession. Accordingly, the transition to normal science isn't generally a sudden event, and so its historical dating can't be attempted with arbitrary precision.[118] But neither is it so gradual as to be congruent with the entire development of a field from its very first beginnings to the emergence of consensual research practice; the transition's temporal order of magnitude is generally in the decades.

The above explanation touches only on the social aspects of the transition to normal science. We must now inquire into the *characteristics* of the "convincing achievements" that trigger the transition and into the *standards* according to which such achievements become so compelling.

The accepted values of a given scientific community function as the standards according to which the community evaluates scientific achievements. In their abstract-universal characterization, these include, above all, accuracy, consistency, scope, simplicity, and fruitfulness.[119] Though the value systems of different scientific communities

114. 1970c, *SSR*, p. 179. —Compare § 5.2.
115. Compare § 5.4.
116. 1959a, *ET* p. 231, with n. 3; *SSR*, p. 163; 1970b, pp. 244–245.
117. *SSR*, pp. 17–19; 1970c, *SSR* p. 178.
118. *SSR*, pp. 21–22; 1970b, p. 245.
119. Compare § 4.3.c.

need not be completely identical, nothing rules out different communities' coming to the same conclusion in evaluating some novel achievement. For one, the various value systems all share a common, atemporal, abstract-universal moment. In addition, different value systems may still reach the same conclusion in individual evaluations, just as individual members of a community may make the same value judgements despite their different slants on communal values.

One of the competing schools must have thus attained an achievement so convincing by these standards that members of other schools begin to defect, and the school is in a position to attract the next generation of scientists. The peculiarly compelling quality of this novel achievement presents itself for three points of characterization.

First, the compelling nature of a novel achievement is always relative to the achievements of the competition.[120] It is thus claimed neither that the novel achievement explains all facts and solves all problems in the relevant field nor that the problems it does solve are solved in every detail. The achievement rather consists in a relatively convincing solution of a few problems whose central position in the relevant field is beyond doubt.

Second, the solution of these problems occurs in such a way as to leave specialists with the impression that certain fundamental issues of the field have thereby been resolved more or less conclusively.[121] Among these fundamental issues are the search for the conceptual system to be used in articulating research problems and their solutions, questions regarding the existence and essential attributes of the basic entities of the field at hand, and questions regarding the domain-specific methods of observation, experiment, and theory needed in the study of these entities. How might the impression arise that such fundamental issues have been resolved more or less conclusively? Three answers present themselves.[122] First, the new problem solutions must allow the successful, concrete prediction of phenomena in the appropriate field.[123] Next, such predictive success must be regular and reliable within a set

120. *SSR*, pp. 10, 17–18, 23–24.

121. *SSR*, pp. 19–20, 21; 1963a, p. 353; 1963b, p. 388; 1970b, p. 247; 1970c, *SSR* p. 178. —I've chosen to use a formulation somewhat more modest than Kuhn's in allowing the fundamental questions to have been *more or less* conclusively resolved. My reason involves the possible individual difference of members of the community with regard to the internal distance each maintains vis-à-vis the relevant tradition of normal science, discussed in § 5.5.a.

122. 1970b, pp. 245.

123. "Prediction" is once again used in the sense common among scientists, referring to a proposition derived from theory. The actual temporal relationship between derivation and derived event is irrelevant.

margin of error, at least for a given subdomain of phenomena. Finally, these predictions must be based on a theory which at the same time justifies the predictions, explains their uncertainty, and points toward ways of improving their accuracy. The theory must also allow some prospects for the expansion of its domain of application.[124]

Third, the novel achievement must sanction that form of scientific activity described by those analogies to puzzle-solving we considered earlier, in short, normal science.[125] To the extent they fulfill this demand, exemplarily solved concrete problem solutions are of a different kind than the paradigmatic solutions otherwise employed by scientific schools.[126]

From these features of the novel achievements which usher in normal science, it follows that the transition to normal science can by no means be forced by social measures alone. Either a field exhibits some paradigmatic achievement in emulation of which research may be conducted in a manner analogous to puzzle-solving, or it doesn't. If the paradigmatic achievements are lacking, they can't be created by decree, nor can the real attraction of alternatives be stifled by social pressure.[127] According to Feyerabend, Kuhn was understood by some social scientists as claiming just the opposite.[128] Kuhn's philosophy of science offers no recipes for determining how a given branch of prenormal science can attain or accelerate the transition to normal science;

> As in individual development, so in the scientific group, maturity comes most surely to those who know how to wait.[129]

The transition to normal science thus can't be arbitrarily decreed but must rather be legitimated in fact. Members of the competing schools join the school whose achievements are better than those of the

124. Convictions regarding the conclusiveness of solutions to fundamental problems can, of course, only be grounded in the available empirical material. In the transition from prenormal to normal science, such material includes relatively easily performed observations and experiments, supplemented over the course of the esoteric practice of normal science by more demanding experimental and theoretical problem solutions (*SSR*, pp. 64–65). Thus unexpected novel phenomena, for example, might shake convictions concerning the conclusiveness of solutions to the fundamental problems.

125. 1970c, *SSR* p. 179; compare § 5.2.

126. 1970c, *SSR* p. 179. —Also compare § 4.2.b.

127. 1970b, pp. 245, 260, 263; also see 1961a, *ET* pp. 200, 221. —In claiming the impossibility of forcing a transition to normal science, Kuhn differs from the so-called Edinburgh school in the sociology of science. For this school's "radical view of the conventional character of knowledge" consists in the doctrine that "Knowledge is conventional *through and through*" (Barnes 1982, p. 30 and p. 27). Also compare § 3.2.

128. Feyerabend 1970, p. 198.

129. 1970b, p. 245, q.v. p. 246.

competition, as measured by scientific values. One qualification should still be made: the transition to normal science is never absolutely compelling, and it is thus not consummated by all members of the various schools, without remainder.[130] This feature of (new) normal science is one reason it is qualified, as hinted at earlier, as "quasi-dogmatic," a qualification we must now discuss.

5.6 The Functional Role of the Quasi-dogmatic Element of Normal Science

There are two reasons why the transition to (new) normal science can't be absolutely compelling.

The feature of scientific value systems that, in § 4.3.c, I called the "underdetermination of scientific value systems" is partially responsible. This underdetermination, we recall, meant that the application of scientific value systems in the comparative evaluation of different theories needn't necessarily produce an unequivocal result. Measured next to the (unreasonable) ideal of a completely determinate comparative evaluation, these value systems are underdetermined. Thus no school's victory over its competitors can be absolutely compelling.

In addition, the phase of normal science always contains certain problems that resist satisfactory solution.[131] In principle, such problems can also be seen as evidence against a current (or future) practice of normal science. Only from a certain perspective can they be seen as merely temporally unsolved, to be set aside for the benefit of problems deemed currently soluble.

Still, normal science is typically characterized by the "assurance that [the present theory] will solve all of its problems," and "that same assurance is what makes normal or puzzle-solving science possible;"[132] the validity of the present theory isn't doubted.[133] Similarly, Kuhn speaks of "faith" in the present theory,[134] of "trust" placed in it,[135] of a theory "taken for granted,"[136] of a feeling of total security toward it,[137] of

130. *SSR*, p. 19.

131. 1959a, *ET* pp. 234, 236; 1961a, *ET* pp. 202, 203–204; *SSR*, pp. 18, 68, 79–80, 81–82, 110, 146–147; 1963b, p. 392; 1964, *ET* p. 262. —Compare § 5.2.d, point 2.

132. *SSR*, pp. 151–152; similarly 1959a, *ET* p. 235.

133. 1959a, *ET* p. 235; 1961a, *ET* p. 193.

134. 1959a, *ET* p. 236.

135. *SSR*, p. 165; 1963a, pp. 356, 363.

136. *SSR*, p. 37; 1970c, *SSR*, p. 178.

137. 1963a, p. 353.

> firm convictions about the phenomena which nature can yield and about the ways in which these may be fitted to theory.[138]

Kuhn characterizes the relationship between scientists and their normal science most often as one of "commitment" (sometimes intensified by such qualifiers as "thorough," "firm," or "deep") where the term is frequently used as a synonym for one of the above expressions.[139]

What emerges from these and similar formulations is the fact that the regulations contained in the paradigmatic achievements that govern normal science aren't open to dispute:[140]

> Normal science . . . is predicated on the assumption that the scientific community knows what the world is like. Much of the success of the enterprise derives from the community's willingness to defend that assumption, if necessary at considerable cost.[141]

But the regulations of normal science can be compellingly justified neither in the transition from prenormal to normal science nor, as we shall see later,[142] in their reestablishment following a scientific revolution. The immunity of the regulations of normal science thus contains a moment of dogmatism. What functional role can the dogmatization of these regulations play? Three to some extent mutually sustaining points might be adduced.

First, the firm commitment to the regulations of normal science makes possible that thoroughness, depth, and accuracy that is lacking in prenormal science and that appears attainable in no other way;

> So long as the tools a paradigm supplies continue to prove capable of solving the problems it defines, science moves fastest and penetrates most deeply through confident employment of those tools.[143]

138. 1963a, p. 348; similarly p. 349.

139. E.g. 1959a, *ET* pp. 235, 236; *SSR*, pp. 40, 41, 42, 43, 60, 100, 101, 144, 151; 1963a, pp. 349, 353, 359, 363, 368, 369; 1963b, pp. 392, 393; 1970a, *ET* p. 268.

140. Compare §§ 5.2.c and 5.2.d. —Compare this with the " 'hard core' of a scientific research programme" discussed by Lakatos in his 1970, section 3a, pp. 133ff (Kuhn notes the congruence of this aspect of his position with that of Lakatos in his 1971a, p. 138). According to Lakatos, this hard core is "conventionally accepted (and thus by provisional decision 'irrefutable')" (Lakatos 1971, § 1d, p. 110). For Lakatos (as for Kuhn), this doesn't exclude the possibility of this core's "crumbling" under pressure from "logical and empirical" sources (Lakatos 1970, § 3a, p. 133ff).

141. *SSR*, p. 5; similarly 1963a, p. 349.

142. See § 7.4.b.

143. *SSR*, p. 76; similarly 1980a, p. 183. —Compare §§ 5.2.d, point 2, and 5.5.b.

The immunity of regulations prevents such difficulties as appear in the course of problem-solving from being laid at the feet of the regulations themselves. And this is a great advantage, for

> Experience shows that, in almost all cases, the reiterated efforts, either of the individual or of the professional group, do at last succeed in producing within the paradigm a solution to even the most stubborn problems.[144]

If the community allowed the abovementioned range of unsolved or unsatisfactorily solved problems to be turned against the regulations of normal science, the potential of this tradition to solve the most difficult problems couldn't be exhausted.

Second, research which holds strictly to the regulations of normal science shows definite progress, so long as these regulations really continue to sanction a scientific practice analogous to puzzle-solving.[145] Such progress is meaningful not only on account of the manifold prospects knowledge of the natural sciences offers for technical exploitation. It is doubtless also an important factor in the social prestige enjoyed by modern science, which in turn proves weighty in recruiting a competent successor generation.[146]

Third, and of overriding importance, is the fact that research moving under the guidance of the regulations of normal science contains a unique dialectic. This dialectic ensures that the binding quality of the regulations of normal science relents, and eventually vanishes entirely, just when it begins to fetter further scientific development. But the fact that, on account of its quasi-dogmatic character, normal science is conducted within the bounds of its regulations for as long as possible explains why scientific revolutions are capable of restructuring existing knowledge as thoroughly as they do.

The dynamic of scientific revolutions alluded to here will be discussed in chapter 7. First, however, Kuhn's concept of a scientific revolution requires thorough exposition.

144. 1963a, p. 363; similarly p. 367; 1961a, *ET* pp. 202–203; *SSR,* pp. 81, 82, 84; 1964, *ET* p. 262; 1970b, p. 248; 1970c, *SSR* p. 186; 1977c, *ET* p. 332.

145. Compare § 5.5; also *SSR,* p. 101.

146. The importance of recruiting the successor generation is especially evident when we consider that the number of scientists has, over the past three centuries, exhibited exponential growth, with a doubling time of roughly fifteen years, and has thus expanded at the expense of other societal domains; on this issue, see Solla-Price 1963.

CHAPTER SIX

The Concept of a Scientific Revolution

THE CONCEPT OF A SCIENTIFIC REVOLUTION as molded by Kuhn has a much broader scope than in usual parlance. In § 6.1 we will discuss in what sense this is so, and why. We will then be ready to disclose the central characteristics of (Kuhnian) scientific revolutions: change of world (§ 6.2), and incommensurability (§ 6.3).

6.1 Kuhn's Extension of the Concept of a Scientific Revolution

It is common practice to call those great events associated with such names as Copernicus, Newton, Lavoisier, Darwin, Bohr, and Einstein scientific revolutions.[1] The episodes of scientific development thus addressed display two features. First, they witnessed the replacement of one fundamental scientific theory by another, resulting in a massive change in both scientific practice and in the attending scientific world view. In addition, these episodes had a transforming effect outside science as well, reaching popular consciousness.

Kuhn sets out to extend the popular notion of a scientific revolution.[2] This extension occurs in two steps. In the first step, the notion of a revolution is extended to include those changes of theory that, though their effects are minimal outside science, or outside the relevant discipline, still have the same sorts of consequences within the disci-

1. 1959a, *ET* p. 226; *SSR,* pp. 6, 66, 92; 1963a, p. 364; 1970b, p. 251; *ET,* p. xvii. —See Cohen 1985 on the history of the concept of a scientific revolution over the past four centuries.

2. On the following, see 1959a, *ET* pp. 226–227; *SSR,* pp. 6–8, 49, 64, 66, 85, 92, 139–140; 1962d, *ET* pp. 176–177; 1963a, p. 364; 1963b, p. 388; 1969c, *ET* p. 350; 1970b, pp. 249–250; 1970c, *SSR* pp. 180–181; *ET,* p. xvii. —Astonishingly, despite Kuhn's explicit pointers to this expansion of the concept of a revolution, there have been some misunderstandings here; see e.g. Grandy 1983, p. 13; Laudan et al. 1986, p. 190; Rodnyi 1973, p. 189; Toulmin 1970, p. 44, and Kuhn's reply in his 1970b, pp. 249–250.

pline as episodes otherwise classed as revolutions. Here Kuhn has in mind the wave-propagation theory of light, the dynamic theory of heat, or Maxwell's theory of electromagnetism. The justification for the first extension of the concept of a revolution lies in the fact that Kuhn's theory is directed at scientific development in the epistemic sense[3] and not (or only subordinately) at an analysis of the mechanisms by which science affects domains external to it. It follows that, for purposes of Kuhn's theory, epistemically similar episodes of scientific development differing in their external effects aren't taken into account.

In the second step of the extension, a certain class of discoveries is termed revolutionary, the class of unexpected discoveries of new phenomena or entities. This extension is justified by the fact that such discoveries, like the great revolutions though less conspicuously, bring about corrections of previous scientific practice and the previous world view. Strictly speaking, this step doesn't extend the meaning of the noun "(scientific) revolution" but only of the adjective "revolutionary." For while, in his writings, Kuhn qualifies discoveries in the appropriate class as "revolutionary," he usually doesn't call them revolutions.

Kuhn is fully aware that "this extension strains customary usage." For it is

> a fundamental thesis of this essay [*SSR*] that they [the defining characteristics exhibited by obvious examples of scientific revolutions] can also be retrieved from the study of many other episodes that were not so obviously revolutionary.[4]

Before we consider these defining characteristics of revolutions in detail, we ought to ask in what respect the three fairly heterogeneous types of episodes in the histories of scientific disciplines may be conceived as belonging to the same class. Their similarity consists in the fact that none of them can be ascribed to normal science. Scientific revolutions, in Kuhn's sense, are all among

> the tradition-shattering complements to the tradition-bound activity of normal science.[5]

The claim that revolutionary scientific development is the "tradition-

3. Compare § 1.1.a.

4. *SSR*, pp. 6–8; similarly 1959a, *ET* pp. 226–227, 234; 1963a, p. 364; 1963b, p. 388. —The claim that scientific revolutions may vary widely in their scope can be found as early as Conant 1947, p. 66.

5. *SSR*, p. 6; similarly 1959a, *ET* pp. 227, 234; *SSR*, pp. 8, 90, 136; 1963a, p. 369; 1970a, *ET* p. 274; 1970b, p. 242; 1970c, *SSR* p. 181; *ET*, p. xvii; 1981, p. 8. —The juxtaposition of normal and revolutionary science may be found in Conant's work; see § 5.1.

shattering complement" of research in normal science has three implications.

First of all, revolutionary scientific development is *different* from development in normal science. And yet it is not the prenormal stage of science[6] but rather a mode of development confined to the mature sciences. Thus revolutionary scientific development must be distinguished above all from normal science. The garnering of knowledge in normal science is cumulative; new knowledge, compatible with previously available knowledge, replaces ignorance.[7] By contrast, in order for a given episode in scientific development to have revolutionary character, it is necessary (though not sufficient) that it not be epistemically cumulative.[8]

Three forms of noncumulative scientific development are conceivable. One consists in the correction of apparent knowledge about a given phenomenal world, where the lexical structure of knowledge remains unchanged over the course of correction.[9] Such correction belongs to the daily life of normal science, for it involves the identification and elimination of error by means of the regulations of normal science.[10]

In a further conceivable form of noncumulative scientific development, knowledge that not only concerns a given phenomenal world but also helps constitute it is destroyed without replacement. According to Kuhn, this variety, the Popperian falsification of basic conjectural knowledge, doesn't actually occur in the history of science.[11]

Finally, it's conceivable that new knowledge, incompatible with previous knowledge in the sense that its acceptance presupposes a modification of the lexical structure of the old, might appear.[12] This form of development is revolutionary in Kuhn's sense. Because they involve conceptual changes alien to normal science, Kuhn often describes scientific revolutions as "conceptual transformations," "fundamental reconceptualizations,"[13] or the like. Because they involve the emer-

6. Compare § 5.5.b.

7. Compare § 5.4.

8. *SSR,* pp. 84–85, 92, 108–109; *ET* p. xvii; 1979c, pp. vii–viii; 1981, pp. 8, 9, 15; similarly 1959a, *ET* p. 226; 1961a, *ET* p. 208; *SSR,* pp. 7, 136; 1962d, *ET* p. 175; 1969c, *ET* p. 350; 1970a, *ET* p. 267; 1970b, p. 252; 1970c, *SSR* p. 181; 1983b, p. 713.

9. Compare §§ 3.6.f and 4.4.a.

10. 1970a, *ET* pp. 278, 280.

11. *SSR,* p. 77; similarly p. 145; 1961a, *ET* p. 211; 1980a, p. 189; 1983a, p. 684 n. 5. —For a more detailed discussion, see § 7.4.a.

12. 1959a, *ET* p. 226; 1961a, *ET* p. 200; *SSR,* pp. 6, 92, 94, 95, 97, 98, 103; 1970a, *ET* p. 267; 1970c, *SSR* pp. 175, 185, 201; 1974a, *ET* p. 304 n. 14; 1980a, p. 189.

13. *SSR,* pp. 90 n. 15, 102, 132; 1964, *ET* pp. 246, 251, 256, 260, 263, 264; 1970b, pp. 249–250; *ET,* p. xiv; 1983b, pp. 715–716; 1989b, p. 49. —Conant already characterizes

gence of novel knowledge together with the modification or abandonment of old knowledge, Kuhn also calls revolutions "destructive-constructive."[14]

Second, the claim that normal and revolutionary science are complementary also implies that, besides these two, *there are no other modes of development for the mature sciences.*

Third, the complementarity of normal science and scientific revolutions suggests that the two processes stand in an (asymmetrical) relation of dependence on one another. For conceptual reasons, there can only be scientific revolutions if there is also normal science, for only an existing scientific tradition can be destroyed.[15] Tradition-bound science, however, theoretically might exist even if there were no scientific revolutions.

The agent of a scientific revolution is, like that of a tradition of normal science, a scientific community.[16] This central thesis of Kuhn's theory is important above all for two reasons. First, an inquiry into the factors swaying theory choice in scientific revolutions amounts to an inquiry into the reasons behind the community's decision.[17] In addition, the question of whether a given episode in scientific development should properly be ascribed to revolution or to normal science can only be answered relative to particular communities.[18] Since some developments have revolutionary character only for the group immedi-

---

revolutions as conceptual changes; his 1947 involves, among other issues, "the evolution of new conceptual schemes as a result of experimentation" (Conant 1947, p. 18). Also compare Reisch 1991 on the substantive parallels between Kuhn and Carnap.

14. *SSR,* p. 66; similarly pp. 77, 97–98, 147; 1961a, *ET* p. 208; 1963b, p. 394.

15. 1959a, *ET* p. 227; 1970b, pp. 233, 242, 249; *ET,* p. xvii. —*SSR* contains another juxtaposition besides that of normal science and scientific revolutions, one which might cause confusion. On p. 91, Kuhn asserts "The proliferation of competing articulations, the willingness to try anything, the expression of explicit discontent, the recourse to philosophy and the debate over fundamentals, all these are symptoms of a transition from normal to extraordinary research. It is upon their existence more than upon that of revolutions that the notion of normal science depends." What Kuhn has in view here is the distinction of the research *activity* of normal science from another research activity, "extraordinary science" (see § 7.3.b below). But the possibility of distinguishing these two forms of research activity does not depend on whether or not extraordinary science results in revolution (*SSR,* p. 90). The juxtaposition of normal science and scientific revolutions, however, concerns the relationship between different *results* of research activity. The results of normal science, but not revolutionary results, are compatible with those of their predecessors.

16. 1959a, *ET* p. 226; *SSR,* pp. 49–50, 52, 61, 92–93; 1962d, *ET* p. 175; 1963a, p. 364; 1970b, pp. 251–253; 1970c, *SSR* pp. 179, 180–181; 1971a, pp. 145–146. —Compare § 1.1.b, the beginning of chap. 3, and §§ 3.4, 4.3.c, and 5.1.

17. See § 7.4.b below.

18. Of the passages cited in n. 16, above, see especially 1970b, pp. 251–253.

ately involved but are cumulative for some more distant groups, this point isn't trivial.[19]

It would be a misunderstanding, and not just on account of this last point, to take the juxtaposition of normal science and scientific revolution as a strict dichotomy. There is rather a fluid transition between these two forms of scientific development, a transition which does not, however, rule out the existence of paradigmatic cases of both forms capable of sustaining the distinction.[20]

Just what Kuhn understands by a scientific revolution or a revolutionary scientific development can, given their complementarity to normal science, be determined by contrast with the central features of that form of scientific practice.

6.2 Change of World

Normal science, as we have seen, works under the guidance of regulations derived from exemplary problem solutions.[21] These exemplary problem solutions have a world-constituting function, in virtue of the similarity relations they contain.[22] Since over the course of normal science the exemplary problem solutions aren't open to dispute, their contribution to the world's constitution remains constant. Consequently, the region of the phenomenal world relevant to the appropriate scientific community's efforts remains stable throughout normal science, and knowledge about this region increases cumulatively.[23] By contrast, as Kuhn often emphasizes, it is a central characteristic of revolutionary scientific development that the world changes over the course of it.[24] Put more precisely, the region of the phenomenal world in which the given community conducts its work changes. Kuhn has primarily three modes of change of world in view.

First, the postrevolutionary (phenomenal) world may contain phenomena and entities which the earlier phenomenal world lacked.[25] But these discoveries must exhibit one further trait in order to count as

19. See § 7.2 below.
20. 1970b, pp. 251–252; see § 7.2 below.
21. Compare § 5.2.a.
22. Compare § 4.4.a.
23. Compare § 5.4.
24. *SSR,* pp. 6, 61, 85–86, 102, 103, 106, 111, 117, 118, 120, 121, 122, 134, 135, 141, 143, 147–148, 150; 1970c, *SSR* pp. 193, 201; 1974a, *ET* p. 309 n. 18; *ET,* p. xxiii; 1979b, pp. 418–419; 1983a, pp. 682–683; 1986, p. 33.
25. 1961a, *ET* pp. 204–206; *SSR,* pp. 7, 52–61, 66, 88–89, 96–97, 103, 111, 116–117; 1962d, *ET* pp. 166, 175–176.

revolutionary, for new phenomena and entities are discovered in normal science, too.[26] The difference between the revolutionary discovery of new phenomena or entities and discovery that occurs in normal science is that the latter case in general leads neither to a revision of prior explicit or implicit knowledge, nor to an alteration of conventional experimental techniques, nor to any correction in ways of interpreting data. Such discovery isn't surprising, and the addition of knowledge about new phenomena and entities to the communal stores is cumulative. By contrast, a revolutionary discovery (by definition) demands certain revisions. An unexpected discovery can force larger changes in explicitly formulated theories if the discovered phenomena or entities can't be brought into harmony with them. Even when an unexpected discovery doesn't call the explicitly formulated portions of a theory into question, it may lead to the revision of the implicit theoretical assumptions at work in the use of instruments or interpretation of data. For conceptions of what's in the world and how it behaves, or more implicit conceptions of what isn't in the world, must necessarily contribute to the employment of instrumentation and the interpretation of data.[27]

Second, there are changes of world in which, though no new phenomena or entities are discovered, "familiar objects are seen in a different light";[28]

> Confronting the same objects as before [the revolution] and knowing that he does so, [the scientist] nevertheless finds them transformed through and through in many of their details.[29]

This transformation of familiar phenomena or objects can come about as a result of their being conceived as instances of different laws after the revolution than they were before.[30] Once again, the acceptance of the objects' new attributes requires the abandonment of some of the old.

Third, certain numerical data may change over the course of a revolution,[31] since new quantitative expectations regarding certain phenomena may replace old expectations or appear where no expectations were before.[32] If these expectations are disappointed, the new normal

26. *SSR*, p. 58; 1962d, *ET* pp. 166–167, 175. Also see § 7.2 below.
27. Compare § 5.2.a.
28. *SSR*, p. 111.
29. *SSR*, p. 122; similarly pp. 7, 102, 103, 134, 150; 1962d, *ET* p. 175.
30. *SSR*, pp. 102, 130–134.
31. *SSR*, pp. 130, 134–135; 1961a, *ET* pp. 194–197.
32. Compare § 5.3.c.

science doesn't give up its guiding regulations but rather attempts to achieve or improve the accordance of theory and observation by theoretical and experimental efforts within its own bounds. In the process, previously hidden influences on measurement may be assessed in light of new theory as sources of error, the correction of which actually changes numerical data.

These three sorts of change may, indeed, not belong in normal science. But does their occurrence really entitle us to talk of "the world changing"? In the case of the second form of change, it's especially obvious that we're dealing with something which happens exclusively in the heads of participating scientists. Wouldn't it thus be more appropriate, instead of talking of a change in the world, to talk only of a noncumulative change in knowledge *about* the world? At the end of the 1950s and the beginning of the 1960s, Kuhn seems somewhat uncertain on this issue. And so we find him speaking, in his 1959a, not of changes in the world but rather only of the fact that, in revolutions, "a scientific community abandons one time-honored way of regarding the world."[33] Similar formulations may also be found in various passages of *SSR* and 1962d.[34] In these passages, Kuhn understands "to see the world differently" in both literal and metaphorical senses.[35] We even find Kuhn giving the tenth chapter of *SSR,* which discusses whether and how revolutions change *the world,*[36] the rather confusing title "Revolutions as Changes of *World View.*" But the description of revolutions as changes of world view seems incompatible with their description as changes in the world; either a revolution changes the world itself, or the world remains constant while our view of it changes.

In *SSR,* Kuhn gives a brief justification for his rather startling talk of change in the world. Of course, so the justification begins, events outside the laboratory continue their usual course even after a scientific revolution.

> Nevertheless, paradigm changes do cause scientists to see the world of their research-engagement differently. In so far as their

33. 1959a, *ET* p. 226.

34. *SSR,* pp. 53, 102, 111, 116–120, 124, 144; 1962d, *ET* p. 175.

35. Compare §§ 3.5, 3.6.a, and 3.6.e.

36. Consider this passage from the first sentence of chap. 10: "the historian of science may be tempted to exclaim that when paradigms change, the world itself changes with them" (*SSR,* p. 111). And in the final sentence of this chapter, Kuhn claims, with reference to the third form of change in the world, the change of the data themselves, "That is the last of the senses in which we may want to say that after a revolution scientists work in a different world" (*SSR,* p. 135).

> only recourse to that world is through what they see and do, we may want to say that after a revolution scientists are responding to a different world.[37]

In the second sentence of this citation, Kuhn hints at the differentiation of two features united in the commonsense world concept (or at least in that of the twentieth century West), whose conflation is what makes talk of the world's changing in a revolution so startling.[38] The first of these features qualifies the world as our real counterpart, as something with purely object-sided being. According to the second, the world is accessible to experience. *Taken together,* the two features are incompatible with the lessons Kuhn draws from the new internal historiography of science.[39] The incompatibility can only be resolved by dropping (at least) one of the two features. If we retain the first, we will talk of scientific revolutions as changes in world *view*. If we retain the second, we will call revolutions changes in the *world*. In the second sentence of the above citation, Kuhn underscores the second feature of the popular world concept; by suppressing the first feature, he makes plausible his talk of change in the world during a revolution.[40] Greater clarity can, of course, be obtained only by distinguishing two separate world concepts.[41] A further reflective step may then reveal the difficulties implicit in the concept of a purely object-sided world, thought to fulfill certain theoretical functions.[42]

Still, Kuhn's vacillation between the two characterizations of revolutions, as changes in the world or changes in world view, shouldn't be entirely attributed to any vagueness in the world concept. For in *SSR,* Kuhn has an additional theoretical interest in characterizing revolutions as changes in world *view*. This characterization makes it possible to describe revolutions as changes in ways of seeing, changes along the lines of visual gestalt shifts. Hanson pointed to similarities between seeing objects in the world and seeing in his *Patterns of Discovery,* a

37. *SSR,* p. 111; also see p. 118.

38. The startling quality of such claims is also found in the title of Fleck's *Genesis and Development of a Scientific Fact* (1935). Precisely this startling quality is what aroused Kuhn's interest in Fleck's book, as Kuhn recalls in his 1979c, p. viii. Here Kuhn also reports Conant's open-mindedness with regard to the "paradox" of Fleck's title.

39. Compare § 1.2.c.

40. This procedure repeats, on the meta-level, what Kuhn's theory described as typical of scientific revolutions on the object-level; the legislative content of a concept, here the popular world concept, is suspended (see § 3.7.a). It also produces what Kuhn calls, in his 1981, p. 21, "the touchstone of revolutionary change," "violation or distortion of a previously unproblematic scientific language."

41. Compare § 2.1.a.

42. Compare §§ 2.2.d, 2.2.e, and 2.3.

suggestion Kuhn pursues, albeit with the qualification that, in some respects, these similarities are misleading.[43] Kuhn achieves two aims by describing revolutions as gestalt switches. First, the indubitable existence of visual gestalt switches in psychological experiments lends credence to the notion that such gestalt switches might also take place in scientific development.[44] At the same time, the view Kuhn attacks, on which scientific development is an entirely cumulative process, becomes less convincing.[45] It further becomes clear how a revolution, for an individual scientist, might be a "relatively sudden and unstructured event" of the sort attested to by scientists' reports.[46]

In his most recent work, however, Kuhn has criticized his earlier talk of gestalt switches in scientific development.[47] To be sure, individuals may experience gestalt switches, as indeed happens during the course of revolutions. But, with regard to the actual agent of a scientific revolution, the scientific community, talk of gestalt switches is inappropriate, for a scientific revolution is a temporally extended process. Though some of the participating individuals may experience gestalt switches, these shifts don't occur simultaneously. Others may experience the requisite change as a series of less discontinuous steps.

> To speak, as I repeatedly have, of a community's undergoing a gestalt switch is to compress an extended process of change into an instant, leaving no room for the microprocesses by which the change is achieved.[48]

A scientific revolution is gradual to a much greater degree than the gestalt switch metaphor suggests, and so, in the 1980s, Kuhn tempers his view of revolutions accordingly.[49]

But even in *SSR* Kuhn is aware that his talk of scientific revolutions as changes in world view or transformations of the world requires further analysis. We thus find him asserting of his claim that the representatives of competing paradigms ply their trades in different worlds that "In a sense" he is "unable to explicate [it] further."[50] Only in the late 1960s does Kuhn become clearer here: what changes above all in

43. Hanson 1958, chap. 1; Kuhn *SSR*, pp. 85, 113; similarly pp. 111, 115–117, 122, 150; and, later, 1976b, p. 196; *ET*, p. xiii; 1983a, p. 669.
44. Compare § 2.1.b.
45. Compare § 1.2.a.
46. *SSR*, p. 122, similarly p. 150; q.v. 1989b, pp. 49–51.
47. Kuhn 1989b, pp. 50–51.
48. Kuhn 1989b, p. 50.
49. 1983b, pp. 714–716; 1989b, p. 49.
50. *SSR*, p. 150; also see p. 121.

scientific revolutions are certain similarity relations.[51] Certain of the features of revolutions attested to in *SSR* thus become intelligible.

First, since similarity relations are coconstitutive of phenomenal worlds,[52] changes in similarity relations may result in changes in the phenomenal worlds they help constitute.[53] Such change is of course compatible with the existence of a world of stimuli, conceived as purely object-sided, which remains untouched by it.

Second, since similarity relations are coconstitutive of perception and of the empirical concepts used in describing the world,[54] the post-revolutionary world is "seen" differently, both in literal and extended senses. Insofar as a phenomenal world is a "view" of a purely object-sidedly conceived world of stimuli, a revolution is a change in world view.

Third, since similarity relations aren't isolated from one another but rather are interwoven,[55] it becomes clear how, for the individual scientist, revolutions can be relatively sudden and unstructured "holistic" experiences.[56] For the network of similarity and dissimilarity relations to be acquired can't be introduced piecemeal into the existing network of similarity relations. The new similarity relations are rather to some extent incompatible with the old, and so if the total network is to remain consistent, either all or none of the new similarity relations must be adopted.

A fourth point, closely related to the last, is that characterizing revolutions as an alteration in the network of similarity relations sanctions the more precise determination of a further central feature of revolutions, incommensurability. But this notion requires discussion in its own right.

6.3 Incommensurability

Incommensurability is a notion by means of which certain central features of scientific revolutions are to be described. This concept was

51. 1970b, pp. 275–276; 1970c, *SSR* pp. 200–201; 1976b, pp. 195–196; 1979b, pp. 416, 417; 1981, pp. 20–21; 1983a, pp. 680, 682–683. —The notion of a similarity relation is already implicitly contained in *SSR*, as noted in § 3.2. Thus, in the cited works, we find Kuhn discussing changes in similarity relations by reference to examples already discussed in *SSR*. These involve the planets before and after Copernicus (*SSR*, pp. 128–129), free fall, pendular and planetary motion before and after Galileo (*SSR*, pp. 118–120, 123–125), and (physical) mixtures, solutions, and chemical compounds before and after Dalton (*SSR*, pp. 130–134).

52. Compare chap. 3, especially § 3.2.

53. Compare 1974a, *ET* p. 309 n. 18.

54. Compare §§ 3.5 and 3.6.

55. Compare §§ 3.2, 3.6.c, and 3.6.e.

56. 1981, pp. 9–10, 11–12, 19; 1983a, pp. 676–677, 682; 1983b, pp. 715–716.

introduced into the philosophy of science by Kuhn in 1962, and also, in the same year, by Feyerabend, with somewhat different emphasis.[57] The incommensurability thesis is one of the most discussed, most controversial parts of Kuhn's theory.[58] Despite the extent of this discussion, a somewhat resigned Kuhn notes in 1982 that "virtually no-one has fully faced the difficulties that led Feyerabend and me to speak of incommensurability."[59] Indeed, the incommensurability discussion exhibits a great diversity of misunderstandings. In Kuhn's own judgement, the source of these misunderstandings lies in part in "the role

57. Feyerabend 1962. —According to Kuhn's admittedly uncertain recollections in 1982, he and Feyerabend introduced the *term* "incommensurability" independently (1983a, p. 669 and 684 n. 2). See Feyerabend 1978a, p. 30, and 1978b, section 10, for Feyerabend's view of the introduction of the *notion* of incommensurability. Here he traces the beginning of the substantive incommensurability debate to Feyerabend 1958, section 6. Interestingly enough, Wieland independently uses "incommensurability" in roughly the same way, at the same time as Kuhn and Feyerabend: "For it can happen that sentences which contradict each other are really completely incommensurable, namely when they seek to answer incommensurable questions" (Wieland 1962, p. 30 [translation mine. —A.T.L.]; similarly, pp. 37, 45).

58. See e.g. Achinstein 1964; Agazzi 1985 and 1990; Andersson 1988a, pp. 110–124, and 1988b, pp. 28–39; Balsiger and Burri 1990; Balzer 1978, 1985a, 1985b, and 1989; Balzer, Moulines, and Sneed 1987, pp. 306–320; Bartels 1990a, 1990b, pp. 108–109; Batens 1983a and 1983b; Bernstein 1983, pp. 79–93 and elsewhere; Biagoli 1990; Bohm 1974; Boros 1990; Boyd 1984, pp. 52–56; Brown 1977, pp. 115–120, 1983, 1983a, pp. 98–99; Burian 1984; Causey 1974; Chen 1990; Churchill 1990, pp. 470–477; Collier 1984; Danneberg 1989, pp. 290–329; Davidson 1974; Devitt 1979, 1984, pp. 151–155; Doppelt 1978; English 1978; Feyerabend 1978b, pp. 178–202, 1987; Field 1973; Fine 1967 and 1975; Flonta 1978; Franklin 1984; Giedymin 1970 and 1971, pp. 45–47; Grandy 1983; Greenwood 1990; Hacking 1982, pp. 58–62, 1983, especially chap. 5; Harper 1978; Hattiangadi 1971; Hautamäki 1983; Hempel 1977, p. 27ff., 1979, pp. 54–55; Hesse 1983; Hintikka 1988; Hung 1987; Jones 1981; Jones 1986, p. 446; Katz 1979; Keita 1988; Kitcher 1978 and 1983; Koertge 1983; Kordig 1971a, 1971b, 1971c and 1971d; Krige 1980, pp. 194–200; Krüger 1974; Lakatos 1970, p. 179 n. 1; Laudan 1976, pp. 593–596, 1977, pp. 139–146, 1990, pp. 121–145; Levin 1979; Lodynski 1982; MacCormac 1971; Malpas 1989; Mandelbaum 1979, pp. 417–423; Martin 1971, 1972 and 1984; Mittelstraß 1984, S. 122–123; Moberg 1979; Musgrave 1971, pp. 295–296 and 1978; Nersessian 1982, 1984 chap. 2, 1989a, and 1989b; Newton-Smith 1981, pp. 9–13 and chap. 7; Nordmann 1986; Oddie 1988 and 1989; Parsons 1971a and 1971b; Pearce 1982, 1986, 1987 (with additional references) and 1988; Pearce and Maynard 1973; Phillips 1975; Porus 1988; Przelecki 1978; Putnam 1981, pp. 113–124; Rabb 1975; Rakover 1989; Rescher 1982, chap. 2; Rorty 1979, pp. 322–333; Sagal 1972; Sankey 1990, 1991a, and 1991b; Scheffler 1967, pp. 81–83 and 1972, pp. 367–368; Scheibe 1988a; Shapere 1964, pp. 390–392, 1966, pp. 67–68, 1969, pp. 106–110, 1971, p. 708, 1984, pp. xv–xvi, and 1989; Siegel 1980, 1987, especially pp. 51–69, pp. 81–85; Stegmüller 1973, p. 167ff., 254, 283–287 and 1986, 298–310; Ströker 1974, pp. 59–63; Suppe 1974, pp. 199–208; Szumilewicz 1977 and 1985; Thagard 1990 and 1991; Tuchanska 1988; Van Bendegem 1983; Van der Veken 1983; Vision 1988, pp. 246–288; Watanabe 1975, pp. 118–122; Watkins 1970, pp. 36–37, Wittich 1981, pp. 147–151; Wong 1989; Zheng 1988.

59. 1983a, p. 669.

played by intuition and metaphor in our [Feyerabend's and Kuhn's] initial discussion."[60] For this reason, Kuhn has modified his initial presentation of incommensurability at least twice.[61] A temporally differentiated discussion of his position thus recommends itself. In § 6.3.a, I will discuss the introduction of the incommensurability concept in *SSR*. Next, I will present its further development over the course of the late 1960s and 1970s (§ 6.3.b) and in the 1980s (§ 6.3.c). Finally, I will discuss two misunderstandings to which Kuhn's incommensurability thesis was subject (§§ 6.3.d and 6.3.e).

a. *The Introduction of the Incommensurability Concept in* SSR

In *SSR*, Kuhn primarily uses "incommensurability" to characterize the relationship between successive traditions of normal science.[62] According to *SSR*, pp. 148–150, the incommensurability which holds between pre- and postrevolutionary traditions has three different aspects.[63]

1. In a scientific revolution, both the set of scientific problems which must necessarily be faced and the set of those which may legitimately be faced change.[64] Problems whose answers were of central importance to the older tradition may be dropped as obsolete or unscientific, while questions which either didn't exist or whose answers were trivial in the older tradition may attain great importance. Along with the questions, the standards which problem solutions must meet in order to be scientifically acceptable often also change.

These two kinds of change are, for the most part, a consequence of a change in phenomenal world, as can be seen in Kuhn's examples.[65]

60. Ibid.

61. As Kuhn himself notes at several points: 1974b, p. 506 n. 4; 1976b, p. 198 n. 11; *ET*, pp. xxii–xxiii; 1983a, p. 684 n. 3.

62. *SSR*, pp. 103, 148, similarly pp. 4, 149, 165. —On pp. 150 and 157 Kuhn also talks of the "incommensurability of paradigms," and on p. 112 (or similarly on p. 4) he claims that, for a scientist after a revolution, "the world of his research will seem, here and there, incommensurable with the one he had inhabited before." Given the close association of a particular tradition of normal science with paradigms and with a phenomenal world, such use of "incommensurability" isn't substantively different. It need hardly be noted that, in subsequent literature, disagreement over which entities are the primary relata of incommensurability has been very widespread.

63. A richer notion of incommensurability is explicated in *SSR* pp. 148–150 than on p. 103, where the concept makes its first important appearance in *SSR*. I will discuss the difference in point (3) of this subsection.

64. On the following, see 1959a, *ET* p. 234; 1961a, *ET* pp. 211–212; *SSR*, pp. 6, 52, 85, 103–309, 140–141, 147–149. —Compare § 5.3.

65. The most important of these examples are the development of dynamics from Aristotle through Descartes and Newton to the eighteenth century, the revolutionary

This point is all too obvious, if both the questions and the sorts of answers to them that are regarded as legitimate within a given discipline depend on which fundamental entities, with which essential attributes, are taken as extant in the appropriate domain. Dependance on the currently accepted ontology is especially obvious for explanations in which elementary processes occur as *explanantia* of more complex phenomena.

It is characteristic of such changes in problem domains and standards as occur over the course of scientific development that they can't be conceived as part of a constant improvement or fine-tuning process, in which error is eliminated and the coarse refined.[66] This characteristic is especially evidenced by the fact that revolutionary changes in problem fields or standards can, in later revolutions, be partially reversed.

2. After a revolution, many procedures and concepts belonging to the previous tradition of normal science may still be employed, albeit in modified ways. Special importance is here ascribed (in *SSR*, and to an even greater extent later) to conceptual change, or change in concept use, so-called meaning change.[67]

The meaning changes Kuhn discusses in *SSR* have an extensional and an intensional aspect.

The *extensional* aspect of meaning change consists in the motion of certain objects out of the extension of one concept and into the exten-

---

transformation of chemistry by Lavoisier, and Maxwell's development of the electromagnetic theory.

66. *SSR*, pp. 108–109. —In this passage Kuhn also describes this state of affairs as the *noncumulative* development of scientific problems and standards. This description is highly misleading. For the counterclaim against which Kuhn seeks to defend his position doesn't assert that the development of scientific problems and standards is *cumulative in the literal sense,* the mere addition of new problems and standards to the old. The counterclaim is rather, in Kuhn's own words, "that the history of science records a continuing increase in the maturity and refinement of man's conception of the nature of science" (*SSR*, p. 108). Such increase is alleged to consist in the rejection of "intrinsically illegitimate" problems, or problems "inherently unscientific or metaphysical in some perjorative sense" (*SSR*, p. 108). What is to be warded off is thus a position on which earlier problems and standards may, indeed, be criticized and abandoned, and which thus, *in this respect,* ascribes no cumulativity to problems and standards. What Kuhn means to attack, the view he describes as "cumulative," is a particular *interpretation* of such criticism and abandonment of problems or standards as occurs over the course of scientific revolutions. On this interpretation, throughout the history of science, only that is corrected or discarded which was never really scientific to begin with. This view is usually defended by those who take the growth of science to be cumulative (see § 1.2.a); hence Kuhn's description of his own position as "noncumulative."

67. Meaning change is cited at *SSR*, pp. 149–150 as an aspect of incommensurability. Meaning change is more or less explicitly treated at *SSR* pp. 64, 101–103, 115, 128–129, 130–134, 142–143, and 149–150. —Cf. § 3.6.a, point 1., on the potential difference between "conceptual change" and "change in concept use," entirely missing in *SSR*.

sion of another (and perhaps vice versa), where the two concepts are extensionally disjoint.[68] Kuhn's favorite example for this sort of extensional change, from *SSR* up to the present day, is the change in the concept of a planet which occurred with the transition from Ptolemaïc to Copernican theories.[69] After this transition, the Earth, for example, becomes a planet, while the sun and moon cease to be called such.

The *intensional* aspect of meaning change consists in a change in the affected concept brought about by a change in the attributes of objects which fall under the concept.

But we won't want to call every case in which the attributes of objects falling under a given concept change a case of meaning change for that concept.[70] Meaning change would require, in addition, that those attributes of the elements of the concept's extension that are contained in the concept's *definition* be among those that change. But this requirement presupposes that the attributes of elements of the concept's extension can be sorted according to whether they pertain to their objects on empirical or definitional grounds. This amounts to an application of the analytic/synthetic distinction, the possibility of which Kuhn denies for empirical concepts.[71] It follows that the legitimacy of Kuhnian pronouncements of meaning change can't (at least immanently) be judged on the basis of whether they offer evidence for an actual change in defining traits. But now Kuhn's diagnosis of meaning change appears largely arbitrary, since there's no criterion for determining which changes in the attributes of elements of a concept's extension lead to meaning change and which don't.

But according to Kuhn, the phase of normal science features certain propositions similar in some respects to analytic sentences, though in other respects their character seems more synthetic. Such propositions are similar to analytic sentences in virtue of their marked resistance to empirical falsification, and so they take on a certain air of necessity.[72] They appear synthetic, however, in that they are by no means the products of arbitrary definitional stipulations. They are rather in part the products of painstaking empirical and theoretical research.[73] Kuhn cites Newton's second law of motion and Dalton's law of fixed proportion as examples:

68. *SSR*, pp. 115, 128–129, 130–134.

69. *SSR*, pp. 115, 128–129; 1970b, p. 275; 1970c, *SSR* p. 200; 1979b, p. 216; 1981, p. 8; 1989a, p. 31; 1990, p. 313.

70. This is the thrust of Shapere's objection in his 1964, p. 390.

71. Compare § 3.6.f.

72. *SSR*, pp. 78, 131–132, 133.

73. *SSR*, p. 78.

> Newton's second law of motion, though it took centuries of difficult factual and theoretical research to achieve, behaves for those committed to Newton's theory very much like a purely logical statement that no amount of observation could refute.[74]

Similarly, Kuhn claims of Dalton that his work

> made the law of constant proportion a tautology . . . A law that experiment could not have established before Dalton's work, became, once that work was accepted, a constitutive principle that no single set of chemical measurements could have upset.[75]

Kuhn later calls propositions of this kind "quasi-analytic."[76] In the 1980s, however, he finds the "analytic" designation inappropriate.[77] Only in his latest work has Kuhn observed that the character of such sentences most closely approximates that of synthetic a priori judgements.[78]

With the help of such sentences we are now in the position to give a criterion for determining under what circumstances changes in the attributes of extension elements must be interpreted as changes in meaning. Changes in the attributes of extension elements of a given set of concepts lead to meaning change just in case sentences with the approximate synthetic a priori status described above, which contain these concepts, now become false. In a certain sense, this criterion is obviously a weakened form of the unfulfillable criterion given above, which required that the concept's defining traits change in order for talk of meaning change to be permissible.

Though the criterion explicated above can't be found in Kuhn's work, it surely captures his intensions. For in later work, Kuhn frequently asserts his suspicion

> that, quite generally, scientific revolutions can be distinguished from normal scientific developments in that the former require, as the latter do not, the modification of generalizations which had previously been regarded as quasi-analytic.[79]

3. In *SSR*, p. 150, Kuhn cites as the "most fundamental aspect of . . . incommensurability" the fact that "the proponents of competing

74. *SSR*, p. 78. —Compare this passage with Kuhn's approach to the two ways of viewing the second law of motion in 1989a and 1990.

75. *SSR*, p. 133.

76. 1974a, *ET* p. 304 n. 14; 1976b, p. 198 n. 9.

77. 1983d, p. 567.

78. Kuhn 1989a, p. 20, and p. 21 n. 19; 1990, p. 306, and p. 317 n. 17.

79. 1974a, *ET* p. 304 n. 14; similarly 1970c, *SSR* pp. 183–184; 1976b, p. 198 n. 9; similarly 1981, p. 21.

paradigms practice their trades in different worlds." This aspect was discussed in detail in § 6.2. It remains for us to observe an element of unclarity in Kuhn's treatment of incommensurability in *SSR* which affects the world-change dimension. For on p. 150, Kuhn explicates a richer notion of incommensurability than that employed on p. 103, the first important occurrence of the term in *SSR*.[80] On p. 103, Kuhn cites only changes in problems and standards as sources of incommensurability between traditions of normal science, characterizing this sort of change as a more subtle effect of paradigm shift[81] than that discussed in the preceding pages,[82] change in concepts and ontological commitments.[83] So while the incommensurability concept presented on p. 103 relates only pairs of pre- and postrevolutionary problems and standards, that utilized on pp. 148–150, in addition to relating concepts or procedures, also introduces the relationship between worlds as the "most fundamental aspect" of incommensurability. The transition to the richer notion seems to occur on p. 112, where Kuhn applies the notion of incommensurability to the relationship between worlds.

b. Further Developments at the End of the 1960s and in the 1970s

Such vagueness in the incommensurability concept of *SSR* is presumably what led Kuhn, in works composed from 1969 on, to change his views in certain respects. To be sure, Kuhn sees these changes only as a refinement in the presentation of his position in *SSR*.[84] I begin by noting two changes which are retained even in papers appearing in the 1980s.

First of all, Kuhn now speaks almost exclusively of the incommensurability of theories, terms, vocabularies, or languages.[85] By 1969 and thereafter, incommensurability characterizes only such differences as result from the differential acquisition of concepts belonging to different theories.[86] The version of the incommensurability concept which

80. The only earlier use of the incommensurability concept is his anticipatory reference on p. 4.

81. *SSR*, p. 104.

82. *SSR*, pp. 99–103.

83. The same contrast is taken up again on p. 109, where Kuhn recalls his examples for change in problems and standards. Here he distinguishes the "cognitive" and "normative functions of paradigms."

84. 1976b, p. 190.

85. 1970b, pp. 267, 268; 1970c, *SSR* p. 198; 1971a, p. 146; 1976b, pp. 190, 191, 198 n. 11; 1979b, p. 416; 1983a, pp. 669, 680; 1983b, p. 714. —Kuhn also talks of "incommensurable standpoints," usually in referring back to *SSR;* 1970c, *SSR* pp. 175, 200; *ET*, p. xxii.

86. 1983a, p. 684 n. 3.

results is substantially narrowed from that of 1962. Where the 1962 notion had three distinct aspects, between which the relations hadn't been worked out, incommensurability is now confined to one of those aspects, that of meaning change. To be sure, the two other aspects, changes in problems and in the standards for their solution, and changes in the phenomenal world, are closely associated with meaning change. Kuhn completes this narrowing of the incommensurability concept with an explicit statement that his incommensurability thesis was never meant to imply that *all* of the concepts employed in both theories change meaning in the transition to a new theory.[87] Kuhn had widely been understood to be claiming precisely this.[88] But what he always had in mind, he claims, is "local incommensurability," in which meaning change affects "only . . . a small subgroup of (usually interdefined) terms."[89]

Second, Kuhn now relies heavily on the notion of (un-) translatability in explicating the "incommensurability of theories." Significant inspiration is here provided by the work of Quine, especially his *Word and Object,* published in 1960, which Kuhn critically adapts.[90] Kuhn's emphasis on this point shifts in the course of the transition from his work in 1969 and the 1970s to that in the 1980s, and so I will make a further temporal distinction in my exposition. In 1969 and the 1970s, two theories are called incommensurable just in case there is no "language into which at least the empirical consequences of both can be translated without loss or change."[91] This language would be a "neutral observation language," at least relative to the two theories in question.[92] But there can be no such language, for

87. This misunderstanding is suggested, for example, by the assertion, in *SSR,* that a revolution is "a displacement of the conceptual network through which scientists view the world" (p. 102). From *SSR* p. 130, for example, or from his discussion of *partial* communication across paradigms (e.g. *SSR,* p. 149), we may infer that Kuhn never meant a revolution to be a displacement of the entire conceptual network (cf. § 7.5.c). In his 1969 papers Kuhn says explicitly, albeit in passing, that only some concepts experience meaning change: "Though most of the same signs are used before and after a revolution . . . *the ways in which some of them* attach to nature has somehow changed (1970b, p. 267, emphasis mine, similarly p. 269; 1970c, *SSR* p. 198; 1977c, *ET* p. 338). This misunderstanding is only emphatically warded off in 1983a, pp. 669–671.

88. E.g. Agazzi 1985, p. 61; Franklin et al. 1989, p. 229; Martin 1971, pp. 17 and 19; Newton-Smith 1981, p. 12; Putnam 1981, p. 115; Shapere 1966, pp. 67–68, 1969, p. 197, and 1984, p. xvi; Suppe 1974a, p. 491; or, indirectly, Toulmin 1970, p. 44.

89. 1983a, pp. 670–671.

90. 1970b, pp. 268–269; 1970c, *SSR* p. 202; 1971a, p. 146; 1976b, p. 191; *ET* p. xxii–xxiii; 1977c, *ET* p. 338; 1979a, p. 126; 1983a, pp. 672–673, 679–681.

91. 1970b, p. 266, similarly p. 268; 1970c, *SSR* p. 201; 1974b, p. 410; 1976b, p. 191; 1979b, p. 416.

92. Only after the publication of *SSR* does Kuhn note that the neutrality of the (hypothetical) observation language need be only relative to the two theories in question,

> In the transition from one theory to the next words change their meanings or conditions of applicability in subtle ways. Though most of the same signs are used before and after a revolution—e.g. force, mass, element, compound, cell—the ways in which some of them attach to nature has somehow changed.[93]

Kuhn thus makes his skepticism with regard to the possibility of a (relatively) neutral observation language the basis of his concept of incommensurability. Though this skepticism was already articulated in *SSR*,[94] its connection with incommensurability wasn't.

How is the incommensurability thesis justified in Kuhn's 1969 papers? In its structure, the argument closely parallels his argument for the nonexistence of any (absolutely) neutral observation language in *SSR*. There Kuhn claimed that efforts at constructing a neutral observation language had so far resulted, at best, in

> a language that—like those employed in the sciences—embodies a host of expectations about nature and fails to function the moment these expectations are violated.[95]

Kuhn's argument in 1969 and immediately thereafter begins with his sketch of the way in which, when learning a theory, a portion of the phenomenal world is constituted by means of immediate similarity relations and of the role the objects and problem situations which must be ostended play in this process.[96] As was already asserted in *SSR*, the system of empirical concepts employed in this process is by no means cognitively neutral but contains knowledge of the appropriate phenomenal world.[97] But it's a central characteristic of scientific revolutions that certain similarity relations change over the course of the revolution.[98] In the process, the phenomenal world changes, along with some of the empirical concepts with which it's described and the knowledge of the phenomenal world implicit in these concepts. The empirical consequences derivable from both theories (given set initial and boundary conditions) are thus formulated in conceptual systems with *mutually*

---

not absolute; 1976b, p. 198 n. 11. Kuhn now makes the claim, stronger than that in *SSR*, that revolutions lack even observation languages of only relative neutrality.

93. 1970b, pp. 266–267; similarly 1970c, *SSR* p. 198.

94. *SSR*, pp. 126–129, 145–146.

95. *SSR*, p. 127.

96. 1974a, *ET* pp. 305–318; 1970b, pp. 270–276; 1970c, *SSR* pp. 200–201. —Compare chaps. 3 and 4.

97. Compare § 3.7.

98. Compare § 6.2, conclusion.

incompatible epistemic claims. There can thus be no conceptual system in which the empirical consequences of both theories can be translated without loss or change.

c. Further Development in the 1980s

Kuhn's alteration and expansion, in the 1980s, of the explication of incommensurability given at the end of the 1960s and in the 1970s in response to popular criticisms of his incommensurability thesis, focuses on the following points.[99]

First of all, the 1969 and 1970s papers relied on the untranslatability of two theories by means of some (relatively) neutral observation language in explicating the concept of incommensurability. But in the 1980s, the recourse to a neutral observation language disappears. The notion of untranslatability is now immediately involved in explicating incommensurability. As Kuhn puts it in a formulation we will shortly take up again, "if two theories are incommensurable, they must be stated in mutually untranslatable languages."[100]

Second, the explication of the concept of incommensurability cited above appears to contradict the 1969 papers. For in those papers Kuhn repeatedly talks of a theory-choice situation's involving the translation of one theory into the language of the other and vice versa, though such translations aren't unproblematic.[101] The contradiction dissolves when we recall that, in the 1980s, Kuhn consciously operates with a different notion of translation than before.

At the end of the 1960s and in the 1970s, Kuhn employs an everyday concept of translation. This concept allows Kuhn to say, for example, that translation

> can present grave difficulties to even the most adept bilingual. He must find the best available compromises between incompatible objectives. Nuances must be preserved but not at the price of sentences so long that communication breaks down. Literalness is desirable but not if it demands introducing too many foreign words which must be separately discussed in a glossary or appendix.[102]

99. In his 1983a, Kuhn cites the criticisms of Davidson 1974; Kitcher 1978; Putnam 1981, pp. 113–124; Scheffler 1967; and Shapere 1966.

100. 1983a, pp. 669–670; similarly 1983b, pp. 713 and 715; 1984, p. 245; 1989a, pp. 10–11, 22, 24; 1990, pp. 299, 308, 315; and 1981, p. 8.

101. 1970b, pp. 267, 268, 269–270; 1970c, *SSR* pp. 202, 203; 1977c, *ET* p. 338. —I will return to this translation situation in § 7.5.d.

102. 1970b, p. 267.

But, as Kuhn claims in 1982, translation in this everyday sense (which also applies to translations of literary texts) consists in two heterogeneous moments, a moment of translation in the narrow, technical sense, and an interpretive moment.[103]

In translation in the narrow, technical sense, single words or word groups in a text of the source language are systematically replaced by single words or word groups in the available target language. Neither source nor target language is changed in the process, and both texts are to be identical in meaning. In this sort of translation, there is no need to compromise between incompatible objectives; the translation of a given text into a given target language is either possible or impossible.

The second moment of translation in the everyday sense is of the interpretive variety. It is especially in evidence when we consider the situations of historians or ethnographers dealing with wholly or partially unintelligible texts. In such cases, the unintelligible must be made intelligible, which requires *learning a more or less significant portion of the language in which the unintelligible material is articulated.* To attain the sought-after degree of understanding, "being able to think in the foreign language," it isn't always enough to be able to identify the referents of previously unintelligible concepts. One must also be able to understand how it is that these referents form the extension of a *single* concept. Toward this end, familiarity with the series of assumptions about the world on which the heretofore unintelligible texts or utterances are based may be necessary.[104]

When the unintelligible has thus been made intelligible, it remains entirely open whether the newly intelligible text can be translated into another language *in the narrow, technical sense of translation* or not. Though the text may resist translation in this sense, it is still translatable in the everyday sense. But such translation requires compromises of the kind mentioned in the passage cited above, in addition to changes in the target language, whether by the addition of new concepts or by the more or less subtle alteration of previously available concepts for purposes of translation.

Two theories are now called incommensurable just in case they are formulated in languages not translatable *in the narrow sense.* As to *why* there might be languages not intertranslatable in this sense, further

103. 1983a, pp. 671–673; also 1989a, pp. 10–12; 1990, pp. 299–301. —In these discussions Kuhn talks of "components" or "processes" rather than moments; compare § 4.3.e. —Also compare Rescher 1982, pp. 32–35.

104. Hesse 1983, p. 707; Kuhn 1983a, pp. 674–676, and 1983b, p. 712. Q.v. the earlier Kuhn 1970b, p. 270.

explanation is required, explanation Kuhn provides in part by a third device.

In the 1980s, Kuhn employs the concept of a lexicon and its structure as a new tool in the explication of incommensurability.[105] According to the position Kuhn takes in these later writings, a speaker applies concepts to nature by determining, with reference to certain criteria, whether or not a given concept pertains to a given object or situation.[106] The criteria any given speaker employs generally aren't universal over the entire speech community; indeed, two speakers might, in principle, employ completely disjoint sets of criteria without differing at all in their concept use. If a concept is to be used unequivocally, the relations the concept bears to other concepts in virtue of a given speaker's set of criteria must be identical to the analogous relations which hold for other members of the community (albeit perhaps in virtue of other criteria). Taken together, the relations between the empirical concepts of a given system, or "lexicon," are called the "structure of the lexicon."

Changes in the criteria for concept employment may even spread throughout the speech community without changing the structure of the lexicon. According to Kuhn, this is just what happens all the time in normal science.[107] By contrast, change in the structure of the lexicon is characteristic of revolutionary linguistic change.[108] A change in the immediate similarity relations constitutive of concepts, and of the world, is responsible here.[109] It brings with it a change in the extensions of concepts, for fundamental taxonomies change in such a way as to place objects formerly belonging to the same extension in the complementary extension, and vice versa.[110] Such change sometimes also results in a change in descriptive vocabulary.[111] At the same time, the knowledge of nature contained in the lexicon also changes,[112] leading to the "violation or distortion of a previously unproblematic scientific language," "the touchstone for revolutionary change."[113]

We are now in a position to explain why pre- and postrevolution-

105. 1983a, pp. 682–683; 1983b, pp. 713–714; 1989a; 1990. —Also compare §§ 3.6.g and 4.4.a.

106. Compare § 3.6.g.

107. 1981, p. 19.

108. 1981, pp. 8, 19–21; 1983a, pp. 682–683; 1983b, pp. 713–714; 1984, p. 246; 1989a, pp. 24, 30–31; 1990, pp. 313, 314–315.

109. 1979b, pp. 416, 417; 1983a, pp. 680, 682–683; 1989a, p. 31.

110. Compare § 6.3.a.

111. 1981, pp. 18–19; 1983a, p. 673; 1984, p. 245.

112. Compare § 3.7.

113. 1981, p. 21; similarly 1983a, p. 683.

ary languages aren't intertranslatable in the narrow sense.[114] Somewhat simplified, the explanation is as follows.[115] A necessary condition for translation in the narrow sense is a systematic mapping of concepts in the source language to concepts in the target language such that each concept is mapped to a concept with the same extension and meaning. If the lexica of source language and target language have different structures, and if the structural difference is such that the extensions of roughly corresponding concepts aren't identical, the necessary condition can't be fulfilled.[116] Incommensurability thus occurs only when the structure of the world, as mediated by the structure of the lexicon, is different from what it was before.[117]

d. The First Misunderstanding: Incommensurability Implies Incomparability

Kuhn's incommensurability thesis triggered a range of misunderstandings, which burden subsequent debate. In § 6.3.b, I discussed the misunderstanding on which Kuhn is thought to hold that *all* of the concepts contained in both theories change meaning over the course of a revolution. I will now treat two further misunderstandings. First, many authors have assumed that the incommensurability of successive theories means, or at least implies, their incomparability. In addition, some have accused the incommensurability thesis of undermining all continuity between successive traditions of normal science (§ 6.3.e).

How might the impression that incommensurable theories can't be compared with regard to their empirical potential come about?[118] The basis for this impression is that such theories appear to bear the same

114. 1983a, p. 683.

115. The simplification glosses over two points. First of all, individual words may also be replaced by expressions containing several words, and vice versa. Second, the replacements may be context-sensitive, so that one or another word in the target language is used depending on context (cf. 1983a, p. 679).

116. Compare § 3.6.g.

117. 1983a, pp. 673, 676, 680, 683. —Compare § 6.2.

118. Among many others, see e.g. Agazzi 1985, pp. 51 and 61; Balzer 1978, pp. 313–314; Bayertz 1981a, p. 23; Devitt 1979, p. 29; Franklin et al. 1989, p. 229; Jones 1981, p. 394; Katz 1979, p. 329; Kitcher 1978, p. 529; Kordig 1981d; Lakatos 1970, p. 179 n. 1; Lee 1984, p. 73; Levin 1979, pp. 407 and 414; Martin 1971 and 1972; Newton-Smith 1981, pp. 9–10, 148, 267; Phillips 1975, pp. 37 and 39; Putnam 1981, p. 118; Radnitzky 1988, pp. 109 and 127; Scheffler 1967, pp. 16–17 and pp. 83–84; Shapere 1966, pp. 67–68, 1969, p. 107 and 1984, p. xxxvii; Stegmüller 1973, p. 167; Szumilewicz-Lachmann 1984, p. 262; Watanabe 1975, pp. 118–119; Wittich 1981, p. 148. But cf. Bernstein 1983, pp. 82 and 86 and Brown 1983 (or, more distantly, Stegmüller 1986, pp. 298, 302; Zheng 1988, p. 233), in which the incomparability charge against incommensurable theories is rejected. Brown cites a number of passages from Feyerabend's work in which the charge is similarly rejected. Kuhn concerns himself with this misunderstanding in 1970b, p. 267; 1976b, pp. 190–191; 1979b, p. 416; 1983a, pp. 670–671.

relation to one another as, for example, theories of the unconscious bear to theories on the stability of globular star clusters. Such theories have no empirical intersection at all, since they address different object domains in which they explore different problems, employing mutually untranslatable vocabularies. Incommensurable theories appear, analogously, to address differently constituted regions of the world with mutually untranslatable lexica and different questions. The empirical potentials of such theories don't appear to be in competition at all.

There is, however, an important disanalogy between incommensurable theories and pairs of theories related as a theory of the unconscious is related to a theory of globular star clusters. The latter theories have totally different object domains, hence totally disparate lexica and research problems. Incommensurable theories, by contrast, target *roughly the same* object domain, *as far as the world-in-itself is concerned,*[119] though this "object domain" isn't graspable in any theory-neutral way, since different lexica must always produce different object domains. But this difference in object domains and the research problems posed within them isn't total, because the incommensurability of the lexica is only local.[120] It follows, as we shall soon see, that the empirical potentials of incommensurable theories can indeed be compared,[121] that such theories have empirical intersections, and that they can be mutually incompatible.[122]

In comparing the potentials of incommensurable theories, three stages of increasing complexity can be distinguished, though transitions between them may prove quite fluid.[123] In the first stage, we find that some of the empirical predictions of incommensurable theories can be compared *immediately,* namely those unaffected by the (merely local) incommensurability of the lexica.[124] Though the incommensurable

119. Such talk of "object domains" obviously can't be taken literally, as this notion properly applies only to regions of the phenomenal world.

120. Compare § 6.3.b.

121. Neither does the mathematical concept of incommensurability, on which that employed in the philosophy of science is modeled, imply incomparability; 1970b, p. 267; 1976b, p. 191; 1983a, p. 670.

122. Compare § 6.1. —In *SSR,* p. 103, Kuhn explicitly claims that incommensurability implies incompatibility: "The normal-scientific tradition that emerges from a scientific revolution is not only incompatible but often actually incommensurable with that which has gone before."

123. I will return to the comparison of incommensurable theories in § 7.4; in particular, in § 7.4.b I will discuss how factors relevant to choice shift *over the course of both theories' development.* The three stages of increasing complexity, which I will discuss without further ado, have to do with an expansion of the comparative basis in which the two theories are taken *as constant.* In truth, the two processes here separated for expository purposes are intermingled.

124. 1970c, *SSR* p. 203; 1977c, *ET* p. 339.

concepts may be central to both theories, many of the theories' predictions may be formulated entirely commensurably. Predictions in which the incommensurable concepts don't figure at all may often be made on the basis of a new theory in the early stages of competition, where they couldn't be made at all (at least not initially) by means of the old theory. Of course, the opposition may not understand or may only partially understand the manner in which these predictions are reached, but this failure poses no obstacle to comparison.

The familiar example of planetary theory illustrates the possibility of immediate comparison. Despite the incommensurability of Ptolemaic and Copernican theories, as evidenced, for example, by the meaning shift in the concept of a planet, predictions of planetary *position* may be immediately juxtaposed. All that's required here, in essence, is the comparison with the appropriate measurements of the predicted angles between the planet and the reference points provided by the fixed stars.

In the second stage, the basis for comparing theories is expanded beyond such immediate comparisons by identifying those portions of the lexicon whose lexical structure has changed. These will usually turn out to be small groups of interrelated concepts, which have to be learned. But even before this learning process has been concluded, new opportunities for comparing the two theories may emerge. For before one becomes capable of applying the new concepts in all situations as well as those thoroughly familiar with the new lexicon might do, there may be *particular situations* in which the referents of the new concepts may be identified by means of the concepts in the old lexicon. Though the two theories describe such situations using incommensurable vocabulary, some of their assertions can thus be compared.

Let us consider, by way of illustration, the relationship between phlogiston theory and its incommensurable successor, oxygen chemistry. In certain types of situations, the referents of the concepts *oxygen* and *hydrogen* can be identified with the help of concepts in the older lexicon, as dephlogisticated air and phlogiston, respectively.[125] Although these identifications by no means capture the concepts of oxygen and hydrogen for someone familiar only with phlogiston theory, empirical assertions about the appropriate class of situations can now be compared.

In the third stage, further opportunities for theory comparison emerge once one has learned the new theory, a process which involves, above all, becoming familiar with the new lexicon.[126] Still, no fully

125. See Kitcher 1983, pp. 691–692, and Koertge 1983, pp. 102–103.

126. Of course, further opportunities for comparison also emerge even before one has *completely* learned the new theory, although some parts of the new theory may then appear odd, if not downright inconsistent.

systematic, "point-by-point" comparison of the two theories is possible.[127] This would involve the dissection of the theories into their component parts, especially their empirical laws, and the separate juxtaposition and comparative evaluation of these parts. It's impossible because a theory isn't, notwithstanding the claims of pre-Kuhnian philosophy of science to the contrary, a set of relatively independent empirical laws.[128] A theory is rather an integrated whole, in which individual laws are conceptually interconnected and, empirically, mutually sustaining.[129] But given incommensurability, the conceptual, and hence also the empirical, connections can be taken to have changed to the point where no unequivocal mapping between the theoretical moments of one side and those of the other is possible.

It follows that any juxtaposition of the two theories must have a holistic character, in the sense that all theoretical moments, hence all differences, must be considered more or less simultaneously. To be sure, the theories differ in their weighting of and solutions to certain problems. To be sure, some facts may be formulated in one theory but not in the other. Yet the holistic comparison of the potentials of the two theories isn't thereby ruled out. Balancing the comparison may, of course, pose substantial difficulties, for weaknesses in one domain may be compensated by strengths in another, where the opposite holds for the other theory.[130] But the fact that such holistic comparison is difficult and, in some situations in theory development, without any unequivocal result,[131] surely doesn't entail that it's *impossible.*

The fact that incommensurable theories can, indeed, be compared in their empirical potentials has important consequences for Kuhn's theory. If incommensurable theories were incomparable, theory choice would perforce be fundamentally irrational, for it could in no way be guided by the aims in pursuit of which theories are used. But Kuhn was, quite against his intentions, understood by many as claiming precisely that the completion of a theory choice always occurs with the help of propagandistic means and isn't swayed by the merits of arguments.[132]

127. 1970b, p. 266; 1976b, pp. 190, 191; 1977c, *ET* p. 338; similarly 1974b, p. 410.

128. 1977b, *ET* pp. 19–20; or 1974a, *ET* p. 299.

129. Kuhn's standard example here is Aristotelian physics, with emphasis on the law that nature abhors a vacuum; 1977b, *ET* p. 20; 1981, pp. 8–12, especially pp. 11–12; or even 1957a, pp. 84–94; 1964, especially pp. 257–259.

130. 1961a, *ET* p. 211 n. 48; also see § 7.6.b below.

131. The manner in which competition between theories develops given the ongoing articulation of the theories, especially of the new one, will be discussed in § 7.4.b.

132. See e.g. Koertge 1988, p. 30; Lakatos 1970, p. 178; Maxwell 1984, p. 99; Mittelstraß 1988, p. 186; Mocek 1988, p. 95; Popper 1970, pp. 56–58; Purtill 1967, pp. 54, 57; Scheffler 1967, p. 18; Shapere 1964, pp. 392–393; 1966, p. 67; 1977, p. 200, and 1984,

e. The Second Misunderstanding: Incommensurability Implies Discontinuity

The misunderstanding on which all continuity between incommensurable theories is denied is related to the misunderstanding discussed above, which took incommensurability to imply incomparability.[133] But Kuhn asserts, as early as *SSR,* that "at least part of that achievement [of normal science] always proves to be permanent"[134] in that it survives revolution. This should be no surprise, since after a revolution

> much of [the scientist's] language and most of his laboratory instruments are the same as they were before. As a result, postrevolutionary science invariably includes many of the same manipulations, performed with the same instruments and described in the same terms as its revolutionary predecessor.[135]

This, in turn, rests on the fact, here mentioned only in anticipation, that one inexorable condition for the success of a new theory is that it

> must promise to preserve a relatively large part of the concrete problem-solving ability that has accrued to science through its predecessors.[136]

Yet Kuhn emphasizes that "recognizing continuity through revolutions has not led historians or anyone else to abandon the notion [of a revolution]."[137] But in his 1976b, we find that Kuhn isn't satisfied with his previous treatment of the continuities persisting through revolutions.[138] The reason for this dissatisfaction is doubtless that, although he attested to these continuities, he didn't analyze them in any depth.[139]

---

p. xvi; Toulmin 1970, p. 44. Contrast these with Austin 1972 and Hempel 1979, especially p. 54. —See § 7.4.b below on Kuhn's version of the grounds for theory choice.

133. See e.g. Fuller 1989, p. 66; Laudan et al. 1986, p. 170; Nersessian 1989b, pp. 324–325; Purtill 1967, pp. 53–54; Shapere 1969, pp. 107–108 and 1984, p. xxxi; Shimony 1976, p. 574; Siegel 1983, p. 76; Stegmüller 1973, p. 167; Toulmin 1967, p. 466 and 1970, p. 43; Watanabe 1975, p. 115; Wittich 1981, p. 150.

134. *SSR,* p. 25.

135. *SSR,* p. 130; similarly p. 149.

136. *SSR,* p. 169; similarly 1961a, *ET* pp. 212–213; 1970a, *ET* p. 289; 1980a, p. 190. —See § 7.4.b.

137. 1970b, p. 250; also 1980a, p. 191.

138. 1976b, p. 185.

139. When the passages cited above are taken together, it seems impossible to agree with Stegmüller's judgment: "We wouldn't be doing justice to Kuhn were we to apply the model of Hegelian dialectic and assert that new theories constitute a form of 'synthesis,' under which the old may be 'subsumed.' No, the relationship between the old and the new is that of sheer opposition" (Stegmüller 1973, p. 167 [My translation. —A.T.L.]). But in fact there are significant parallels between Kuhn's exposition and Hegel's *Phenomenology of Spirit,* though this isn't the proper place to pursue them.

CHAPTER SEVEN

The Dynamic of Scientific Revolutions

SCIENTIFIC REVOLUTIONS AREN'T, according to Kuhn, coincidental events. They are rather, in a certain sense, the necessary product of normal science. For normal science produces the significant anomalies which serve as the concrete point of departure for revolutions (§ 7.1). Significant anomalies may lead to the relatively isolated, unexpected discovery of new phenomena or entities (§ 7.2), or they may trigger a great revolution of the kind in which theories aren't just modified but replaced (§ 7.3). A choice between two (or more) theories is then forced, and we must ask what reasons weigh in to determine the outcome of such choice (§ 7.4). The discourse between proponents of the different theories exhibits certain special features, which we must discuss next (§ 7.5). In conclusion, one might ask what sort of progress results from scientific revolutions (§ 7.6).

7.1 The Dialectic of Normal Science: The Production of Significant Anomalies

The practice of normal science strives to expand and improve knowledge of its particular phenomenal world on the basis of previously available knowledge. As such prior knowledge of the world also encompasses (to some extent implicit) knowledge about the sorts of objects there are and the fundamental attributes they have,[1] it places substantive expectations on experimental and theoretical results whose details have yet to be worked out.[2] Such concrete expectations concern

1. Compare § 3.7.
2. 1959a, *ET* pp. 227, 234, 237; *SSR*, pp. 5–6, 24, 34, 52–53, 57, 64–65, 100–101, 152, 166; 1962d, *ET* pp. 173–174; 1963a, pp. 349, 364–365, 368; 1970a, *ET* pp. 272, 277; 1970b, p. 247.

the solubility of appropriately selected problems,[3] as well as many details, including quantitative details, of the expected solutions.[4]

But it's not the case that such expectations regarding the behavior of nature (that is, of the current phenomenal world), when it comes to the actual performance of instrumentation or achievements of theory, are always fulfilled. On the contrary, normal science repeatedly witnesses situations which contradict expectations. Kuhn calls observations or experimental and theoretical findings that are surprising relative to normal-scientific expectations, and appear to contradict them, *anomalies*.[5] Accordingly, problems that, by the standards of normal science, really ought to be soluble, but whose solution poses unexpected difficulties, are called "anomalous" or "extraordinary" problems,[6] or anomalies as well.[7] Normal science is always confronted with anomalies and anomalous problems.[8] As Kuhn so forcefully puts it in *SSR*, using "counterinstance" instead of "anomaly," "there is no such thing as research without counterinstances."[9]

But anomalies and anomalous problems aren't normally taken as refuting counterexamples to the current governing regulations. Anomalies may come about when apparatus doesn't function the way predicted, or when approximations are made unconsciously, or when their accuracy is falsely assessed, or the like.[10] For the same sorts of reasons, problems may unexpectedly turn out not to be satisfactorily soluble or not to be soluble at all. And so normal science often deals with anomalies by setting them aside rather than scrutinizing them, until they have somehow demonstrated their conformity to regulations.[11] The reason behind such treatment is the expectation that anomalies and anomalous problems would dissolve under closer analysis but that such analysis isn't worthwhile, since it promises no interesting epistemic gain. And, indeed, it happens again and again that anomalies simply dissolve "by themselves," without any special expenditure of energy on their behalf. If all problems that, contrary to expectations, resisted solution were taken as falsifying instances for the appropriate normal-scientific prac-

3. Compare § 5.2.b.
4. Compare § 5.2.c.
5. 1961a, *ET* p. 221; *SSR*, pp. ix, 5–6, 52–53, 57, 65, 96–97, 101; 1962d, *ET* pp. 173–174; 1963a, pp. 364–365, 368; 1964, *ET* p. 263; 1971b, *ET* p. 28; *ET*, p. xvii; 1980a, pp. 183, 191; also 1959, *ET* p. 235.
6. E.g. *SSR*, p. 34.
7. E.g. *SSR*, pp. 5–6.
8. Compare § 5.2.d point 2, and § 5.6.
9. *SSR*, p. 79, similarly pp. 17–18, 81, 110, 146, 147.
10. 1961a, *ET* p. 202; *SSR*, pp. 81–82.
11. Compare § 5.2.d.

tice, their ubiquity would mean that "all theories ought to be rejected at all times."[12]

Still, it's not the case that anomalies never cast the slightest doubt on the regulations of normal science. Kuhn calls an anomaly with the ability to call the regulations of normal science into question "more than just an anomaly"; and he similarly speaks of "serious," "meaningful," "troublesome," "especially compelling," "admittedly fundamental," "crisis-provoking," or "significant" anomalies.[13] Kuhn sometimes calls this kind of anomaly just an "anomaly," in situations in which the critical character of the anomaly is clear from the context.[14] Now, what distinguishes a "mere" anomaly from a "significant" anomaly?

This question amounts to the demand for a criterion or defining characteristic capable of specifying the difference between mere and significant anomalies. But as such, for Kuhn, it isn't answerable. For being a significant anomaly isn't a (relational) property of any given experimental or theoretical finding or problem. To be sure, failure to harmonize with the expectations of a given community is a (relational) property of findings or problems, and accordingly, the *identification* of anomalies is largely unanimous in scientific communities.[15] But assessing whether a given anomaly calls the governing regulations into question or not involves a judgment or *evaluation* on which members of the relevant community may have differences of opinion. For

> there are no rules for distinguishing an essential anomaly from mere failure. . . . I was concerned not to find a methodological rule for individual scientists.[16]

Still, something may be said about the circumstances which may prompt, though not force, the evaluation of an anomaly as a significant anomaly. With no claims to completeness, Kuhn lists four relevant factors.[17] First, an anomaly may appear to call the accepted, interpreted

12. *SSR,* p. 146.

13. 1961a, *ET* pp. 204, 205, 209, 211; *SSR* pp. 77, 81, 82, 86, 97 (similarly p. 169); 1963a, pp. 365, 368; 1963b, p. 392; 1964, *ET* pp. 262, 263; 1970b, p. 248; 1970c, *SSR* p. 186; 1980a, p. 191.

14. E.g. 1961a, *ET* pp. 191, 211; *SSR,* pp. 6, 67, 97, 101; 1962a, *ET* p. 173; 1962d, *ET* pp. 173–174; 1963a, p. 365.

15. 1980a, p. 183.

16. 1963b, p. 392; similarly *SSR,* pp. 79–80; 1970b, pp. 248–249; 1980a, p. 183. The same position may also be found in Kuhn's presentation of scientific values, where he claims that scientific value judgements generally aren't the application of algorithmic procedures. Compare § 4.3.c.

17. 1961a, *ET* pp. 202, 208–211, 221; *SSR,* pp. x, 69, 75, 82, 97, 153; 1964, *ET* p. 262; *ET,* p. xvii; 1980a, p. 190 n.

symbolic generalizations immediately into question. This occurs when the problem at issue, though of a kind these symbolic generalizations are supposed to be able to handle, still gives every appearance of refusing to be subsumed under them. A quantitative discrepancy between theoretical prediction and experimental or observational findings is especially wont to trigger the reevaluation of a mere anomaly as a significant anomaly. Second, an anomaly initially set aside as of little significance may, on grounds of developments elsewhere in normal science, become an indicator for the doubtfulness of the governing regulations. Third, an anomaly may become significant when, over a long period of time, its solution resists efforts by even the best specialists in the field. The reasons behind such protracted engagement with the problem may in this case be external to science, as when certain applications of theory are especially important on extrascientific grounds but are hindered by anomalies.[18] Fourth, one and the same anomaly may appear repeatedly in different laboratories, or different anomalies may appear to have the same root. Such anomalies, or groups of anomalies, can then no longer be pushed aside as "noise," unworthy of analysis.

As suggested above, such reasons for reevaluating a mere anomaly as a significant anomaly aren't necessarily compelling. Accordingly, such reevaluation may be subject to difference of opinion within a scientific community. It follows that the *universal* recognition of an anomaly as significant usually occurs, if at all, as the result of a gradual process and not a sudden event.

Now, it's a historical fact that the practice of normal science has given rise to significant anomalies again and again. This is understandable, for anomalies may only exist where there are particular expectations. But the presence of such expectations is only a necessary condition for the appearance of those significant anomalies that concern not just inconsistencies within a theory, or between theories, but discrepancies between theoretical expectation and empirical fact. For them to

18. Here it becomes clear that Kuhn's position can't be unproblematically classed as either externalistic or internalistic. As far as the universal phase model, or "structure," of scientific development is concerned (compare § 1.3), extrascientific factors can be largely neglected, as Kuhn asserts in *SSR* (*SSR,* pp. x with n. 4, 69; q.v. *ET,* p. xv; 1983c. See 1971a, pp. 140–141; 1979a, p. 128 on Kuhn's distinction between "external" and "internal"). But accounts of scientific development that go beyond this structural description must also take factors external to science into consideration, especially as regards their accelerating or dilatory effects, but also as they affect the substantive content of science; on this issue, see *SSR,* pp. x with n. 4, 68–69, 75, 123–124; 1968a, *ET* pp. 118–120; 1971c, *ET* pp. 148–150, 159–160; 1976a, *ET* p. 62; *ET* p. xv; 1977c, *ET* p. 333 n. 8; 1979a, p. 122; 1983c.

occur, the phenomenal world under study must also exhibit a certain independence from theoretical expectations; otherwise such discrepancies between theory and experience couldn't exist. The phenomenal world must assert such resistance despite the fact that similarity relations are coconstitutive *both* of the concepts occurring in the relevant theory *and* of the corresponding region of the phenomenal world.[19] The possibility of discrepancies between theory and experience is explained by the fact that similarity relations are only coconstitutive, and not *entirely* constitutive, of the appropriate region of the phenomenal world. The resistance of the world-in-itself (or of stimuli)[20] may, to some extent, penetrate the network of similarity relations, despite the fact that these similarity relations are, in a certain respect, already attuned to such resistance by virtue of their incorporation of genetically object-sided moments.[21] The repeated occurrence of significant anomalies in the history of science can, in fact, prove one motive for assuming the existence of a theory-independent, and by its own determinate, proprietary features resistant, world-in-itself.[22]

In the occurrence of significant anomalies lies the dialectic (or the irony!) of normal science. For it's precisely because normal science is conducted with the expectation that its chosen problems admit of solution in accordance with regulations[23] that, historically, it repeatedly gives rise to significant anomalies. Precisely because it attacks ever more esoteric problems at ever greater instrumental and theoretical expense,[24] still trusting to its regulations, do the chances rise that these regulations will prove insufficient. Precisely because training in normal science is narrow and rigid, conducted with confidence in the governing regulations,[25] are those thus trained so extraordinarily suited to diagnosing real failure on the part of these same regulations.[26] In this way, normal science has, time and time again, led to its own cessation. "The ultimate effect of this tradition-bound work has invariably been to change the tradition,"[27]

19. Compare chap. 3, especially § 3.6.
20. Compare §§ 2.1.a and 2.2.b.
21. Compare § 3.2.
22. For other such motives, compare § 2.2.c.
23. Compare § 5.2.b.
24. Compare § 5.2.d.
25. Compare § 5.5.a.
26. *SSR,* p. 166.
27. 1959a, *ET* p. 234; similarly *SSR,* pp. 5, 64–65, 122; 1963a, pp. 349, 365; 1964, *ET* p. 262; 1970b, p. 247. —Kuhn's thesis shouldn't be misunderstood as claiming that normal science *necessarily* leads to the production of significant anomalies, as Kuhn appears to assert when he says that the practice of normal, puzzle-solving science "can and

for no other sort of work is nearly so well-suited to isolate for continuing and concentrated attention those loci of trouble or causes of crisis upon whose recognition the most fundamental advances in basic science depend.[28]

What follows when normal science has given rise to one or more significant anomalies? These indicate that the practice of normal science may no longer proceed as before; significant anomalies reveal "that something has gone wrong in ways that may prove consequential."[29] Some significant anomalies lead to the relatively isolated, but unexpected, discovery of new phenomena or entities (§ 7.2). Others necessitate the replacement of a previously accepted theory by a novel theory incompatible with the older one (§ 7.3).

To be sure, as Kuhn notes in *SSR,* the distinction between the unexpected discovery of new phenomena or entities and revolutions in theory is artificial,[30] for two reasons. First, superficial appearances to the contrary, *both* cases involve changes on the theory level. Second, the discovery of new facts may herald the replacement of a previously accepted theory, and conversely, the replacement of a previously accepted theory may lead to or be accompanied by unexpected discoveries.[31]

7.2 Unexpected Discoveries

In normal science, too, new phenomena and entities are discovered, but these discoveries aren't unexpected; they may rather be largely anticipated with the help of accepted theory.[32] By contrast, the discoveries resulting from significant anomalies are such that the newly unearthed phenomena or entities "could not be predicted from accepted theory and . . . therefore caught the assembled profession by surprise."[33] But this apparently unequivocal characterization of unexpected discoveries calls for three points of qualification.

---

inevitably does lead to the isolation and recognition of anomaly" (1963a, p. 365). But his claim here is rather that normal science necessarily runs into significant anomalies only where there are actually some to be discovered.

28. 1959a, *ET* p. 234; similarly *SSR,* p. 52.

29. 1962d, *ET* p. 173; similarly 1959a, *ET* p. 235; 1961a, *ET* p. 203; *SSR,* pp. 5–6, 57, 166; 1963a, pp. 366, 368.

30. *SSR,* pp. 52, 66.

31. See *SSR,* chap. 6, and 1962d for examples.

32. Compare § 5.2.c.

33. 1962d, *ET* p. 166; similarly *SSR,* p. 52; 1963a, p. 365.

First of all, if nonpredictability by accepted theory is cited as one feature of unexpected discoveries, this should not be taken to suggest that the discovered phenomena can't, in principle, be subsumed under the *symbolic generalizations* of accepted theory.[34] "Accepted theory" rather refers to these symbolic generalizations *along with* the meanings of concepts mediated by exemplary problem solutions, as well as to theoretical and experimental standard operational procedures.[35] Accordingly, a discovery may be unexpected on the grounds that it conflicts with expectations implicitly contained in current experimental procedures.[36] All experimental procedures must obviously contain a great number of explicit and implicit expectations about the attributes of entities that exist, and about the nonexistence of others. And so if a discovery collides with the expectations of normal science, it is revolutionary in Kuhn's sense, for it demands the correction of prior knowledge and isn't simply the cumulative acquisition of new.[37] Such correction may—as happened with the discovery of X-rays—affect only the standard experimental procedures, without necessitating any change in the *formal* portion of accepted theory, the symbolic generalizations.[38]

Second, talk of surprising the "assembled profession" primarily concerns the community of specialists actually working in the relevant field. For members of other communities, whether neighboring communities or others belonging to the larger community which includes that of the specialists, the discovery need not be revolutionary.[39] Where specialists, given their particular experimental procedures, have certain expectations, members of other communities may have none. For the outsider, then, the discovery so revolutionary for the specialist may, in fact, be cumulative.

Third, first impressions to the contrary, the distinction between expected (normal-scientific) and unexpected (revolutionary) discoveries isn't a strict dichotomy.[40] For one, there's a fluid transition between the expected and the surprising. Furthermore, the epistemic state of members may vary so much over the community that what was expected for one surprises another.

Unexpected discoveries begin with an anomaly's taking on the

34. Compare § 4.3.a.

35. *SSR,* pp. 58–61, 96–97; 1962d, *ET* pp. 173–177.

36. Compare § 5.2.a.

37. Compare § 6.1.

38. *SSR,* pp. 7, 58–59; 1962d, *ET* pp. 175–176.

39. *SSR,* pp. 49–50, 52, 61, 92–93; 1962d, *ET* p. 175; 1963a, p. 364; 1970b, pp. 251–253; 1970c, *SSR* pp. 180–181.

40. *SSR,* p. 62; 1962d, *ET* p. 167 n. 3.

character of a significant anomaly.[41] Following its discovery, the significant anomaly must be understood, which is only possible when the expectations that made the relevant phenomenon into an anomaly have been abandoned.[42] Gaining understanding is a process that necessarily takes a certain amount of time, time in which to reflect over precisely which expectations of normal science must be given up and which new expectations ought to take their place. For this reason, unexpected discoveries can't be precisely dated, nor do they often have any individual discoverer.[43] This latter condition holds when the analysis of an anomaly is carried out by more than one scientist.

Once an understanding of the anomaly has been gained, and the former anomaly has become expected, a revolutionary episode of scientific development, in Kuhn's sense, is immanent.[44] For what has occurred isn't some normal-scientific addition to the prior store of knowledge but rather a replacement of part of this store with new, incompatible knowledge.

7.3 The Triggering of Revolutions in Theory

Like unexpected discoveries, revolutions in theory are triggered by anomalies. These anomalies here lead to a state of affairs markedly different from normal science: crisis (§ 7.3.a). Research practice in times of crisis is called extraordinary science, to be discussed in § 7.3.b.

a. Anomalies and Crisis

Revolutions of the sort in which a theory isn't just modified but is rather overturned and replaced by a new one begin, like unexpected discoveries, with the appearance of significant anomalies.[45] These

41. On this point, Kuhn seems to vacillate a bit in his writings. While he usually claims that unexpected discoveries *always* start with anomalies (*SSR,* pp. 52–53, 62, 64, 96; 1962d, *ET* p. 174; 1963a, pp. 365–366), he sometimes leaves open the possibility that they might occur, albeit rarely, without a positive expectation's first being disappointed. For example, he asserts, "Cumulative acquisition of unanticipated novelties proves to be an *almost* non-existent exception to the rule of scientific development" (*SSR,* p. 96, emphasis mine). But Kuhn appears to retract this statement at the bottom of p. 96. —Also cf. 1963a, p. 349.

42. 1959a, *ET* pp. 226–227; *SSR,* pp. 7, 53, 62, 64, 66; 1962d, pp. 174–177.

43. *SSR,* pp. 2, 52–53, 54–56, 57–58; 1962d, *ET* pp. 166, 171, 173, 174, 176–177.

44. *SSR,* pp. 53, 66.

45. 1959a, *ET* pp. 234–235; 1961a, *ET* pp. 203–204, 208–211, 221; *SSR,* pp. ix, 67–68, 79, 97, 144, 145, 152, 153; 1963a, pp. 366–368; 1964, *ET* p. 262; 1970a, *ET* pp. 267, 273.

anomalies call the foundations of accepted theory into question. Distinguishing this phase from the state of affairs in normal science, Kuhn calls this situation the "abnormal situation," "crisis state," or, most often, "crisis."[46] This situation is characteristically a time "when research projects go consistently astray and when no usual techniques seem quite to restore them."[47] In this situation, it's clear that normal science can't continue as before. An innovation is needed,[48] for "the traditional rules of normal science . . . no longer define a playable game."[49] The natural inference, then, is that "something has gone wrong with existing knowledge and beliefs."[50] This admission produces "pronounced professional insecurity"[51] in the scientific community. Such insecurity is sometimes revealed by scientists' open dissatisfaction with previously accepted theory, dissatisfaction grounded in the theory's inability to further guide productive research.[52]

In order to ward off misunderstandings, it should be insisted that three claims, in particular, are not implied in this characterization of the crisis state. First, it's by no means claimed that every generally recognized anomaly necessarily leads to a crisis.[53] Even when such an anomaly casts doubt on the governing regulations, it need not trigger a crisis so long as it poses no obstacle to the treatment of other sorts of problems within normal science. Second, it's by no means claimed that all members of a community will perceive a crisis in the same way. On the contrary, it may happen that two scientists

> reach different judgements in concrete cases, one man seeing a cause of crisis where another sees only evidence of limited talent for research.[54]

46. 1961a, *ET* pp. 202–204, 206–211, 221–222; *SSR,* pp. ix, 61, chap. 7, and chap. 8; 1963a, pp. 367–368.

47. 1961a, *ET* p. 202, similarly p. 221; *SSR,* pp. ix, 5–6, 67–68, 69, 74–75, 145, 153; 1963a, p. 367; 1970a, *ET* pp. 267, 273; 1970b, p. 257.

48. 1969c, *ET* p. 350.

49. *SSR,* p. 90, similarly pp. 76, 85–86; 1963a, pp. 349, 365; 1970a, *ET* p. 281; 1970b, p. 247; 1980a, p. 190.

50. 1959a, *ET* p. 235; similarly 1961a, *ET* p. 203; *SSR,* pp. 6, 92, 93, 114, 153, 166; 1963a, pp. 367–368; 1970a, p. 273; 1970b, p. 247; 1970c, *SSR,* p. 181; 1980a, p. 190 n. 1.

51. *SSR,* pp. 67–68.

52. *SSR,* pp. 83–84; 1964, *ET* pp. 262–263.

53. *SSR,* pp. 81–82. —The examples Kuhn uses here, the Newtonian theoretical values for the speed of sound and the precession of Mercury's perihelion, are, however, differently and incompatibly interpreted at 1961a, *ET* pp. 203–204. I will return to this issue in § 7.3.b, where I discuss the ways in which crises may come to an end.

54. 1970b, p. 248; similarly 1961a, *ET* pp. 202, 206; 1970c, *SSR,* p. 185. —Kuhn appears to contradict this statement in 1970c, *SSR* p. 181, where he speaks of crisis as

The reason for such disagreement is that the attestation of a crisis is a form of theory evaluation, and given the peculiarity of scientific value systems, theory evaluations need not be unanimous.[55] Third, nothing in the concept of a crisis implies that the crisis state itself need be entirely transparent to members of the community.[56] What may occur instead, Kuhn suggests, is that a community drifts into a crisis state that, though an external observer might recognize it as such by its characteristic symptoms, only gradually penetrates the consciousness of the participants.

Before I turn, in the next subsection, to the research practice typical of the crisis state, one controversial thesis of Kuhn's, put forth in *SSR* and other work composed around the same time, ought to be discussed. This thesis asserts that all revolutions in theory are initiated by crises, crises triggered, in turn, by the appearance of significant anomalies in the relevant fields.[57] The validity of this general claim is thought to be evidenced, for one, by the history of science. In addition, the fact that, in the course of a revolution in theory, a theory that served as the basis for a (more or less extended) period of normal science must be abandoned is also thought to substantiate this claim. For how might it happen that confidence in this older theory wanes and vanishes, making room for confidence in some new theory? This process can only be set in motion if those reasons that argued for confidence in the older theory[58] cease to be convincing; "Nature itself must first undermine professional security by making prior achievements seem problematic."[59] We recall that the reasons for confidence in the older theory leaned heavily on its sanctioning of puzzle-solving science, an ability it no longer has in times of crisis.

The thesis that all revolutions in theory are initiated by crises, triggered in turn by the appearance of significant anomalies in the relevant field, has, subsequent to being met with criticism,[60] been somewhat weakened.[61] Kuhn remains, as before, convinced that crises are usually the prelude to revolution, but he acknowledges that revolutions

---

"the *common* awareness that something has gone wrong" (emphasis mine). This strikes me as merely a somewhat negligent, abbreviated formulation.

55. Compare § 4.3.c.

56. *SSR*, pp. 84, 86.

57. 1959a, *ET* pp. 234–235; 1961a, *ET* pp. 206–209; *SSR*, pp. 67–68, 74–76, 77, 85–86, 92, 145, 158, 169; 1963a, pp. 349, 365; 1964, *ET* p. 263.

58. Compare §§ 5.1, 5.5.b, 5.6, and 7.4.b.

59. *SSR*, p. 169.

60. E.g. Curd 1984, pp. 9–10; Cohen 1985, p. 27; Hörz 1988, p. 86; Mandelbaum 1977, pp. 446–447; Watkins 1970, pp. 30–31.

61. 1970c, *SSR*, p. 181; *ET* p. xvii; 1980a, p. 190 n. 1.

might also, albeit rarely, get started in other ways. A further modification of the thesis consists in the concession that a crisis need not be triggered by anomalies *within* the field at issue. Crisis might rather be imported through the results of other communities.

b. Research in Times of Crisis: Extraordinary Science

Crises may vary significantly both in the number of scientists they affect and in their duration. In one limiting case a crisis might affect only a single scientist's work, while in the other it affects the entire professional community. Most often, crises appear in particular specializations and occupy the relevant specialists.[62] Similarly, crises may vary along a time scale, ending quickly or lasting a long time.[63] When a crisis affects a given field of research, research practice in that field changes progressively by comparison with normal science. Kuhn calls this altered practice "extraordinary science" or "science in the crisis state."[64]

The chief research topics of extraordinary science, naturally enough, are the significant anomalies which triggered the crisis.[65] But where normal science presupposes certain regulations in problem solving, prior regulations, and especially the former governing theory, are open to dispute in extraordinary science. This doesn't mean that these regulations are completely suspended from one day to the next. But because the prior regulations have led to crisis, their collective validity can no longer be taken for granted. The task of extraordinary research is then to change the regulations in such a way as to preserve as many previously attained problem solutions as possible, on the one hand, while dissolving those stubborn anomalies which triggered the crisis, on the other.

Extraordinary science may be identified above all by the following four symptoms, which need not all occur together.[66] The first symptom (cited in the last subsection) is openly voiced dissatisfaction with the previous governing theory's aptitude. The second symptom is the continued use of old regulations for problem solving but, as the crisis situation persists, in increasingly modified and supplemented forms.[67]

62. 1961a, *ET* p. 203.

63. I will discuss the question of how crises end later in this section.

64. 1961a, *ET* p. 202; *SSR,* pp. 6, 82, 86, 87, 89, 91, 101, 154; 1970a, *ET* p. 272.

65. *SSR,* pp. 82–83, 86–87, 88, 152; 1964, *ET* pp. 262–263; also 1961, *ET* pp. 209–211.

66. *SSR,* p. 91; 1964, *ET* p. 263.

67. *SSR,* pp. 61, 70–71, 74, 75, 80, 83, 84, 86, 86–87, 89, 154; 1963a, p. 367; 1970b, p. 257; 1970c, *SSR,* p. 181.

The readiness to soften regulations which have held up to now naturally arises out of the awareness that something is wrong with them. Some of these modifications and additions—and the longer the crisis, the more of them there are—are ad hoc; they are geared only toward a single problem and thus appear rather arbitrary. In addition, new, speculative theories are often coined in hopes of finding the way to a new normal science. These speculative theories are tested on their capacity to solve the known anomalies. Throughout the period, there is no consensus over either the proposed modifications of and additions to the old theory or the proposed replacement theories.

A third symptom of extraordinary science is the scientists' willingness to try many things whose results can be predicted only vaguely, if at all.[68] In particular, experiments are carried out without any precise expectations regarding their results. Such experiments serve the purpose of gathering data toward the more precise localization of the sources of anomalies. One frequent consequence of this strategy is new discoveries that can't be made compatible with previous theory and are thus important in the development of further theories.

Fourth, extraordinary science may be recognized by its recourse to philosophical analyses of the foundations of the previous research tradition. This strategy amounts, above all, to an attempt to make previously implicit regulations explicit for purposes of testing them.[69] Thought experiments may play a great role in the conceptual penetration of a given theory, since they are capable of teasing out the theory's implications with greater clarity than is possible in actual experimental studies.

All of these symptoms make it obvious that extraordinary science exhibits certain *similarities* with prenormal science.[70] The fact that extraordinary science, too, may witness the formation of schools with competing approaches also contributes to this similarity. But the *differences* between prenormal and extraordinary science, differences deriving from the fact that extraordinary science is preceded by normal science, also deserve notice. The prior existence of normal science means that there are already broad domains of specialized knowledge, including the requisite vocabulary and established techniques of the most diverse varieties. But above all else, in extraordinary science, by contrast with prenormal science, it's clear which central problems must be solved: the significant anomalies which brought on the crisis in the first place. These provide a relatively well defined focus for research activities and the clash of the schools. By comparison, prenormal sci-

68. 1961a, *ET* p. 203; *SSR,* pp. 61, 87, 88–89; 1963a, p. 367.
69. *SSR,* p.88; 1964, *ET* pp. 262–265.
70. *SSR,* pp. 61, 72, 84, 101, 112, 149. —Compare § 5.5.b.

ence is substantially more diffuse. For in this phase there is little specialized or universally accepted knowledge and no sure guide to the fundamental problems which must be solved.

Here we might ask how the substantial consensus of normal science might ever come to dissolve in extraordinary science. The reason is that, even in normal science, *consensus isn't total*. There are rather significant epistemic differences between members of a community engaging in normal science. Despite their unequivocal use of central scientific concepts, there may be individual differences in the criteria members apply in identifying referents and nonreferents of these concepts.[71] Despite general agreement over scientific values, such values remain open to individual shaping.[72] And despite the communal practice of normal science, the degree of inner distance from the governing regulations is also subject to individual difference.[73]

All such individual differences are practically invisible throughout the phase of normal science. But once significant anomalies appear, unexpected situations that seem impossible to surmount within the bounds of previously successful practice, these differences rise to the surface. Different criteria for concept use may make the current difficulties appear differently caused;[74] differently molded values may lead to different evaluations of competing theories; and differential identification with the previously reigning theory may result in differential readiness to abandon it. Dissent in extraordinary science is thus the manifestation of differences that exist in normal science, but only in latent form.

How may crises end, or, posed differently, under what circumstances may a community find its way back to the consensual work of normal science? This can only happen in one of three ways.[75] First, concerted efforts may, notwithstanding expectations by some members of the community to the contrary, cause an anomaly to vanish within the bounds of valid regulations. The second possible way of ending a crisis doesn't occur too often, but may also be found in the history of science. It consists in acknowledging the anomalies as such and returning to the work of normal science.[76] This possibility is open only

71. Compare § 3.6.d.

72. Compare § 4.3.c, along with §§ 7.4.b and 7.4.c below.

73. Compare § 5.5.a.

74. 1989a, p. 21; 1990, p. 307.

75. 1961a, *ET* pp. 203–204; *SSR*, p. 84. In the first edition of *SSR*, Kuhn cites revolution in theory as the only way of ending a crisis (p. 84), while in the second edition this is only one of three possibilities.

76. The examples Kuhn cites at 1961a, *ET* pp. 203–204 for this way of ending a crisis (none are cited in *SSR*), the Newtonian theoretical values for the speed of sound and the precession of Mercury's perihelion are used in *SSR*, p. 81, as examples of

when the known anomaly (or anomalies) doesn't (or don't) pose an obstacle to normal-scientific work on problems other than the anomalous ones. The third way of ending a crisis consists in a revolution in theory, in which the theory which led to the crisis of normal science is replaced by a new one.

But where does this new theory come from? In many cases, the foundations of the new theory were laid before the crisis but received little or no attention.[77] Only a crisis which shows that something is fundamentally amiss with the established theory suffices to awaken interest in such alternatives. In other cases, the new theory is developed during the crisis as a potential avenue of escape,[78] thus providing evidence for the "creative function" of crises.[79] The innovators are often people new to the field on account of either their youth or some change in their field of study.[80] Such people are more inclined to tread new paths than scientists whose reputations are based on achievements within the bounds of the old theory.

7.4 Theory Comparison and Theory Choice

For several reasons, the process of theory replacement is not without problems. In order to explicate this point, we will begin by discussing why a new theory must succeed in competition with other theories (§ 7.4.a). Next we must ask what sorts of reasons play a role in theory choice (§ 7.4.b). Finally, we will consider Kuhn's attack on the distinction between the context of discovery and the context of justification (§ 7.4.c).

a. Theory Comparison, Not Theory Falsification

The need for deciding between (at least)[81] two different theories instead of simply choosing or refusing a theory on the basis of its potential

---

anomalies that *didn't* trigger crises, because "apparently neither had seemed sufficiently fundamental to evoke the malaise that goes with crisis" (*SSR*, pp. 81–82). This vacillation in Kuhn's judgement of whether these cases made crisis immanent suggests that difference of opinion over the presence of a crisis in a given field isn't limited to participating scientists but also affects the historians who examine such episodes.

77. *SSR*, pp. 75–76, 86, 97.

78. *SSR*, pp. 75, 154.

79. 1970b, p 258.

80. 1961a, *ET* p. 208 n. 44; *SSR*, pp. 90, 133, 144, 151, 166.

81. For simplicity's sake alone, I will limit myself in the following discussion to the situation in which a community faces a choice between only two theories. Of course it's conceivable for the choice to involve more than two candidates.

arises, according to Kuhn, from the fact that a theory in crisis doesn't count simply as falsified.[82] As Kuhn understands it, the falsification of a given theory means that, on grounds of its demonstrated empirical falsity, the theory is no longer used in scientific practice. Crisis may, of course, lead to a situation in which the scientists in the affected community "begin to lose faith [in the previously governing theory] and then to consider alternatives."[83] But this doesn't mean that all work is henceforth conducted without the theory in crisis or that a new theory arises as the basis of a new phase of normal science given only the results of its comparison with empirical findings. The contrary is shown, so Kuhn claims, by the history of science:

> No process yet disclosed by the historical study of scientific development at all resembles the methodological stereotype of falsification by direct comparison with nature.[84]

But we mustn't remain content with historical fact. The historical absence of any theory falsification (in the sense given above) can rather be explained in a way that lends it an inexorable character.[85] Here Kuhn offers two arguments.

The first argument begins with a thesis which itself requires further justification. By the time a theory that was once the foundation for a tradition of normal science has reached a crisis, it's impossible to do without it in the phase of extraordinary science. A scientist who absolutely refused to continue using a theory on account of the significant anomalies that plague it would be forced, for two reasons, to stop doing science.[86]

The first reason is that the theory in crisis must still be used in identifying the problems that extraordinary research must face. Its store of previously solved problems places decisive demands on possible competing theories;[87] significant anomalies, the very focus of extraordinary research, only exist relative to the theory in crisis.[88] To stop using a falsified theory in times of crisis would mean a return to the state

82. *SSR*, p. 77; similarly pp. 145, 147; 1961a, *ET* p. 211; 1980a, p. 189; 1983a, p. 684 n. 5. —This thought was already quite clearly expressed by Conant: "We can put it down as one of the principles learned from the history of science that a theory is only overthrown by a better theory, never merely by contradictory facts" (Conant 1947, p. 36, similarly p. 84).

83. *SSR*, p. 77.

84. Ibid.

85. *SSR*, pp. 77–78.

86. *SSR*, pp. 78–79.

87. See § 7.4.b below.

88. Compare §§ 7.1 and 7.3.b.

science was in before the theory was established, a return which doesn't happen in extraordinary science.[89]

In addition, the world-constitutive function of a theory in crisis necessitates the theory's further use;

> Once a first paradigm through which to view nature has been found, there is no such thing as research in the absence of any paradigm.[90]

Kuhn here appears to be operating under the assumption that an ordering of some region of the world gained by means of similarity relations can't simply be abandoned without replacement.[91] While crisis may relax the injunction to use this particular ordering, crisis can't lead to its rejection without some alternative that may be used in its place.

But the inexorable need to continue using a theory in crisis doesn't immediately entail Kuhn's claim that theory comparison is unavoidable. Theoretically, it might be the case that, though the retention of the old theory is necessary for the research practice of extraordinary science, the evaluation of new theory candidates need not involve comparison. But in fact, in order to gain universal acceptance by the community, a new theory must be measured according to the achievements of the old; it must exceed them. This is the second argument against theory falsification, and its argumentative weight derives from the particular kinds of reasons which play a role in theory choice. I will return to these reasons in § 7.4.b.

First, however, I shall attempt to explain the difference between Popper's and Kuhn's positions on theory falsification. Kuhn diagnoses the difference between himself and Popper on this issue as residing in the distinction between normal and extraordinary science, a distinction that Popper doesn't make.[92] This diagnosis is entirely correct, but its correctness derives in turn from a difference in Popper's and Kuhn's basic epistemological-ontological assumptions. While, for Popper, absolute or purely object-sided reality is, in principle, knowable (in a

89. Put in Hegelian terms, the new theory must be a *determinate* negation of the old. The old is thus indispensable until the new is actually in hand (see Hegel, *Phenomenology of Spirit*, Hegel's *Works*, vol. 3, p. 74).

90. *SSR*, p. 79.

91. This assumption is most easily made plausible for the case of gestalt perception. Once one has learned to see certain gestalts in a given domain, it is only with difficulty, if at all, that one can return to the mode of seeing in which these gestalts weren't visible. By contrast, it's quite conceivable that one might learn to see different gestalts in this same domain (under certain conditions). Still, the illustration of gestalt perception shouldn't be taken as the last word on this assumption of Kuhn's.

92. *SSR*, pp. 146–147.

certain sense), for Kuhn, the reality accessible to us is always steeped in genetically subject-sided, essentially variable moments, moments that may in no way be (reflectively) subtracted.[93] Consequently, for Popper, a theory's task is to capture and represent purely object-sided reality. Though we may never hope to accomplish this task conclusively, we may still continually improve our attempts. A theory must be improved or discarded as soon as its differences with (empirically captured) reality reveal themselves. For Kuhn, by contrast, a theory's task can't consist in the representation of purely object-sided reality, as this task is unfulfillable. Instead, theories participate in the constitution of what the given community takes to be, and experiences as, reality. But the constitution of reality, or of the world, is most definitely not arbitrary, since the object-sided moments which contribute to a phenomenal world offer resistance (whose nature is indeterminable by us). And so a theory, though coconstitutive of the same phenomenal world it represents, may still run afoul of empirical difficulties.[94] When this happens, the theory can't simply be abandoned, for this would amount to the (impossible) abandonment without replacement of the ordering of the phenomenal world gleaned from this theory. Instead, those similarity relations which refuse to harmonize with the resistance of the purely object-sided must be corrected or replaced. This correction or replacement is precisely what occurs in a scientific revolution.[95]

b. The Reasons behind Theory Choice

An inquiry into the reasons underlying theory choice must distinguish three questions—the neglect of this distinction has led to serious misunderstandings of Kuhn's position.[96] First, we might ask after the reasons relevant to the very first adherents of a new theory. Second, we might ask what sorts of reasons are generated while working on the new theory, such that they eventually bring more and more members of the community to adopt it. And third, we might ask which reasons prove decisive in the choice of an entire community. After addressing each of these questions, I will turn, in a fourth point, to consider the argumentative weight of the reasons canvassed.

1. What reasons guide the very first adherents of a new theory in their decision to articulate it and flesh it out? One negative reason is

93. Compare part 2.

94. Compare § 7.1.

95. Compare § 6.2.

96. *SSR*, p. 153, Kuhn draws special attention to the importance of the distinction between the first two questions and the third.

the crisis of the older theory.[97] But in positive terms, especially as regards its demonstrated problem-solving capacity, any new theory must initially be entirely inferior to the old. In its early stages, the new theory may not even succeed in coping with the significant anomalies that led to the crisis of the old.[98] In this case, the individual decision to work with the new theory can only be made on the basis of "faith." For someone facing this decision

> must . . . have faith that the new paradigm will succeed with the many large problems that confront it, knowing only that the older paradigm has failed with a few.[99]

As for the nature of such faith,

> There must also be a basis, though it need be neither rational nor ultimately correct, for faith in the particular candidate chosen.[100]

This is one of the few passages in Kuhn's work where the possible absence of rationality in theory choice is explicitly discussed, and it is thus doubtless one source for the common accusation that, for Kuhn, theory choice, hence science as a whole, is an irrational enterprise. Kuhn has emphatically rejected this charge,[101] a charge for which the passage cited above really offers no evidence. For this passage explicitly refers to the motives of the very first adherents of the new theory, not to the reasons that weigh in the actual decision by an entire scientific community. Yet it is these latter reasons that determine the rationality of theory choice. The reasons that prompt the first supporters of a new theory to make their choice may, indeed, be of dubious rationality.[102] But whether a new theory is capable of attracting further adherents, and eventually convincing the entire community, depends on the arguments these first adherents produce in fleshing out the theory.

The work done by a new theory's first supporters—whatever their motives—is, of course, of potentially great importance to scientific development. For any theory that is eventually to gain universal acceptance must find its first adherents, even if its later success isn't yet apparent.

97. *SSR,* p. 158.

98. *SSR,* p. 154.

99. *SSR,* p. 158.

100. *SSR,* p. 158; q.v. 152–153.

101. 1970b, pp. 234–235, 263–264, 275; 1970c, *SSR,* pp. 175, 186, 191, 199; 1971a, pp. 139, 143–146; 1976b, p. 196; 1977c, *ET* p. 321; 1983d, p. 563. —I will return once more to the issue of "the rationality of science" in § 7.4.c, point 5.

102. Q. v. *SSR,* pp. 156, 159.

2. The reasons that motivate a new theory's first adherents may be distinguished, albeit not rigidly, from those that adherents to the theory develop and adduce in the course of extraordinary science. Such reasons are coconstituted by the scientific values that form a substantial component of consensus within the community.[103] For it's on the basis of such values that the various candidates for the position of future research-governing theory are comparatively evaluated.

Theories are, for the most part, evaluated on their suitedness to the task for which they are intended, the solution of as many research problems as possible, as accurately as possible.[104] This evaluation has three dimensions. First, and most importantly, the candidate must be able to cope with the problems that brought on the crisis. If a new theory accomplishes this task with substantially greater accuracy than its older competitor, this success argues strongly for the selection of the new theory. Second, a new theory must be able to solve at least a large portion of the problems solved by the older theory with comparable (or greater) accuracy. Its paths to these solutions may, however, be substantially different from those previously accepted. What also argues for the problem-solving capacity of a new theory is, third, its ability to predict phenomena that, from the perspective of the older theory, are unexpected.

In addition to this evaluation of problem-solving capacity, other values also play a role, described by Kuhn as that of "more subjective and aesthetic considerations"[105]—"the new theory is said to be 'neater,' 'more suitable,' or 'simpler' than the old."[106] The role played by these values is significant, especially because the problem solutions *already attained* by a theory aren't sufficient to compel a positive decision.[107] For the theory is also supposed to guide *future* research in a productive way. The problems a new theory has already solved naturally *also* argue for its future capabilities. But, in addition, aesthetic considerations may help nurture hope for the future fruitfulness of the new theory.

Throughout the phase of extraordinary science, competing theories are differently evaluated by different members of the community, despite the set of common scientific values that govern all such decisions.[108] This difference has two primary sources. First, in evaluating the problem-solving capacities of competing theories, the difficulty

103. 1970b, pp. 262–263; 1970c, *SSR*, pp. 199–200; compare § 4.3.c.

104. *SSR*, pp. 147, 153–155, 169–170, 173; 1970a, *ET* pp. 289, 290; 1970c, *SSR*, pp. 185, 205; 1977c, *ET* pp. 320, 339; 1980a, pp. 190–191; 1983d, pp. 563–564.

105. *SSR*, p. 156.

106. *SSR*, p. 155.

107. *SSR*, pp. 155–158.

108. Compare § 4.3.c.

arises that neither the catalog of problems to be solved, nor their importance, nor the standards for their successful solution can be evaluated in a manner entirely independent of the theories at issue.[109] Accordingly, assessments of the problem-solving capacities of both theories may vary, depending on which theory serves as the point of departure for this assessment.[110] Furthermore, individually variable moments always contribute to the evaluation of theories by means of scientific values; for values may be differently interpreted and differently weighted.[111]

3. In time, dissent on theory choice may dissipate, as the number and variety of arguments in favor of the new theory steadily increases. However the individual community member shapes communal values, the new theory now cuts a better figure in comparative theory evaluation. And so, after a time, most if not all members of the community will engage in normal science under the guidance of the new regulations. In light of this conclusion, we must consider which reasons ultimately compelled the community's decision.[112]

To begin with, those aesthetic factors which may, as described above, play a role in an individual decision in favor of a new theory don't weigh in the community's decision. As early as *SSR,* Kuhn is quite unequivocal here;

> [The aesthetic appeal of theories to individuals] is not to suggest that new paradigms triumph ultimately through some mystical aesthetic. On the contrary, very few men desert a theory for these reasons alone.[113]

Instead, arguments based on the new theory's problem-solving capacity are what prove compelling.[114] Only if the new theory can cope with at least some of the significant anomalies which brought on the crisis, and only if it can also solve a large proportion of the problems the old theory solved in its own way, can the new theory pervade the entire community.

But arguments for the problem-solving capacity of the new theory are based on the values accepted in the relevant community. To be sure, each member of the community shapes communal values in his or her own way, and so the individual grounds for choice aren't uniform. But if (more or less) all members of the community still make the

109. Compare § 6.3.a, point 1.
110. I will return to this circularity in § 7.5.b.
111. Compare § 4.3.c.
112. Compare § 4.3.c.
113. *SSR,* p. 158, also cited approvingly at 1970b, p. 261.
114. *SSR,* pp. 169–170; 1970a, *ET* pp. 289, 290; 1970c, *SSR,* p. 205.

same choice in the end, despite individual differences in their shaping of values, then these communal values are what determine the outcome of the community's choice.

4. Even if (more or less) the entire community decides in favor of the new theory, this choice still isn't, on the individual level, as rigorously justifiable as a mathematical proof.[115] The rigorous proof of a given proposition is only possible given the availability of a number of premises and of inference rules sufficient for the deduction of the desired conclusion from the premises by means of them. What would correspond to the satisfaction of this condition in comparative theory choice? Here, above all, the successful applications of theory serve as premises, and scientific values sanction the inference to a conclusion, the preferential choice of one theory over the other. Let us examine these premises and inference "rules" a little more closely.

Despite the two theories' incommensurability, some of their empirical applications, those in which the incommensurable concepts don't occur, can be compared relatively unproblematically.[116] Such applications of theory are the premises shared by both parties; they include, for example, statements of the form: Where theory T_1 predicts value e_1 for quantity E, theory T_2 predicts value e_2 (or fails to make any prediction at all). Such statements are the unproblematic premises of comparative theory evaluation; if both parties assess quantitative accuracy in the same way, and if, for example, e_2 is, empirically, the more accurate value, then both parties will agree to take this as evidence for T_2. But there may also be problems that don't occur in T_2 at all, whose solution by T_1 is thus of no interest from the perspective of T_2. In addition, given incommensurability, there are, for example, applications of T_1 that can't be accurately articulated in the vocabulary of T_2. While, for someone seeking to defend T_2, such propositions aren't relevant premises for theory evaluation, the proponent of T_1 will indeed take them as relevant. So the set of propositions used as premises for theory evaluation by the proponent of T_1 won't be identical with the set of premises used by the proponent of T_2. There is thus no agreement on which premises should be used in theory evaluation.

The inference "rules" of comparative theory evaluation are also substantially different from those employed in mathematics or logic, for they don't sanction any absolutely compelling inferences. On the contrary, different individuals may, with respect to their shared prem-

115. *SSR,* pp. 94, 148, 150, 151, 152, 155, 158; 1970a, *ET* p. 280; 1970b, pp. 234, 260–261, 266; 1970c, *SSR,* pp. 198–199; 1971a, pp. 144–145; 1977c, *ET* p. 320; 1979b, p. 416.

116. Compare § 6.3.d.

ises, differ in their concrete theory evaluations, since individual shaping plays a role in their application of communal values.

When Kuhn expresses the dissimilarity between comparative theory evaluation and rigorous proof by claiming that theory choice "can never be unequivocally settled by logic and experiment alone,"[117] he by no means wishes to assert that logic and empirical findings are irrelevant to theory choice.[118] Of course logic and experiment play an important role, but they don't have the power to determine a decision rigorously.

If a theory cannot be subject to strict refutation, an adherent of the older theory may thus retain "the assurance that the older paradigm will ultimately solve all of its problems."[119] Resistance to theoretical innovation—a phenomenon very familiar from the history of science[120]—thus can't be interpreted simply as a human failing.[121] The same faith in the governing theory is rather what makes normal science possible, along with the thoroughness, depth, and accuracy of that practice and the unequivocal progress which results.[122] Insistence on the old theory also leads to the close inspection of its entire potential, so that the old theory is only given up if it really copes with anomalies less well than competing theories do.

But this interpretation of resistance to innovation shouldn't be taken to imply that insistence on an old theory is just as reasonable as the decision in favor of a new one. The set of empirical arguments in favor of the new theory may become so overwhelming that further insistence on the old theory can only be described as "stubborn," "pigheaded," or "unreasonable."[123] But since the arguments in favor of a new theory may gain force more or less continuously without ever becoming absolutely compelling, there is no specific "point at which resistance becomes illogical or unscientific."[124] It follows that, like unexpected discoveries, revolutions in which one theory is replaced by a

117. *SSR,* p. 94; similarly 1970a, *ET* p. 292; 1970b, p. 234; 1971a, p. 144.

118. 1970b, pp. 234, 261; also 1970a, *ET* p. 267.

119. *SSR,* p. 151.

120. *SSR,* pp. 150–151; 1963a, p. 348; 1974a, *ET* p. 304 n. 14; 1981, p. 15.

121. *SSR,* pp. 64–65, 151–152.

122. Compare § 5.6.

123. *SSR,* pp. 152, 159; 1970c, *SSR* p. 204.

124. The full passage is as follows: "Though the historian can always find men—Priestley, for instance—who were unreasonable to resist [the new theory] for as long as they did, he will not find a point at which resistance becomes illogical or unscientific" (*SSR,* p. 159, approvingly cited at 1970b, p. 260, and 1977c, *ET* p. 320; similarly 1971a, p. 145). This passage provides a good example of just how negligently Kuhn has often been read. We find Laudan asserting in his 1984, pp. 72–73 that Kuhn claimed that it was *perfectly reasonable* for Priestley to insist on the phlogiston theory. Cf. Hoyningen-Huene 1985 and, on this entire problem-complex, Worall 1990.

novel one aren't instantaneous events in which the scientific community changes theories in one fell swoop;

> Rather than a single group conversion, what occurs [in revolutions] is an increasing shift in the distribution of professional allegiances.[125]

c. Context of Discovery and Context of Justification

Kuhn's analysis of the reasons behind theory choice apparently stands in sharp contrast to that entailed by a distinction fundamental to pre-Kuhnian philosophy of science, the distinction between context of discovery and context of justification.[126] According to the context distinction, as I shall call it, while a diverse range of subjective factors is relevant to scientific development, their relevance involves only the actual conditions of discovery. As regards the justification of scientific knowledge, however, only such strictly intersubjective means as deductive or inductive logic and certain intersubjective "elementary observation sentences" are legitimate.

By *SSR,* Kuhn is well aware that his theory is incompatible with the then current context distinction.[127] But his explicit and implicit criticism of the distinction meets with marked opposition.[128] Kuhn later renews and broadens his attack on the context distinction, attempting to disarm his opponents.[129] His critique of the context distinction may be reconstructed as a four-phased attack.

1. In *SSR,* Kuhn begins his attack on the context distinction by noting that his attempts

> to apply [it], even *grosso modo,* to the actual situations in which knowledge is gained, accepted, and assimilated have made [the context distinction] seem extraordinarily problematic.[130]

The suggestion is that the context distinction "does not fit observations of scientific life,"[131] in which scientists facing a choice of theories may be seen to decide on the basis of individually shaped but communal scientific values.[132] It follows that individually variable factors play a

125. *SSR,* p. 158. —I will return to the notion of conversion in § 7.5.e.
126. See Hoyningen-Huene 1987b and 1987a for an analysis of the distinction.
127. *SSR,* pp. 8–9.
128. See especially Scheffler 1967, chap. 4; or Watanabe 1975, p. 126.
129. 1977c, *ET* pp. 325–330.
130. *SSR,* p. 9.
131. 1977c, *ET* pp. 327.
132. Compare § 7.4.b.

role even in the (comparative) justification of theories, and not just in the factual conditions of discovery, contrary claims on behalf of the context distinction notwithstanding.

But this stage of Kuhn's attack on the context distinction may be dismissed as somewhat beside the point. To be sure, scientists' decisions in theory-choice situations may actually work as Kuhn describes them. But if so, this fact is irrelevant to the soundness of the context distinction.[133] For what matters in the context of justification is whether decisions are *justified.* And this question can't be answered by recourse to the history of science, by the description of actual processes of discovery.

Still, even if this argument is conceded, the fact that actual science doesn't proceed in the manner deemed reasonable in philosophical reflections ought to give us pause. We find Kuhn asking, rhetorically, at the end of his introduction to *SSR,*

> How could history of science fail to be a source of phenomena to which theories about knowledge may legitimately be asked to apply?[134]

2. In the second stage of his attack on the context distinction, Kuhn assumes that the actual presence of individually variable factors in the theory choice situation has been conceded. Given this concession, one might ask

> whether or not the invocation of the distinction between contexts of discovery and of justification provides even a plausible and useful idealization.[135]

The context distinction would provide a "plausible and useful idealization" only if the individually variable factors in theory choice were "eliminable imperfections."[136] Kuhn denies that they are and traces the origin of this erroneous conviction to the following sources.[137]

One such source lies in the inadequacies of the old internal historiography of science.[138] First of all, the examples by which this historiographic tradition attempts to illustrate the strict intersubjectivity of reasons for theory choice, such as the "crucial experiment," usually aren't presented in a historically adequate way. The actual, historically

133. This line is taken, for example, by Siegel in his 1980, pp. 369–372, and 1980a, pp. 309–313. See Hoyningen-Huene 1987b, § 4.d, for a critique of Siegel's presentation.

134. *SSR,* p. 9.

135. 1977c, *ET* p. 327.

136. 1977c, *ET* p. 330.

137. 1977c, *ET* pp. 327–329.

138. Compare § 1.2.a.

effective arguments were, in part, shaped by individual factors. Furthermore, the arguments in favor of the theory which, in the end, was ruled inferior are usually left out of these later presentations. This omission results in an unrealistic simplification of actual theory choice situations.

There is a further, more subtle source for the erroneous assumption that theory choice criteria are strictly intersubjective. It begins with the observation that many theory-choice controversies have, in the end, an unequivocal outcome. But the intersubjectivity of the *reasons* behind such choices can't be inferred from the intersubjectivity of their outcomes. The fact that (more or less) all members of the community reach the same decision is entirely compatible with their doing so for different reasons.[139]

Once the influence of such factors is recognized, the insistence of Kuhn's critics that the context distinction, though not the most strictly observed norm in scientific practice, is still an idealized description of scientific behavior becomes intelligible. The persistent neglect by the philosophy of science of the role played by individually variable factors in theory choice is thus explained.

3. But, as Kuhn is well aware, the empirical demonstration of the presence of subjective factors in theory choice (point 1), together with a genetic explanation of the minimal attention paid them in the philosophy of science (point 2), isn't sufficient for an acceptance or rejection of the context distinction. What matters here is rather the *justified evaluation* of the presence of individually variable factors in theory choice. The question is thus whether "these facts of scientific life have *philosophic* import"[140] or whether such factors are "an index only of human weakness, not at all of the *nature of scientific knowledge*."[141] These factors have "philosophic import" just in case they aren't simply imperfections eliminable in principle but are rather something essential to scientific knowledge. What's at issue is an *evaluation* of the individually variable factors, a decision as to whether these factors *are allowed to* or perhaps even *should* participate in theory choice. Kuhn will answer this question affirmatively.[142]

But from the standpoint of the context distinction, the very question seems to indicate massive confusion, an impression Kuhn anticipates, in *SSR*, that many readers will be left with.[143] For, according to

139. Compare § 7.4.b, point 3
140. 1977c, *ET* p. 325, emphasis mine, similarly pp. 326–327.
141. 1977c, *ET* p. 326, emphasis mine.
142. See point 4 of this subsection.
143. *SSR*, p. 9.

this distinction, only intersubjective means may be employed in the context of justification, simply because what matters in this context is the *justification* (or critical testing) of knowledge *with claims to intersubjectivity*. Anyone who poses the above question appears not to understand what the justification of epistemic claims amounts to. In fact, Kuhn's (and other authors') critique of the context distinction has caused some to shrug in puzzlement. Herbert Feigl, for example, was "surprised" that such brilliant and knowledgeable scholars as N. R. Hanson, Thomas Kuhn, Michel Polanyi, Paul Feyerabend, Sigmund Koch, and others hold the distinction of being invalid or at least misleading.[144]

So in order even to be able to ask after the legitimacy or illegitimacy of individually variable factors in theory choice, Kuhn must first neutralize the opposing implications of the context distinction. Toward this end, his strategy is to expose the context distinction as *not philosophically neutral,* hence *needing justification.* And so he claims of this and similar distinctions,

> Rather than being elementary logical or methodological distinctions, which would thus be prior to the analysis of scientific knowledge, they now seem integral parts of a traditional set of substantive answers to the very questions upon which they have been deployed.[145]

In other words, the context distinction doesn't allow the differentiation of different perspectives from which to pose meaningful questions about scientific knowledge but is rather already part of a particular theory of knowledge. The distinction is thus not a condition for the possibility of epistemology and the philosophy of science but rather *part of a particular position in epistemology and the philosophy of science.* This position and, along with it, the context distinction must be subjected to the same strict tests as theories in other fields.[146]

Though Kuhn never explicitly carries out this testing program in *SSR* or in later work, he appears to be aiming at the following conclusions.

The context distinction, in the form familiar to analytic philosophy, goes back to Popper and Reichenbach.[147] And in fact it is not simply a philosophically neutral distinction between different perspectives from which to regard knowledge. In truth, the context distinction is an agglomeration (or identification) of at least four different distinc-

144. Feigl 1974, p. 2.
145. *SSR,* p. 9.
146. Ibid.
147. Popper 1934, chap. 1, § 2; Reichenbach 1938, pp. 6–7.

tions.[148] Of primary relevance to us is the context distinction's implicit identification of the *distinction between descriptive and normative* with the *distinction between empirical and logical.*

For our purposes, the distinction between descriptive and normative amounts to a distinction between two different (metascientific) perspectives from which to consider epistemic claims. On the one hand, epistemic claims may be described with an eye toward their genesis and other accompanying circumstances, while on the other they may be evaluated with an eye toward their justification.

Two points should be noted immediately. First of all, nothing in this distinction implies the strict separation of descriptive and normative. For example, the correct maintenance of the descriptive perspective requires that certain norms be followed, namely those which govern accurate description. The descriptive perspective may itself be directed at norms, as in attempts to describe them. But following or describing norms is still different from evaluating them. Second, the distinction between normative and descriptive perspectives does nothing to determine which means are appropriate to the description or evaluation of epistemic claims. The distinction simply amounts to the claim that either of the two perspectives may be adopted with respect to knowledge, and it leaves entirely open what, precisely, epistemic claims consist in or by what criteria they may be evaluated.

Now, the current context distinction identifies the distinction between descriptive and normative with the distinction between empirical and logical. This identification is evidenced by the fact that, according to the context distinction, the evaluation of theoretical epistemic claims has only the methods of formal logic at its disposal (with "elementary observation sentences" of some kind as premises), while only empirical methods are permissible in the procurement of descriptive facts. But, in consequence, a commitment to the current context distinction also involves a commitment to a certain basic position on the permissible means for the description, and especially the evaluation, of theoretical epistemic claims. This basic position is the common ground between the two most influential pre-Kuhnian schools in the philosophy of science, logical empiricism and critical rationalism.[149]

Kuhn's attack on the context distinction amounts to an attack on just this basic position. For Kuhn claims that formal methods are fundamentally insufficient to determine theory choice and that *individually*

148. On the following, see Hoyningen-Huene 1987b and 1987a.

149. The presence in the history of philosophy of alternatives to this basic position shows that the position is by no means self-evident. Consider, for example, Kant's transcendental logic, or Hegel's science of logic.

shaped communal values *ought* also to play a role (as they in fact do). His attack may thus not be *warded off* with the simple observation that it violates the context distinction, for Kuhn never attacks the seemingly philosophically neutral distinction between descriptive and normative. What's at issue is precisely the identification of this distinction with that between the empirical and the logical.

To recapitulate: Kuhn's project of justifying the presence of individual factors in theory choice is not, as it appears from the perspective of the context distinction, self-refuting and confused. Kuhn is rather attempting to formulate and argue a thesis which stands in opposition to a particular theory, implied by the context distinction, of the legitimate means for justifying epistemic claims. Whether Kuhn's thesis or the theory implied by the context distinction is right must be decided by argument. How, then, does Kuhn argue his normative claim that individually variable factors don't just participate factually in theory choice but also *ought* to participate?

4. The normative consequences of Kuhn's philosophy of science are fundamentally grounded in the following argument form:

> scientists behave in the following ways; these modes of behaviour have (here theory enters) the following essential functions; in the absence of an alternate mode *that would serve similar functions,* scientists should behave essentially as they do if their concern is to improve scientific knowledge.[150]

In order to apply this argument form to our case, we must ask after the essential functions of the individually variable factors in theory choice.[151] According to Kuhn, their function is as a "means of spreading the risk which the introduction or support of novelty always entails."[152] By thus spreading the risk, they are "assuring the long-term success of [the community's] enterprise."[153] The function they fulfill thus appears "indispensable."[154] If there were no individually variable factors to supplement communal values but rather just binding algorithms for theory choice, all scientists in a given theory choice situation would be forced to reach the same decision. The competition of theories would thus be ruled out, along with all scientific controversies, where these are understood as controversies between reasonable people. Scientific communities would unanimously opt for a new theory,

150. 1970b, p. 237, original emphasis; similarly 1970c, *SSR,* pp. 207–208.
151. Compare § 4.3.c.
152. 1977c, *ET* p. 332; similarly 1970b, pp. 241, 248–249, 262; 1970c, *SSR,* p. 186.
153. 1970c, *SSR,* p. 186.
154. 1977c, *ET* p. 332.

or for the old one, and would thus be in constant danger of failing adequately to develop the potential of one theory or the other.

Kuhn thus evaluates the individually variable factors in theory choice by comparing the consequences of their presence with those of their absence, or specifically, with a situation in which theory choice is instead determined by algorithms. The result of this evaluation is that, with reference to a higher-order goal, the further development of scientific knowledge, the presence of individually variable factors in theory choice is preferable (if not downright unavoidable).

5. Should Kuhn's attack on the context distinction thus be convincing? I suspect that some of the distinction's defenders won't be convinced. To be sure, one might object, scientists must, for pragmatic reasons, behave in theory choice situations just as they actually do, as Kuhn so convincingly demonstrates. But there's a difference between the justification of norms for *practically purposeful behavior,* on the one hand, and a *justification or critical test of claims to truth* on the other. As regards the latter, it's still not clear how individually variable factors could or ought to play a role, for veridical claims are, fundamentally, claims to *inter*subjectivity. And it's precisely on these grounds, the objection concludes, that the context distinction identifies the logical with the normative, with good reason.

This objection is obviously grounded in a different understanding of the notion of justification than that which guides Kuhn. Kuhn has in view the form of justification appealed to by scientists (defensibly, Kuhn thinks) in their decisions in theory choice situations: better correspondence with the facts, successful coping with anomalies, greater consistency, and the like. Our objection, by contrast, employs a standard of justification appropriate to mathematics and logic: the apodictic licensing of claims to truth, by means of shared premises and uncontroversial inference rules. Yet this form of justification is impossible to apply to theory choice in the empirical sciences.[155] But if this understanding of "justification" is presupposed, to ask whether a given community's choice of theories in a given historical situation was justified or not is to ask the wrong question. For under these conditions, no theory choice was ever justified, and science as a whole becomes an arbitrary enterprise devoid of any inner rationality. But Kuhn is far from branding science an irrational enterprise—on the contrary,

> I do not for a moment believe that science is an intrinsically irrational enterprise . . . I take this assertion not as a matter of fact,

155. Compare § 7.4.b, point 4.

> but rather of principle. Scientific behavior, taken as a whole, is the best example we have of rationality.[156]

And so if, from a certain perspective, in this case one which assumes a certain concept of justification, science appears irrational, this is simply an indicator that something is wrong with the perspective.[157]

7.5 The Discourse of Theory Choice

The distinctiveness of the reasons that figure in theory choice, along with the incommensurability of the theories under discussion, shapes the discourse carried on between proponents of competing theories in a very special way. The typical features of this discourse include its persuasive character (§ 7.5.a), a certain circularity in its arguments (§ 7.5.b), the partial character of communication (§ 7.5.c), the necessity of translation (§ 7.5.d), and finally the fact that the resulting theory choice is really a matter not of choice but of conversion (§ 7.5.e).

a. Persuasion

In considering how scientists are brought to attach themselves to a new theory, Kuhn talks, in *SSR,* of "the techniques of persuasive argumentation," or simply the "techniques of persuasion" he sees as playing a role in this process.[158] This expression has fueled the most grotesque misunderstandings of Kuhn's theory. These misunderstandings were further encouraged among Kuhn's German-speaking readers by the frequent rendering, in German translations, of "persuasion" and "persuasive arguments" by "*Überredung*" and "*überredende Argumentation,*" respectively.[159] When we consider the context in which "persuasion" occurs in Kuhn's work, it becomes clear that "*Überredung,*" which suggests the absence of any good reasons, hence an arbitrary or haphazard nature, is inappropriate. Neither is this misreading of "persuasion" limited to the German-speaking world.

It's not as if Kuhn introduces his use of "persuasion" in *SSR* without any explanation. On the contrary, immediately after introducing

156. 1971a, pp. 143–144.

157. 1970b, pp. 235, 264; 1971a, p. 144; 1976b, p. 196.

158. *SSR,* pp. 93, 152.

159. Examples of this rendering may be found in the German edition of *SSR,* pp. 106, 163 (twice); in the German edition of 1970b, pp. 252, 253 (twice); in the German translation of the postscript to *SSR* (1970c), German edition pp. 210, 211; and in the German translation of 1977c, German edition of *ET* p. 421.

the term, he makes it clear what's meant by talking about "techniques of persuasion, or about argument and counterargument in a situation in which there can be no proof."[160] The issue here is explicitly one of "argument and counterargument," and accordingly, in later work, Kuhn has every right to insist that

> To name persuasion as the scientist's recourse is not to suggest that there are not many good reasons for choosing one theory rather than another.[161]

Kuhn's notion of persuasion has very often been understood as implying the absence of any good reasons for theory choice.[162] There can be no question of the inappropriateness of this reading.[163]

"Persuasion" thus refers to a form of discourse which, though argumentative, can't be likened to a form of proof. It lacks the requisite number of both common premises and sufficiently strong shared inference rules needed for proof.[164]

b. Circularity

Closely associated with its status as a form of persuasion, theory choice discourse typically manifests certain circularities.[165] These circularities rose to the surface in our discussion of the argumentative weight of various reasons for theory choice.[166] At the heart of this circularity is the fact that the importance of certain problems and the legitimacy of certain solutions can't be evaluated in isolation from the theories under discussion.[167] A problem that may appear extremely important from the standpoint of one theory may, from the standpoint of another, appear to be of subordinate importance, perhaps even a pseudoprob-

160. *SSR,* p. 152 (also cited approvingly at 1970b, p. 260, and 1977c, *ET* p. 320); a few lines below the cited passage, we once again find the opposition of "persuasion" and "proof." —Kuhn is in error, however, when he assumes that the inquiry into the techniques of persuasion "is a new one, demanding a sort of study that has not previously been undertaken" (*SSR,* p. 152). For Artistotle's *Topics* and *Rhetoric* address these problems, as do contemporary rhetorical studies grounded in Aristotle's work, which renew this tradition; see Perelman and Olbrechts-Tyteka 1958.

161. 1970b, p. 261, similarly pp. 234, 235; 1970c, *SSR,* pp. 199, 204; 1974b, p. 509; 1977c, *ET* pp. 320–321.

162. E.g. Mittelstraß 1988, p. 186; Scheffler 1967, p. 81; Toulmin 1970, p. 44; van der Veken 1983, pp. 43–44.

163. Compare §§ 7.4.b and 7.4.c.

164. Compare § 7.4.b, point 4.

165. *SSR,* pp. 94, 109–110, 148.

166. See § 7.4.b, point 4.

167. Compare § 6.3.a, point 1.

lem. The analogous claim holds of problem solutions. Accordingly, concrete problem solutions may be quite differently evaluated by two parties holding two differing theories, as may the problem-solving capacities of their respective theories; "Each group uses its own paradigm to argue in that paradigm's defense."[168]

But such circularity doesn't make the arguments in favor of a theory unsound, let alone ineffective.[169] For by such means a proponent of a given theory may, in fact, demonstrate what it means to conduct research on the basis of the regulations derived from this theory. To be sure, such demonstration presupposes that its audience is willing and able to entertain the theory, along with the corresponding phenomenal world.

It is easy to see that such willingness to entertain a new theory is found especially among young people, who aren't yet committed to conducting science in either one of the two candidate ways.[170] By contrast with those familiar with the old theory, they lack the almost irresistible tendency to see the new theory in light of the old, from which perspective the new theory's potential must appear inferior. Novices in any field are thus more easily convinced of the fruitfulness of a new theory, even if this fruitfulness is problematic from the perspective of the old theory. Consequently, young people are overrepresented both among the pioneering innovators and among the very first adherents of new theories.[171]

c. Partial Communication

A further characteristic of theory-choice discourse closely associated with the previous two has also led to numerous misunderstandings: "Communication across the revolutionary divide is inevitably partial."[172] As regards the thrust of this thesis, two points must be emphasized. First, the issue in this passage of *SSR* is explicitly the *partial* nature of comunication between representatives of different theories, not the complete breakdown of communication some of Kuhn's critics sought to attribute to him.[173] Instead, misunderstandings crop

168. *SSR,* p. 94.

169. Ibid.

170. 1961a, *ET* p. 208, n. 44; *SSR,* p. 151; 1970c, *SSR,* p. 203.

171. Compare § 7.3.b.

172. *SSR,* p. 149, similarly (but indirectly) p. 150; 1970b, pp. 231–232, 250, 267, 276–277; 1970c, *SSR,* pp. 198–199, 201; 1976b, p. 190; *ET* pp. xxii–xxiii; 1977c, *ET* pp. 338–339; 1983b, p. 713; 1989b, p. 50.

173. E. g. Koertge 1983, p. 97; Newton-Smith 1981, p. 12; Popper 1970, pp. 55, 57; Scheffler 1967, pp. 16, 19; Toulmin 1970, pp. 43–44. —Still, one may find passages in

up,[174] opponents "talk through each other,"[175] and "logical contact" between parties to the discourse is incomplete.[176]

In addition, the above passage's description of communication across the revolutionary divide as "*inevitably* partial" requires interpretation. Does this mean that the obstacles to communication can't, in principle, be overcome? Kuhn is by no means of this opinion. Problems in communication are rather inevitable *if and for so long as* both parties to the conversation continue to argue from incommensurable standpoints and fail partly or entirely to understand the standpoint of their counterparts. But such obstacles to communication can, in principle, be overcome.[177]

What sorts of obstacles to comunication arise out of the incommensurability of the standpoints involved? To begin with, a difference in problem fields and standards for solution may pose difficulties for communication when it comes to the assessment of the problem-solving capacities of the two theories, since the arguments for one or the other theory's superiority are, in part, circular.[178] Further obstacles to communication also arise out of meaning change, which is so characteristic of revolutions, and out of the attending alteration of that knowledge of nature implicitly contained in the changed concepts.[179] If the parties to communication fail to take note of meaning change, misunderstandings naturally follow. But both parties will first have trouble localizing these misunderstandings, then diagnosing and finally curing them, for the following reasons:

First, many of the changed concepts retain the same names as before.[180] In such cases, changes in the meanings of concepts aren't marked by the employment of new words. And so parties to communication will initially attribute their own respective understandings of such concepts to their counterparts, erroneously.

Second, the misunderstandings which result from meaning change can't be eliminated by recourse to the shared definitions of problematic

---

SSR which, taken in isolation, appear to suggest the complete breakdown of communication. For example, Kuhn asserts of Berthollet and Proust that they "were as fundamentally at cross-purposes as Galileo and Artistole" (*SSR*, p. 132).

174. "[to be] slightly at cross-purposes": *SSR,* pp. 112, 148, and 149. In the latter passage, though, Kuhn asserts that the term "misunderstanding" isn't quite right.

175. *SSR,* pp. 109, 132, 148.

176. *SSR,* pp. 110, 148.

177. On this issue, see the following subsection, 7.5.d.

178. *SSR,* pp. 109–110; compare § 7.5.b.

179. Compare §§ 6.3.a, point 2, 6.3.b, and 6.3.c.

180. E.g. 1970b, pp. 266–267, 269, 275; 1970c, *SSR,* p. 198; 1977c, *ET* p. 338; 1989a, p. 31.

concepts.[181] For, to begin with, there is no language in which such explications might be articulated which is neutral with respect to both conceptual systems and thus intelligible to both parties; this is assured by the incommensurability thesis. Furthermore, there are in general no truly adequate explicit definitions of empirical concepts, even within the languages employed by the relevant parties.[182]

And third, the knowledge implicit in each conceptual system is, at the very least, extremely difficult to explicate accurately.[183] First of all, such knowledge is more or less taken for granted by the user of a given language, and it forms an integral, if hardly conscious, portion of his or her world view. In addition, bringing such knowledge to consciousness calls for decisions regarding its precise scope and other claims, decisions that never had to be made before. Finally, the knowledge implicitly contained in language vacillates between analytic and synthetic status, yet another reason why furnishing one's opponent with a reasonably convincing justification of such epistemic claims proves problematic.

So if, as should by no means be taken for granted, the proponents of different theories want to get beyond their merely partial communication, they must somehow attempt to understand and appropriate each other's languages. The problems which arise in this process may be understood, in a first step, as problems of translation.

d. Translation

We begin by noting that the concept of translation relevant to our current purposes isn't the narrow, technical concept, on which translation involves the systematic replacement of words or word groups in the source language by words or word groups in the target language.[184] For incommensurable theories, this kind of translation is, by definition, impossible. What matters here is rather the everyday notion of translation, the kind of translation in which an interpretive moment always participates and in which the target language is more or less subtly changed over the course of translation.

The participating parties must begin by localizing the concept groups and manners of speaking in which obstacles to communication arise.[185] I move directly to concept groups because, in general, it isn't

181. 1970b, p. 276; 1970c, *SSR*, p. 201.
182. Compare § 3.6.f.
183. Compare § 3.7.b.
184. Compare § 6.3.c.
185. On the following, see 1970b, pp. 267–270, 276–277; 1970c, *SSR*, pp. 175,

individual concepts that cause problems; a revolution changes the network of similarity relations responsible for the local holism of language.[186] Once both parties know which concepts are problematic, they may call upon their shared everyday and scientific vocabulary for help in explaining them. In particular, they may demonstrate the situations in which they employ these concepts to their counterparts. To be sure, learning currently unintelligible concepts is made more difficult by the local holism of language, since some concepts can only be understood together. But in time, the members of one group will be able to translate (in the everyday sense of "translate") portions of their counterparts' theory, along with many of its empirical consequences, into their own language. Such translations may be very complicated, involve a strong interpretive element, and change the target language by introducing new concepts and more or less subtly changing the old. But they permit access to further empirical results, to the different sorts of explanation sanctioned by the other theory, and, more generally, to the alien phenomenal world. The comparison of theories thus acquires a broader basis than that provided by those concrete empirical results communicable without recourse to translation between theories.[187]

e. Conversion, Not Choice

For someone who finds the arguments in favor of the new theory convincing but still only uses the new theory by translating it into the vocabulary of the old, the individual transition to the new theory is not yet complete. What's still missing is the unmediated employment of the new theory, without the bridge of translation—employment like that of one's own native language. In his work since 1969, Kuhn has characterized this final transition by means of the notion of a conversion.[188] Conversion can be neither accomplished nor refused voluntarily; instead, it simply happens over time—or fails to happen, as in the most advanced stages of learning a foreign language. Kuhn also asserts of the transition to the unmediated employment of the new

202–204; 1974b, p. 505; 1976b, p. 191; *ET*, pp. xxii–xxiii; 1977c, *ET* pp. 338–339; 1983a, pp. 671–673; 1983b, p. 713.

186. Compare §§ 3.6.c, 3.6.e, and 3.6.g.

187. Compare § 6.3.d.

188. 1970b, p. 277; 1970c, *SSR*, pp. 198, 202–204; 1977c, *ET* pp. 338–339. —In his 1970c, Kuhn explicitly asserts that his understanding of conversion has become more precise since the composition of *SSR*. "The two experiences [persuasion and conversion] are not the same, an important distinction that I have only recently fully recognized" (1970c, *SSR* p. 203). I will shortly return to Kuhn's use of "conversion" in *SSR*.

theory that the notion of theory *choice* (or the notion of a decision) is inapplicable.[189] For these notions imply a voluntary moment inappropriate for the phenomenon in question.

Yet Kuhn uses the notion of conversion much less specifically in *SSR*. Here "conversion" refers to the transition from one way of seeing science and the world to another, or to the entire process of (individually) experiencing a scientific revolution.[190] This more general use of "conversion," together with Kuhn's characterizations of conversions in *SSR,* have led to many misunderstandings. We find Kuhn claiming, for example, that a conversion experience "cannot be made a step at a time, forced by logic and neutral experience."[191] And yet, at the same time, we find that in regard to

> personal and inarticulate aesthetic considerations . . . Men have been converted by them at times when most of the articulable technical arguments pointed the other way.[192]

Such characterizations appear to place scientific revolutions quite close to instantaneous religious conversions, in which rational control is largely or completely absent.[193] In the German-speaking world, the apparently quasi-religious character of scientific revolutions became almost unavoidable with the translation of "conversion," or "to convert" by "*Bekehrung*" and "*bekehren,*" respectively.[194] But this proximity to the cliché of religious conversion was never really part of Kuhn's theory; instead, as explained above, revolutionary change is brought on by identifiable reasons.[195]

7.6 Scientific Progress through Revolutions

On the traditional view, science is usually ascribed cumulative epistemic progress.[196] Kuhn by no means entirely denies the presence in

189. 1970b, pp. 238, 277; 1970c, *SSR* p. 204; 1977c, *ET* pp. 338–339.

190. *SSR,* pp. 144, 148, 150 (twice), 151 (twice), 152 (four times), 153, 155, 158 (twice), 159 (twice). —Conant also uses "conversion" in this way. In discussing Priestley's resistance to oxygen, he writes, for example, "But Priestley died in 1804 without ever being converted to the new doctrine" (Conant 1947, p. 80).

191. *SSR,* p. 150.

192. *SSR,* p. 158.

193. See e.g. Shapere 1971, p. 707, and 1984a, p. 162; Vollmer 1988, p. 205.

194. In the German edition of *SSR,* pp. 161, 169; the German translation of 1970b, pp. 268, 269; the German translation of 1970c, in *SSR,* German edition, p. 214; and the German edition of 1977c, in *ET* German edition pp. 443, 444.

195. Compare §§ 7.4.b and 7.4.c.

196. Compare § 1.2.a.

science of this kind of progress, though he confines it to the phase of normal science.[197] Scientific revolutions, however, are events in which there is no mere cumulation of knowledge; they are constructive-destructive.[198] In what sort of progress, if any, do scientific revolutions result? In § 7.6.a, we will discuss Kuhn's basic conception of revolutionary epistemic progress, a conception which will be qualified in §§ 7.6.b–7.6.d.

a. Scientific Progress as Increasing Problem-solving Capacity

Following unexpected discoveries, the theories accepted in the practice of normal science are somewhat modified.[199] Subsequent to this modification, scientists may confront a larger domain of problems or treat familiar problems with improved accuracy.[200]

Any discussion of progress through revolutions in which new theories triumph must begin with the reasons which weigh in the decision in favor of a new theory.[201] For the manner in which a particular field develops over the course of revolutions in theory depends on the nature of these reasons. A scientific community's decision in favor of the new theory is based on scientific values and especially on arguments for the superior problem-solving capacity of the new theory.[202] The new theory must be able both to solve a large portion of the problems dispatched by means of the old theory and to cope with the anomalies which brought on the crisis state.

The progress of some thing involves its positively valued gain in some respect important to it.[203] Both unexpected discoveries and revolutions in theory thus result in progress, as problem-solving capacity is both positively valued in and important to science, and it is this problem-solving capacity which increases. In this sense, "scientific development is, like biological, a unidirectional and irreversible process."[204]

This doctrine allows Kuhn to reject indiscriminate charges of rela-

197. Compare § 5.4.

198. Compare § 6.1.

199. Compare § 7.2.

200. *SSR*, p. 66.

201. On the following, see *SSR*, pp. 8, 167–170; 1970a, *ET* pp. 289–290; 1970b, p. 264; 1970c, *SSR*, pp. 205–206; 1971b, *ET* p. 30; 1979b, p. 418.

202. Compare § 7.4.b.

203. Compare § 5.4.

204. 1970c, *SSR*, p. 206; similarly *SSR*, pp. 146, 172; 1970b, p. 264; 1974b, p. 508. —The parallel to biological evolution extends to a further dimension, which I will discuss in § 7.6.d.

tivism.[205] He is especially justified in rejecting the charge that his theory falls prey to the kind of relativism on which successive scientific theories don't differ in quality. But this characterization of the scientific progress attained through revolutions in theory calls for three important qualifications.

b. First Qualification: "Losses" in Revolutions

Kuhn often emphasizes the fact that along with a revolution—and the associated gain in problem-solving capacity—generally come certain losses. Among these are losses in the ability to explain certain phenomena whose authenticity continues to be recognized, losses of scientific problems or the narrowing of the field of research, and, relatedly, increased specialization and increased difficulty in communicating with outsiders.[206] And so, for Kuhn, the progress which comes with a revolution appears to have been bought at the price of a certain recession, albeit one quickly forgotten along with the articles and textbooks in which the conquered theory, in its historical form, is contained.[207]

But an exposition of the losses attendant on scientific revolutions requires some differentiation. For Kuhn talks not simply of the "relinquishment" of explanatory power, problems, and such but rather of their "loss," where "loss" implies a negatively valued relinquishment. Yet the evaluation of the relinquishment of explanatory power, problems, and such in a scientific revolution isn't necessarily unequivocal. This evaluation can (though it doesn't have to) depend on whether it's conducted from the perspective of the conquered theory or that of the revolutionary theory. And so the relinquishment of certain problems can be seen as the emancipation from pseudoproblems and the relinquishment of explanations as a liberation from pseudoexplanations and thus both may be positively evaluated.

The examples by which Kuhn illustrates revolutionary "losses" strongly suggest the possibility of such differential evaluation. We find him claiming, for example, that

205. 1970b, pp. 234, 264–265; 1970c, *SSR,* pp. 175, 205–207; 1974b, p. 508; 1989a, p. 23 n. 24; in 1990, p. 317 n. 21. —The charge of relativism is raised, for example, by Popper 1970, pp. 55–56; Shapere 1964, pp. 392–393, 1966, pp. 66–69, 1971, pp. 708, 709, 1974, p. 507, 1977, pp. 184–185, 200, 1984, pp. xvi–xvii, xx, 1984a, pp. 162–165; Siegel 1980, pp. 368–369. On this entire problem-complex, see especially Doppelt 1978, Holcomb 1987, Mandelbaum 1979, Siegel 1987, as well as the other papers reprinted in Meiland and Krausz 1982.

206. 1961a, *ET* pp. 211–213; *SSR,* pp. 103–109, 148–149, 167, 169, 170; 1970a, *ET* p. 289; 1976b, p. 192.

207. 1959a, *ET* p. 230; *SSR,* pp. ix, 140, 144, 167.

> the seventeenth century's new commitment to mechanico-corpuscular explanation proved immensely fruitful for a number of sciences, *ridding them of problems that had defied generally accepted solution.*[208]

In this case, from the viewpoint of the "new science," it would be inappropriate to describe the relinquishment of such problems as a loss. The relinquishment of explanations attained under a theory now overcome may behave similarly. For Newtonians from the middle of the eighteenth century on, the relinquishment of any ability to explain gravitation, which Cartesians had retained, was no loss (though it was for Newton himself, and for his immediate successors).[209] Instead, the relinquishment of all explanations of gravitation was an immediate consequence of its status as one of the "physically irreducible primary properties of matter."[210]

Regardless of how the victorious community evaluates its relinquishment of problems and explanations, the revolution will, from its perspective, represent progress—just as normal science perforce exhibits its progress for those who practice it.[211] For with the accomplished fact of its choice, based on scientific values, the scientific community bears witness to its preferential evaluation of the new theory, thus committing itself to the view that its decision is progressive.

c. Second Qualification: The Devaluation of the Loser's Perspective

To be sure, the revolution will be differently evaluated from the perspective of the adherents of a theory abandoned by most community members. Sensitivity to their perspective isn't without consequence, in part because, in many revolutions in theory, some scientists, especially the elderly, never defect to the new theory,[212] in part because their counterarguments against the new theory are never compellingly refuted.[213] Just as the adherents of the new theory are compelled to see the revolution as progressive, these scientists see it as a step backward. For one, they negatively evaluate the relinquishment of problems and explanations discussed above, assessing it as a real loss. In addition, they may find the kind of explanation associated with the new theory generally unacceptable. In the history of physics it has repeatedly hap-

208. *SSR,* p. 104, emphasis mine.
209. 1961a, *ET* pp. 211–212; *SSR,* pp. 103–106, 108, 148.
210. *SSR,* p. 106.
211. *SSR,* pp. 166–167; compare § 5.4.
212. *SSR,* pp. 150–152, 159; 1963a, p. 348; 1970b, p. 260; 1977c, *ET* p. 320.
213. Compare § 7.4.b, point 4.

pened that proponents of the old theory deny any explanatory power to the new one, though they concede its markedly improved predictive success.[214]

But regardless of how negatively the new theory may be evaluated from the perspective of adherents of the old, the victory of the new theory leads to the disappearance of this perspective. For succeeding generations of scientists attach themselves to the new theory, and the proponents of the old, in time, die out. Consequently, only the perspective of the victorious party is preserved, and in this respect—*and only in this respect*

> the member of a mature scientific community is, like the typical character of Orwell's *1984,* the victim of a history rewritten by the powers that be.[215]

But in the exposition immediately following this passage, Kuhn explains why an expansion of the set of solved problems as well as an increase in the accuracy of solutions is practically guaranteed by revolution, given the application of scientific values.[216] "Solved problems" here evidently refers primarily to theoretical predictions of empirical data—in other words, to that which Kuhn also describes as the institutional goal of normal science.[217] As far as these solved problems are concerned, the progress of science is *objective,* or independent of the different possible perspectives.

d. Third Qualification: No "Drawing Closer to the Truth"

But, says Kuhn, in a certain sense scientific development doesn't exhibit any progress, namely where progress is taken as a drawing closer to the truth. This amounts to the rejection of Peircean realism.[218] In *SSR,* Kuhn claims only that it's not necessary to conceive of scientific development as a drawing closer to the truth; once this conception is abandoned, he asserts, a variety of problems disappear.[219] Like phylogenesis, as presented by Darwinian evolutionary theory, scientific development isn't a process tending toward some fixed goal set in advance. Instead it's a development in which the specialization and articulation of scientific knowledge increase.

214. 1971b, *ET* pp. 28–29.
215. *SSR,* p. 167.
216. *SSR,* pp. 167–170.
217. Compare § 5.2.d, point 2.
218. Compare §§ 2.2.e, 3.8.
219. *SSR,* pp. 170–173; also see 1974b, p. 508; 1984, p. 244; 1989a, pp. 23–24, 31–32.

In later work, however, Kuhn offers two arguments supporting the claim that the conception of scientific development as a drawing closer to the truth is not only just unnecessary but actually untenable. These arguments are directed, above all, against the Popperian school in the philosophy of science, but the conception at issue is quite common among philosophers (especially in the English-speaking world) and scientists.[220]

The first argument is historical.[221] The claim that successive theories draw closer to the truth primarily asserts that the ontologies of later theories are better approximations of that which exists absolutely—the purely object-sided—than those of earlier theories. In Kuhn's own words,

> One often hears that successive theories grow ever closer to, or approximate more and more closely to, the truth. Apparently generalizations like that refer not to the puzzle-solutions and the concrete predictions derived from a theory but rather to its ontology, to the match, that is, between the entities with which the theory populates nature and what is "really there."[222]

Claims of increasing proximity to the truth thus, first of all, are *distinct* from the claim that problem-solving capacity increases over successive theories, and, second of all, they can't be *justified* by reference to this increasing problem-solving capacity alone.[223] But according to Kuhn, when we consider an actual historical succession of theories, such as the series composed of Aristotelian physics, Newtonian mechanics, and Einsteinian general relativity, we find no hint of any ontological convergence. On the contrary, in a certain sense the ontology of relativity theory is closer to that of Aristotle than to that of Newton. As far as the theories themselves are concerned, since there is no indication of ontological convergence, there is no evidence for any drawing closer to the truth.

The second argument is epistemological;[224] it proceeds from the assumption that it's essentially meaningless to talk of what there really

220. In his 1979b, for example, Kuhn argues against Boyd 1979. —See Shimony 1976 for a critical encounter with Kuhn's arguments; see Laudan 1984, chap. 5, for a detailed discussion of "convergent epistemological realism," as Laudan calls Peircean realism.

221. 1970a, *ET* pp. 288–289; 1970b, pp. 264–265; 1970c, *SSR* pp. 206–207; 1979b, pp. 417–418.

222. 1970c, *SSR* p. 206; also 1970b, p. 265.

223. For an improved problem-solving capacity would be, at most, a *consequence* of the later theory's closer proximity to the truth.

224. 1970c, *SSR* pp. 206; also 1970b, pp. 265–266 and 1979b, p. 417.

is, beyond (or outside) of all theory. If this insight is correct, it's impossible to see how talk of a "match" between theories and absolute, or theory-free, purely object-sided reality could have any discernable meaning. How could the (qualitative) assertion of a match, or the (comparative) assertion of a better match, be assessed? The two pieces asserted to match each other more or less would have to be accessible independently of one another, where one of the pieces is absolute reality. But if we had access to absolute reality—and here we can only return to our initial premise—what interest would we have in *theories* about it?

Scientific progress thus must be interpreted not as a progressive approximation of the truth but rather only as an improvement, in the instrumental sense, of scientific knowledge;

> Conceived as a set of instruments for solving technical puzzles in selected areas, science clearly gains in precision and scope with the passage of time. As an instrument, science undoubtedly does progress.[225]

225. 1979b, p. 418; similarly *SSR,* p. 173; 1970c, *SSR,* p. 206.

SUMMARY OF PART III

There are two alternating phases in the development of the mature sciences: normal and revolutionary scientific development. Normal science is carried by a broad-based consensus in the relevant scientific community, knowledge about a particular phenomenal world grows cumulatively, and research practice exhibits marked similarities with puzzle-solving. But normal science repeatedly produces anomalies, anomalies which compel more or less thoroughgoing revisions of its guiding regulations. These destructive-constructive episodes in scientific development are Kuhnian scientific revolutions. Revolutions vary substantially in their extent. Unexpected discoveries may sometimes be assimilated by small changes in (often merely implicit) expectations, without necessitating any changes in the formal part of the accepted theory. In greater revolutions, theories are entirely discarded and new theories take their place.

The alteration of world-constitutive similarity relations is characteristic of revolutions. Such alterations simultaneously involve changes in the phenomenal world, changes in the empirical concepts with which the world is described, and changes in the knowledge implicitly contained in these concepts. These changes, treated together under the rubric of incommensurability, make revolutions complex processes in which continuity and discontinuity are knotted together. On Kuhn's view, revolutions by no means proceed irrationally. On the contrary, there are phases of reasonable, justified difference of opinion which ultimately, following a closer examination of the relevant field from different standpoints, make way, under the pressure of arguments, for a new consensus, hence a new phase of normal science. Revolutions result in scientific progress in the affected disciplines, progress in the accuracy and number of problem solutions attained. However, the conception on which scientific knowledge progressively draws closer, in an ontological sense, to "the truth" must be abandoned.

EPILOGUE

Reality as Understood by Kuhn's Philosophy of Science

IN CONCLUSION, we must inquire into the understanding of reality presupposed by or following from Kuhn's philosophy of science. The fundamental point of departure for Kuhn's understanding of reality is the distinction between world-in-itself and phenomenal world.[1] According to Kuhn, reality, as it is generally referred to in everyday and scientific contexts, is *a* phenomenal world, not *the* (only possible) phenomenal world, and certainly not the world-in-itself. A phenomenal world contains genetically subject-sided moments, its appearance from the "natural" standpoint notwithstanding. But it's by no means a whimsical construction, an arbitrary invention of consciousness. Instead, on Kuhn's understanding, genetically object-sided moments also enter into every phenomenal world. Before I further discuss these characteristics of phenomenal worlds, the notion of the world-in-itself deserves further attention.

The concept of the world-in-itself is the product of a conceptual process of subtraction, the subtraction from a phenomenal world of all genetically subject-sided moments. Four points deserve special attention here. This process of subtraction is, first of all, *possible* in the sense that, since phenomenal worlds contain genetically object-sided moments in addition to their genetically subject-sided moments, there will be something left after subtraction. In addition, the result of this process of subtraction is *unequivocal* in the sense that, regardless of which phenomenal world it starts out with, the same world-in-itself results.[2] Furthermore, we *can't say very much* about the product of the

1. Though Kuhn sometimes holds the concept of a world-in-itself to be eliminable; cf. § 2.3. —On the following, see above all §§ 2.2, 3.2, and 7.1.

2. This claim must presumably be formulated somewhat more precisely in order to avoid the obvious objections; the object-sided dimension of a phenomenal world in which, for geographical reasons, there are no kangaroos (and in which little travel takes place), must surely be distinct from that of a phenomenal world in which there are kangaroos.

subtraction process, for the world-in-itself is accessible to neither mundane nor scientific channels. We may only infer some of its most general characteristics on the basis of the function we've assigned it, that of discernibly contributing to the constitution of phenomenal worlds. Finally, the theoretical efficacy of this postulated world is, at least within the bounds of a theory like Kuhn's, highly *questionable.*

But however things stand with regard to the world-in-itself, it remains of subordinate importance to the issue of Kuhn's theory's understanding of reality. For reality is always a phenomenal world, whether or not a theoretically useful world-in-itself may be gleaned from it by subtraction. Kuhn's theory thus appears to be an idealist one, or at least one with strong idealist leanings; for a phenomenal world also contains genetically subject-sided moments.[3] But this characterization—which in some circles amounts to utter derision—calls for more precise qualification, as its appropriateness stands or falls depending on what we understand by "idealism."

The following caricature of idealism is widespread. On this caricature, idealism claims that reality is, in all of its aspects, the product of consciousness through and through, where "consciousness" is understood as individual consciousness. Nothing truly distinguishes reality from the things which occur in dreams; as regards both their material existence and all their properties, they are entirely at the whim of the subject who imagines them. Reality, in a sense, is the invention of consciousness; it is the object of a special kind of dream.

But this kind of idealism isn't at all one that applies to the understanding of reality in Kuhn's theory. I shall elaborate on three points of difference. First of all, for Kuhn, reality, that is, a particular phenomenal world, is indeed object-sided, independent of all influence by subjects, in its *substantiality.* It's by no means the case that the substance of real objects is the same as that of a winged horse consciously imagined or dreamed up by some individual. As their substance is purely subject-sided, such objects vanish completely, without remainder, once they cease being imagined. By contrast, for Kuhn, a phenomenal world is a particular reshaping of the world-in-itself, of that which is in itself. Kuhn's is thus hardly an immaterialist position. So the world's substantiality isn't, in opposition to the popular version of idealism, entirely at the whim of the individual representing subject. Neither are the world's *properties.* But, though Kuhn rejects the view on which all of the world's properties are placed on the side of the subject, he doesn't endorse the position at the opposite extreme either. Such a position

3. Scheffler, for example, call's Kuhn's theory "extravagant idealism"; Scheffler 1967, p. 19; Loeck refers to its "historicist idealism"; Loeck 1987, p. 206.

would be a version of realism on which the properties of reality aren't at the whim of the subject because they attach to the world-in-itself. According to this version of realism, the properties of reality are, in origin, entirely object-sided; they are given of whole cloth to epistemic subjects and aren't open to dispute. Kuhn's position is located somewhere between these coarse, popular forms of idealism and realism, respectively.

In order to situate his position more precisely, let us consider the network of similarity and dissimilarity relations which serves as the foundation both of the organization of a phenomenal world and of the language geared toward this phenomenal world. A network of similarity relations of this kind is the product of a historical process, that of the development of a given language, together with the development of the particular understanding of reality implicit in this language. This historical process generally has no driving agent, though it does have a reference point in its corresponding linguistic community. Accordingly, the network of similarity and dissimilarity relations is, at most, the "property" of this community, and, as such, isn't open to free dispute by individuals. Just as an individual can't *change* grammatical rules—though, to be sure, the individual might, by systematic violation of these rules, *leave* the linguistic community—neither can any individual change the structuring of the world inherent in a scientific or an everyday language. An individual might, of course, *lend impetus toward* linguistic change, but the success of such change in the appropriate community is an essentially social process. And so the element of idealism in Kuhn's understanding of reality, whatever it turns out to be, is of a *social,* not an *individual,* nature.

Third, given all that's been said so far, Kuhn's understanding of reality might still, in some rough sense, be idealistic, if it allowed the particular ordering of any phenomenal world to be entirely independent of all object-sided influence. Though this wouldn't be *individual* idealism, it would still be a rough form of idealism to the extent that it failed either to admit a world-in-itself, or properties attached to the world-in-itself, or any capacity on behalf of the world-in-itself or its properties to affect our actual understanding of reality.

But according to Kuhn, any net of similarity and dissimilarity relations is *also,* in part, determined by genetically object-sided influences. For on Kuhn's conception, the world-in-itself offers resistance, resistance which makes it impossible to impose just any network of similarity relations. To be sure, such resistance isn't of the kind which would *uniquely* determine which network of similarity relations will be imposed; for we recall that *all* of the conceptual systems accepted at one point or another in the history of science were successful, in a given domain of phenomena, for a certain amount of time. The proprietary resistance of the world-in-itself makes itself felt more indirectly, not by unique determination. Its presence in scientific development reveals itself especially in the following situations.

First of all, it reveals itself when, over the course of the community-sanctioned research activity of normal science, significant anomalies appear, situations in which something happens which violates the expectations permitted by the network of similarity relations. In this case the consensus which reigned up to this point is shattered in a manner not explicable by social causes alone. For those forms of resistance which, though experienced in the perception and thought of many individuals, result from purely *social* circumstances may be a prerequisite of uniformity, but they can't destroy uniformity. Consequently, the proprietary resistance of the world-in-itself must be a participant in the production of significant anomalies.

In addition, the proprietary resistance of the world-in-itself reveals itself with particular clarity when, in an effort to cope with anomalies, an attempt is made to modify a previously successful network of similarity and dissimilarity relations in such a way as both to preserve the classifications worth preserving *and* to transform anomalies into the scientifically expected. The demands this effort places on revolutionary science are extraordinarily difficult to meet. But scientists don't experience this difficulty only as the difficulty involved in seeking the social acceptance of one's own opinion. In addition to the problem of defending favored candidate solutions socially, there is also the problem of finding solutions which meet one's own standards.[4] In solving this problem, one must grasp the world's unexpected behavior conceptually, where the socially accepted regulations which reigned up to now no longer offer any (more or less) unequivocal guidance toward this end. This situation, therefore, can't be understood as a confrontation between the individual and purely social constraints. Instead, the individual must cope with something which, to all appearances, may only be conceived as the resistance of the world-in-itself.

In the network of similarity and dissimilarity relations coconstitutive of a given phenomenal world, genetically object-sided and genetically social, subject-sided moments are united. And so when we examine a given such network, we can't separate these moments from each other; there is no way of telling where, precisely, either genetically

4. Of course such standards also have a social moment. But if it were only a matter of satisfying the purely social moment, then all innovations whose eventual acceptance was assured would—as never actually happens—gain immediate acceptance in the community, since their salient quality is, *ex hypothesi,* conformity with the social.

subject-sided or genetically object-sided moments are located. Accordingly, we can never subtract the genetically subject-sided from a phenomenal world in such a way as to permit our undistorted view of the purely object-sided, of absolute reality, or of the world-in-itself. The concrete properties of the world-in-itself are, rather, inaccessible; though we feel the effects of these properties in the resistance the world offers to our epistemic efforts, we aren't in a position to grasp this resistance as it is in itself.

B I B L I O G R A P H Y

Works of Thomas S. Kuhn

Reprints cited include only those in *The Essential Tension* (1977a).

1945 Subjective View [on: General Education in a Free Society]. *Harvard Alumni Bulletin* 2 (22 Sept. 1945): 29–30.

1949 The Cohesive Energy of Monovalent Metals as a Function of Their Atomic Quantum Defects. Ph.D. diss. Harvard, Cambridge.

1950a (with J.H. van Vleck): A Simplified Method of Computing the Cohesive Energies of Monovalent Metals. *Physical Review* 79:382–388.

1950b An Application of the W.K.B. Method to the Cohesive Energy of Monovalent Metals. *Physical Review* 79:515–519.

1951a A Convenient General Solution of the Confluent Hypergeometric Equation, Analytic and Numerical Development. *Quarterly of Applied Mathematics* 9:1–16.

1951b Newton's "31st Query" and the Degradation of Gold. *Isis* 42:296–298.

1952a Robert Boyle and Structural Chemistry in the Seventeenth Century. *Isis* 43:12–36.

1952b Reply to M. Boas: Newton and the Theory of Chemical Solutions. *Isis* 43:123–124.

1952c The Independence of Density and Pore-size in Newton's Theory of Matter. *Isis* 43:364–365.

1953a Review of *Ballistics in the Seventeenth Century, A Study in the Relations of Science and War with Reference Principally to England,* by A.R. Hall. *Isis* 44:284–285.

1953b Review of *The Scientific Work of René Descartes,* by J.F. Scott, and of *Descartes and the Modern Mind,* by A.G. Balz. *Isis* 44:285–287.

1953c Review of *The Scientific Adventure: Essays in the History and Philosophy of Science,* by H. Dingle. *Speculum* 28:879–880.

1954a Review of *Main Currents of Western Thought. Readings in Western European Intellectual History from the Middle Ages to the Present,* ed. F.L. Baumer. *Isis* 45:100.

1954b Review of *Galileo Galilei: Dialogue on the Great World Systems,* ed. G. de Santillana, and of *Galileo Galilei: Dialogue Concerning the Two Chief World Systems—Ptolemeic & Copernican,* trans. S. Drake. *Science* 119:546–547.

1955a Carnot's Version of "Carnot's Cycle." *American Journal of Physics* 23:91–95.

1955b La Mer's Version of "Carnot's Cycle." *American Journal of Physics* 23:387–389.

1955c Review of *New Studies in the Philosophy of Descartes* and *Descartes' Philosophical Writings,* ed. N.K. Smith, and *The Method of Descartes. A Study of the Regulae,* by L.J. Beck. *Isis* 46:377–380.

1957a *The Copernican Revolution. Planetary Astronomy in the Development of Western Thought.* Cambridge: Harvard University Press.

1957b Review of *A Documentary History of the Problem of Fall from Kepler to Newton. De Motu Gravium Naturaliter Cadentium in Hypothesi Terrae Motae,* by A. Koyré. *Isis* 48:91–93.

1958a The Caloric Theory of Adiabatic Compression. *Isis* 49:132–140.

1958b Newton's Optical Papers. In *Isaac Newton's Papers and Letters on Natural Philosophy, and Related Documents,* ed. I.B. Cohen, 27–45. Cambridge: Cambridge University Press. pp. 27–45.

1958c Review of *From the Closed World to the Infinite Universe,* by A. Koyré. *Science* 127:641.

1958d Review of *Copernicus: The founder of modern astronomy,* by A. Armitage. *Science* 127:972.

1959a The Essential Tension: Tradition and Innovation in Scientific Research. In *The Third (1959) University of Utah Research Conference on the Identification of Scientific Talent,* ed. C.W. Taylor, 162–174. Salt Lake City: University of Utah Press. In *ET,* 225–239.

1959b (with S. Parnes and N. Kaplan.) Committee Report on Environmental Conditions and Educational Methods Affecting Creativity. In *The Third (1959) University of Utah Research Conference on the Identification of Scientific Talent,* ed. C.W. Taylor, 313–316. Salt Lake City: University of Utah Press.

1959c Energy Conservation as an Example of Simultaneous Discovery. In *Critical Problems in the History of Science,* ed. M. Clagett, 321–356. Madison: University of Wisconsin Press. In *ET,* 66–104.

1959d Review of *A History of Magic and Experimental Science, Vols. VII and VIII: The Seventeenth Century,* by L. Thorndike. *Manuscripta* 3:53–57.

1959e Review of *The Tao of Science: An Essay on Western Knowledge and Eastern Wisdom,* by R.G.H. Siu. *The Journal of Asian Studies* 18:284–285.

1959f Review of *Sir Christopher Wren,* by J. Summerson. *Scripta mathematica* 24:158–159.

1960 Engineering Precedent for the Work of Sadi Carnot. In *Actes du IXe Congrès d'Histoire des Sciences,* 530–535. Barcelona: Asociación para la historia de la ciencia española, 1980.

1961a The Function of Measurement in Modern Physical Science. *Isis* 52:161–193. In *ET,* 178–224.

1961b Sadi Carnot and the Cagnard Engine. *Isis* 52:567–574.

1962a *The Structure of Scientific Revolutions,* Chicago: University of Chicago Press, 2d ed. 1970.

1962b Comment [on *Intellect and Motive in Scientific Inventors: Implications for Supply*], by D.W. MacKinnon. In *The Rate and Direction of Inventive Activity: Economic and Social Factors,* 379–384. Princeton: Princeton University Press.

1962c Comment [on *Scientific Discovery and the Rate and Direction of Invention*], by I.H. Siegel. In *The Rate and Direction of Inventive Activity: Economic and Social Factors,* 450–457. Princeton: Princeton University Press.

1962d The Historical Structure of Scientific Discovery. *Science* 136:760–764. In *ET,* 165–177.

1962e Review of *Forces and Fields,* by M.B. Hesse. *American Scientist* 50:442A–443A.

1963a The Function of Dogma in Scientific Research. In *Scientific Change. Historical Studies in the Intellectual, Social and Technical Conditions for Scientific Discovery and Technical Invention, from Antiquity to the Present,* ed. A.C. Crombie, 347–369. London: Heinemann.

1963b Discussion [on 1963a]. *Scientific Change. Historical Studies in the Intellectual, Social and Technical Conditions for Scientific Discovery and Technical Invention, from Antiquity to the Present,* ed. A.C. Crombie, 381–395 passim. London: Heinemann.

1964 A Function for Thought Experiments. In *L'aventure de la science. Mélanges Alexandre Koyré* vol. 2., 307–334. Paris: Herman. In *ET,* 240–265.

1966 Review of *Towards an Historiography of Science,* by J. Agassi. *British Journal for the Philosophy of Science* 17:256–258.

1967a (with J.L. Heilborn, P. Forman and L. Allen). *Sources for the History of Quantum Physics. An Inventory and Report.* Philadelphia: American Philosophical Society.

1967b Review of *The Questioners: Physicists and the Quantum Theory,* by. B.L. Cline. *Thirty Years that Shook Physics: The Story of Quantum Theory,* G. Gamow; *The Conceptual Development of Quantum Mechanics,* by M. Jammer; *Korrespondenz, Individualität, und Komplementarität: eine Studie zur Geistesgeschichte der Quantentheorie in den Beiträgen Niels Bohrs,* by K.M. Meyer-Abich; *Niels Bohr: The Man, His Science, and the World They Changed,* by R. Moore; and *Sources of Quantum Mechanics,* ed. B.L. van der Waerden. *Isis* 58:409–419.

1967c Review of *The Discovery of Time,* by S. Toulmin and J. Goodfield. *American Historical Review* 72:925–926.

1967d Review of *Michael Faraday: A Biography,* by L. Peirce Williams. *British Journal for the Philosophy of Science* 18:148–161.

1967e Review of *Niels Bohr: His Life & Work As Seen By His Friends & Colleagues,* ed. S. Rozental. *American Scientist* 55:339A–340A.

1968a The History of Science. In *International Encyclopedia of the Social Sciences.* vol. 14, 74–83. New York: Crowell. In *ET,* 105–126.

1968b Review of *The Old Quantum Theory,* ed. D. ter Haar. *British Journal for the History of Science* 98:80–81.

1969a (with J.L. Heilbron.) The Genesis of the Bohr Atom. *Historical Studies in the Physical Sciences* 1:211–290.

1969b Contributions [to the discussion of "New Trends in History"]. *Daedalus* 98:896–897, 928, 943, 944, 969, 971–976 passim.

1969c Comment [on The Relations of Science and Art]. *Comparative Studies in Society and History* 11:403–412. In *ET,* 340–351.

1969d Comment [on *The Principle of Acceleration: A Non-Dialectical Theory of Progress* by F. Dovring]. *Comparative Studies in Society and History* 11:426–430.

1970a Logic of Discovery or Psychology of Research? In *Criticism and the Growth of Knowledge,* ed. I. Lakatos and A. Musgrave, 1–20. Cambridge: Cambridge University Press. 1970. In *ET,* 266–292.

1970b Reflections on My Critics. In *Criticism and the Growth of Knowledge,* ed. I. Lakatos and A. Musgrave 231–278. Cambridge: Cambridge University Press, 1970.

1970c Postscript—1969. In *The Structure of Scientific Revolutions.* 2d ed., 174–210.

1970d Comment [on *Uneasy Fitful Reflections on Fits of Easy Transmission,* by R.S. Westfall. In *The Annus Mirabilis of Sir Isaac Newton 1666–1966,* ed. R. Palter, 105–108. Cambridge: MIT Press.

1970e Alexandre Koyré and the History of Science. On an Intellectual Revolution. *Encounter* 34:67–69.

1971a Notes on Lakatos. In *PSA 1970. In Memory of Rudolph Carnap.* ed. R.C. Buck and R.S. Cohen, 137–146. *Boston Studies in the Philosophy of Science.* vol. 8. Dordrecht: Reidel.

1971b Les notions de causalité dans le developpement de la physique. *Etudes d'épistémologie génétique* 25:7–18. English translation in *ET,* 21–30.

1971c The Relations between History and the History of Science. *Daedalus* 100:271–304. In *ET,* 127–161.

1972a Scientific Growth: Reflections on Ben-David's "Scientific Role." *Minerva* 10:166–178.

1972b Review of *Paul Ehrenfest, Vol. 1: The Making of a Theoretical Physicist,* by M.J. Klein. *American Scientist* 60:98.

1974a Second Thoughts on Paradigms. In *The Structure of Scientific Theories,* ed. F. Suppe, 459–482. Urbana: University of Illinois Press. In *ET,* 293–319.

1974b Discussion [on 1974a, and other papers given at the conference]. In F. Suppe (ed.): *The Structure of Scientific Theories,* ed. F. Suppe, 295–297, 369–370, 373, 409–412, 454–455, 500–517 passim. Urbana: University of Illinois Press.

1975 The Quantum Theory of Specific Heats: A Problem in Professional Recognition. In *Proceedings of the XIV International Congress for the History of Science 1974,* vol. 1, 170–182. vol. 4, 207. Tokyo: Science Council of Japan.

1976a Mathematical versus Experimental Traditions in the Development of Physical Science. *The Journal of Interdisciplinary History* 7:1–31. In *ET,* 31–65.

1976b Theory Change as Structure-Change: Comments on the Sneed Formalism. *Erkenntnis* 10:179–199.

1976c Review of *The Compton Effect: Turning Point in Physics,* by R.H. Stuewer. *American Journal of Physics* 44:1231–1232.

1977a *Die Entstehung des Neuen. Studien zur Struktur der Wissenschaftsgeschichte,* ed. L. Krüger. Frankfurt: Suhrkamp. English edition: *The Essential Tension. Selected Studies in Scientific Tradition and Change.* Chicago: University of Chicago Press, 1977.

1977b The Relations between the History and the Philosophy of Science. In *ET,* 3–20.

1977c Objectivity, Value Judgement, and Theory Choice. In *ET,* 320–339.

1978 *Black Body Theory and the Quantum Discontinuity, 1894–1912.* Oxford: Clarendon.

1979a History of Science. In *Current Research in Philosophy of Science,* ed. P.D. Asquith and H.E. Kyburg, 121–128. Ann Arbor: Edwards.

1979b Metaphor in Science. In *Metaphor and Thought,* ed. A. Ortony, 409–419. Cambridge: Cambridge University Press.

1979c Foreword to *Genesis and Development of a Scientific Fact,* by L. Fleck, ed. T. J. Trenn and R. Merton, vii–xii. Chicago: University of Chicago Press.

1980a The Halt and the Blind: Philosophy and History of Science. *British Journal for the Philosophy of Science* 31:181–192.

1980b Einstein's Critique of Planck. In *Some Strangeness in the Proportion: A Centennial Symposium to Celebrate the Achievements of Albert Einstein,* ed. H. Woolf, 186–191. Reading: Addison-Wesley.

1980c Open Discussion Following Papers by M.J. Klein and T.S. Kuhn. In *Some Strangeness in the Proportion: A Centennial Symposium to Celebrate the Achievements of Albert Einstein,* ed. H Woolf, 192–196 passim. Reading: Addison-Wesley.

1981 What are Scientific Revolutions? Originally published as Occasional Paper #18: Center for Cognitive Science, MIT. In *The Probabilistic Revolution.* Vol. 1, *Ideas in History,* ed. L. Krüger, L.J. Daston, and M. Heidelberger, 7–22. Cambridge: MIT Press, 1987.

1983a Commensurability, Comparability, Communicability. In *PSA 1982. Proceedings of the 1982 Biennial Meeting of the Philosophy of Science Association,* ed. P.D. Asquith and T. Nickles, 669–688. East Lansing: Philosophy of Science Association.

1983b Response to Commentaries [on 1983a]. In *PSA 1982. Proceedings of the 1982 Biennial Meeting of the Philosophy of Science Association,* ed.P.D. Asquith and T. Nickles, 712–716. East Lansing: Philosophy of Science Association.

1983c Reflections on Receiving the John Desmond Bernal Award. *4S Review: Journal of the Society for Social Studies of Science* 1:26–30.

1983d Rationality and Theory Choice. *Journal of Philosophy* 80:563–570.

1983e Foreword to *The tiger and the shark. Empirical roots of wave-particle dualism,* by B.R. Wheaton, ix–xiii. Cambridge: Cambridge University Press.

1984 Revisiting Planck. *Historical Studies in the Physical Sciences* 14:231–252.

1984a Professionalization Recollected in Tranquility. *Isis* 75:29–32.

1985 Panel Discussion on Specialization and Professionalism within the University. *The American Council of Learned Societies Newsletter* 36(Nos. 3 & 4): 23–27.

1986 The Histories of Science: Diverse Worlds for Diverse Audiences. *Academe. Bulletin of the American Association of University Professors* 72(4): 29–33.

1989a Possible Worlds in History of Science. In *Possible Worlds in Humanities, Arts, and Sciences,* ed. S. Allén, 9–32. Berlin: de Gruyter.

1989b Response to Commentators [to 1989a]. In *Possible Worlds in Humanities, Arts, and Sciences,* ed. S. Allén, 49–51. Berlin: de Gruyter.

1990 Dubbing and Redubbing: the Vulnerability of Rigid Designation. In *Scientific Theories. Minnesota Studies in Philosophy of Science 14,* ed. C.W. Savage, 298–318. Minneapolis: University of Minnesota Press.

For works published in 1991 and 1992, see page 302 below.

Secondary Literature

Bibliographical listings on the debate over Kuhn's theory may be found in Burrichter 1979, Gutting 1980, Hacking 1981, Kisiel and Johnson 1974, Laudan 1984, Laudan et al. 1986, Spiegel-Rösing 1973, and Stegmüller 1973.

Achinstein, P. 1964. On the Meaning of Scientific Terms. *Journal of Philosophy* 61:497–509.

Agassi, J. 1963. Towards an Historiography of Science. *History and Theory,* Beiheft 2. The Hague.

Agassi, J., and R.S. Cohen, eds. 1981. *Scientific Philosophy Today: Essays in Honor of Mario Bunge*. Dordrecht: Reidel.

Agazzi, E. 1985. Commensurability, Incommensurability, and Cumulativity in Scientific Knowledge. *Erkenntnis* 22:51–77.

———. ed. 1988. *Die Objektivität in den verschiedenen Wissenschaften.* Fribourg: Editions Universitaires.

———. ed. 1989. *Die Vergleichbarkeit wissenschaftlicher Theorien.* Freiburg: Editions Universitaires, 1990.

Andersson, G., ed. 1985. *Rationality in Science and Politics.* Boston Studies in the Philosophy of Science, vol. 79. Dordrecht: Reidel.

———. 1988a. *Kritik und Wissenschaftsgeschichte. Kuhns, Lakatos' und Feyerabends Kritik des kritischen Rationalismus.* Tübingen: Mohr.

———. 1988b. Die Wissenschaft als objektive Erkenntnis. In Agazzi 1988, 27–39.

Asquith, P. D., and P. Kitcher, eds. 1984. *PSA 1984. Proceedings of the 1984 Biennial Meeting of the Philosophy of Science Association.* Vol. 1. East Lansing: Philosophy of Science Association.

———. 1985. *PSA 1984. Proceedings of the 1984 Biennial Meeting of the Philosophy of Science Association.* Vol. 2. East Lansing: Philosophy of Science Association.

Asquith, P. D., and H. E. Kyburg, eds. 1979. *Current Research in Philosophy of Science.* Ann Arbor: Edwards.

Asquith, P. D., and I. Nickles, eds. 1983. *PSA 1982. Proceedings of the 1982 Biennial Meeting of the Philosophy of Science Association.* East Lansing: Philosophy of Science Association.

Austin, W. H. 1972. Paradigms, Rationality, and Partial Communication. *Zeitschrift für allgemeine Wissenschafstheorie* 3:203–218.

Balsiger, F., and A. Burri. 1990. Sind die klassische Mechanik und die spezielle Relativitätstheorie kommensurabel? *Journal of General Philosophy of Science* 21:157–162.

Balzer, W. 1978. Incommensurability and Reduction. *Acta Philosophica Fennica* 30:313–335.

———. 1985a. Was ist Inkommensurabilität? *Kant-Studien* 76:196–213.

———. 1985b. Incommensurability, Reduction, and Translation. *Erkenntnis* 23:255–267.

———. 1989. On Incommensurability. In Gavroglu, Goudaroulis, and Nicolacopoulos, 1989, 287–304.

Balzer, W., C. U. Moulines, and J. D. Sneed. 1987. *An Architectonic for Science. The Structuralist Program.* Dordrecht: Reidel.

Barbour, I. 1974. Paradigms in Science and Religion. In Gutting, 1980, 223–245.

Barker, P. 1988. Wittgenstein and the Historicist Project in Philosophy of Science. In Weingartner and Schurz 1989, 243–246.

Barnes, B. 1982. *T. S. Kuhn and Social Science.* London: McMillan.

Barnes, S. B., and R. G. A. Dolby. 1970. The Scientific Ethos: A Deviant Viewpoint. *Archives européennes de sociologie* 9:3–25.

Barnett, R. E. 1977. Restitution: A New Paradigm of Criminal Justice. *Ethics* 87:279–301.

Bartels, A. 1990. Mario Bunge's Realist Semantics. An Antidote against Incommensurability? In Weingartner and Dorn. 1990, 39–58.

———. 1990b. Review of *Wissenschaftsphilosophie Thomas S. Kuhns,* by P. Hoyningen-Huene. *Conceptus* 24:104–109.

Bastide, R., ed. 1962. *Sens et usage du terme structure dans les sciences humaines et sociales.* 's-Gravenhage: Mouton.

Batens, D. 1983a. The Relevance of Theory-Ladenness and Incommensurability, and a Survey of the Contributions to this Issue. *Philosophica* 31:3–6.

———. 1983b. Incommensurability is not a Threat to the Rationality of Science or to the Anti-Dogmatic Tradition. *Philosophica* 32:117–132.

Bayertz, K. 1980. *Wissenschaft als historischer Prozess. Die antipositivistische Wende in der Wissenschaftstheorie.* München: Fink.

———. ed. 1981. *Wissenschaftsgeschichte und wissenschaftliche Revolution.* Köln: Pahl-Rugenstein.

———. 1981a. Über Begriff und Problem der wissenschaftlichen Revolution. In Bayertz, 1981, 11–28.

———. 1981b. *Wissenschaftstheorie und Paradigmenbegriff.* Stuttgart: Metzler.

Berding, H. 1978. Revolution als Prozess. In Faber and Meier 1978, 266–289.

Bernstein, R. J. 1983. *Beyond Objectivism and Relativism: Science, Hermeneutics, and Praxis.* Philadelphia: University of Pennsylvania Press.

Biagioli, M. 1990. The Anthropology of Incommensurability. *Studies in History and Philosophy of Science* 21:183–209.

Blaug, M. 1975. Kuhn versus Lakatos, or Paradigms versus Research Programs in History of Economics. *History of Political Economy* 7:399–433.

Bluhm, W. T., ed. 1982. *The Paradigm Problem in Political Science. Perspectives from Philosophy and from Practice.* Durham: Carolina Academic Press.

Blum, W., H.-P. Dürr, and H. Rechenberg, eds. 1984. *Werner Heisenberg: Gesammelte Werke. Abteiling C. Allgemeinverständliche Schriften, Band I: Physik und Erkenntnis 1927–1955.* München: Piper.

Blumenberg, H. 1971. Paradigma, grammatisch. In Blumenberg, 1981, 157–162.

———. 1981. *Wirklichkeiten in denen wir leben.* Stuttgart: Reclam.

Böhler, D. 1972. Paradigmawechsel in analytischer Wissenschaftstheorie? Wissenschaftsgeschichtliche und Wissenschaftstheoretische Aufgaben der Philosophie. *Zeitschrift für allgemeine Wissenschaftstheorie* 3:219–242.

Bohm, D. 1974. Science as Perception-Communication. In Suppe, 1977, 374–391.

———. 1980. *Wholeness and the Implicate Order.* London: Routledge & Kegan Paul.

Bollnow, O. F., ed. 1969. *Erziehung in anthropologischer Sicht.* Zurich: Morgarten.

Boltzmann, L. 1905. *Populäre Schriften.* Leipzig: Barth.

Boos, B., and K. Krickeberg, eds. 1977. *Mathematisierung der Einzelwissenschaften.* Basel: Birkhäuser.

Boros, J. 1990. Probleme der Inkommensurabilität. In Agazzi, 1990, 125–132.

Boyd, R. N. 1979. Metaphor and Theory Change: What is "Metaphor" a Metaphor for? In Ortony, 1979, 356–408.

———. 1984. The Current Status of Scientific Realism. In Leplin, 1984, 41–82.

Brand, M. 1979. Causality. In Asquith and Kyburg, 1979, 252–281.

Briggs, J.P., and F. D. Peat. 1984. *Looking Glass Universe. The Emerging Science of Wholeness*. Glasgow: Fontana.

Briskman, L. B. 1972. Is a Kuhnian Analysis Applicable to Psychology? *Science Studies* 2:87–97.

Brown, H. I. 1977. *Perception, Theory and Commitment. The New Philosophy of Science*. Chicago: University of Chicago Press. 2d ed., 1979.

———. 1983. Incommensurability. *Inquiry* 26:3–29.

———. 1983a. Response to Siegel. *Synthese* 56:91–103.

Bryant, C. G. A. 1975. Kuhn, Paradigms and Sociology. *British Journal of Sociology* 26:354–359.

Buchdahl, G. 1965. A Revolution in Historiography of Science. *History of Science* 4:55–69.

———. 1969. *Metaphysics and the Philosophy of Science. The Classical Origins: Descartes to Kant*. Cambridge: MIT Press.

Buck, R. C., and R. S. Cohen, eds. 1971. *PSA 1970. In Memory of Rudolph Carnap, Boston Studies in the Philosophy of Science*. Vol. 8. Dordrecht: Reidel.

Bunge, M., ed. 1964. *The Critical Approach to Science and Philosophy*. New York: Free Press.

Burian, R. M. 1975. Conceptual Change, Cross-Theoretical Explanation, and the Unity of Science. *Synthese* 32:1–28.

———. 1984. Scientific Realism and Incommensurability: Some Criticisms of Kuhn and Feyerabend. In Cohen and Wartofsky, 1984, 1–31.

Burr, R. L., and B. B. Brown. 1988. Wittgenstein and Kuhn: Paradigms and the Standard Metre. Paper presented at the 13th International Wittgenstein Symposium, 14–18 August, Kirchberg.

Burrichter, C., ed. 1979. *Grundlegung der historischen Wissenschaftsforschung*. Basel: Schwabe.

Butterfield, H. 1931. *The Whig Interpretation of History*. London: Bell.

Cassirer, E. 1910. *Substanzbegriff und Funktionsbegriff. Untersuchungen über die Grundfragen der Erkenntniskritik*, Darmstadt: Wissenschaftliche Buchgesellschaft. 5th ed., 1980.

Causey, R. L. 1974. Professor Bohm's View of the Structure and Development of Theories. In Suppe, 1977, 392–401.

Cedarbaum, D. G. 1983. Paradigms. *Studies in History and Philosophy of Science* 14:173–213.

Chen, X. 1990. Local Incommensurability and Communicability. In Fine, Forbes, and Wessels, 1990, 67–76.

Churchill, J. 1990. The Bellman's Map: Does Antifoundationalism Entail Incommensurability and Relativism? *Southern Journal of Philosophy* 28:469–484.

Cohen, I. B. 1974. History and the Philosopher of Science. In Suppe, 1977, pp. 308–349.

———. 1985. *Revolution in Science.* Cambridge: Harvard University Press.

Cohen, R. S., P. K. Feyerabend, and M. W. Wartofsky, eds. 1976. *Essays in Memory of Imre Lakatos.* Boston Studies in the Philosophy of Science, vol. 34. Dordrecht: Reidel.

Cohen, R. S., and M. W. Wartofsky, eds. 1974. *Methodological and Historical Essays in the Natural and Social Sciences.* Boston Studies in the Philosophy of Science, vol. 14. Dordrecht: Reidel.

———. 1984. *Methodology, Metaphysics and the History of Science; in Memory of Benjamin Nelson,* Boston Studies in the Philosophy of Science, vol. 84. Dordrecht: Reidel.

Colemann, S. R., and R. Salamon. 1988. Kuhn's *Structure of Scientific Revolutions* in the Psychological Journal Literature, 1969–1983: A Descriptive Study. *Journal of Mind and Behavior* 9:415–445.

Collier, J. 1984. Pragmatic Incommensurability. In Asquith and Kitcher, 1984, 146–157.

Colodny, R., ed. 1966. *Mind and Cosmos. Essays in Contemporary Science and Philosophy.* Pittsburgh: University of Pittsburgh Press.

Conant, J. B. 1947. *On Understanding Science. An Historical Approach.* New Haven: Yale University Press.

Crane, D. 1980. An Exploratory Study of Kuhnian Paradigms in Theoretical High-Energy Physics. *Social Studies of Science* 10:23–54.

———. 1980a. Exemplars and Analogies. A Comment on Crane's Study of Kuhnian High Energy Physics. Reply. *Social Studies of Science* 10: 502–506.

Crombie, A. C., ed. 1963. *Scientific Change: Historical Studies in the Intellectual, Social and Technical Conditions for Scientific Discovery and Technical Invention, from Antiquity to the Present.* London: Heinemann.

Crowe, M. J. 1986. *The Extraterrestrial Life Debate 1750–1900. The Idea of a Plurality of Worlds.* Cambridge: Cambridge University Press.

Curd, M. V. 1984. Kuhn, Scientific Revolutions and the Copernican Revolution. *Nature and System* 6:1–14.

Csonka, P. L. 1969. Advanced Effects in Particle Physics. *Physical Review* 180:1266–1281.

Danneberg, L. 1989. *Methodologien. Struktur, Aufbau und Evaluation.* Berlin: Duncker and Humblot.

Danto, A. C. 1965. Analytical Philosophy of History. Cambridge: Cambridge University Press.

Davidson, D. 1974. The Very Idea of a Conceptual Scheme. *Proceedings & Addresses of the American Philosophical Association* 47:5–20. In Meiland and Krausz, 1982, 66–80.

Davidson, D., and G. Harman, eds. 1972. *The Semantics of Natural Languages*. Dordrecht: Reidel.

Devitt, M. 1979. Against Incommensurability, *Australasian Journal of Philosophy* 57:29–50.

———. 1984. *Realism and Truth*. Oxford: Blackwell.

Dick, S. J. 1984. *Plurality of Worlds. The Origins of the Extraterrestrial Life Debate from Democritus to Kant*. Cambridge: Cambridge University Press.

Diederich, W., ed. 1974. *Theorien der Wissenschaftsgeschichte. Beiträge zur diachronischen Wissenschaftstheorie*. Frankfurt: Suhrkamp.

Diemer, A. 1970. Zur Grundlegung eines allgemeinen Wissenschaftsbegriffs. Zeitschrift für allgemeine Wissenschaftstheorie 1:209–227.

———. ed. 1977. *Die Struktur wissenschaftlicher Revolutionen und die Geschichte der Wissenschaften*. Meisenheim: Hain.

Domin, G. 1978. *Wissenschaftskonzeptionen. Eine Auswahl von Beiträgen sowjetischer Wissenschaftshistoriker zur Geschichte der Ideen über die Wissenschaft*. Berlin: Akademie-Verlag.

Doppelt, G. 1978. Kuhn's Epistemological Relativism: An Interpretation and Defense. *Inquiry* 21:33–86. Excerpted in Meiland and Krausz, 1982, 113–146.

Dreyfus, H. L. 1979. *What Computers Can't Do. The Limits of Artificial Intelligence*. rev. ed. New York: Harper.

Dürr, H-P. 1988. *Das Netz des Physikers. Naturwissenschaftliche Erkenntnis in der Verantwortung*. München: Hanser.

Eimas, P. D. 1985. The Perception of Speech in Early Infancy, *Scientific American* 252 (Jan.):34–40.

Einem, H. v. et al. 1973. *Der Strukturbegriff in den Geisteswissenschaften*. Mainz: Akademie der Wissenschaften and der Literatur.

Embree, L. 1981. The History and Phenomenology of Science is Possible. In Skousgaard, 1981, 215–228.

Englehardt, W.v. 1977. Das Erdmodell der Plattentektonik—ein Beispiel für Theorienwandel in der neueren Geowissenschaft. In Diemer, 1977, 91–109.

English, J. 1978. Partial Interpretation and Meaning Change. *Journal of Philosophy* 75:57–76.

Erpenbeck, J., and U. Röseberg. 1981. Dialektik der Wissenschaftsentwicklung. In Hörz and Röseberg, 1981, 434–447.

Eysenck, H. J. 1983. Is There a Paradigm in Personality Research? *Journal of Research in Personality* 17:369–397.

Faber, K.-G., and C. Meier, eds. 1978. *Historische Prozesse. Theorie der Geschichte, Beiträge zur Historik*. Bd. 2. München: dtv.

Feigl, H. 1964. What Hume Might Have Said to Kant (And a few questions about induction and meaning). In Bunge, 1964, 45–51.

———. 1974. Empiricism at Bay? Revisions and a New Defense. In Cohen and Wartofsky, 1974, 1–20.

Feigl, H., and G. Maxwell, eds. 1962. *Scientific Explanation, Space, and Time. Minnesota Studies in the Philosophy of Science*. Vol. 3. Minneapolis: University of Minnesota Press.

Fetzer, J. H. 1991. *Epistemology and Cognition*. Dordrecht: Kluwer.

Feyerabend, P. K. 1958. An Attempt at a Realistic Interpretation of Experience. *Proceedings of the Aristotelian Society* 58:143–170. In Feyerabend, 1981, 17–36.

———. 1962. Explanation, Reduction, and Empiricism. In Feigl and Maxwell, 1962, 28–97. Also in Feyerabend, 1981, 44–96.

———. 1970. Consolations for the Specialist. In Lakatos and Musgrave, 1970, 197–230.

———. 1976. *Wider den Methodenzwang. Skizze einer anarchistischen Erkenntnistheorie*. Frankfurt: Suhrkamp. *Against Method*. rev. English ed. London: Verso, 1988.

———. 1978a. *Der wissenschaftstheoretische Realismus und die Autorität der Wissenschaften. Ausgewählte Schriften*. Bd. 1. Braunschweig: Vieweg.

———. 1978b. Kuhns Struktur wissenschaftlicher Revolutionen. Ein Trostbüchlein für Spezialisten? In Feyerabend, 1978a, 153–204.

———. 1981. *Realism, Rationalism, and Scientific Method. Philosophical Papers*. Vol. 1. Cambridge: Cambridge University Press.

———. 1981a. *Probleme des Empirismus. Schriften zur Theorie der Erklärung der Quantentheorie und der Wissenschaftsgeschichte. Ausgewählte Schriften*, Bd. 2. Braunschweig: Vieweg.

———. 1981b. More clothes from the Emperor's Bargain Basement. *British Journal for the Philosophy of Science* 32:57–71.

———. 1984. Mach's Theory of Research and its Relation to Einstein. *Studies in History and Philosophy of Science* 15:1–22.

———. 1987. Putnam on Incommensurability. *British Journal for the Philosophy of Science* 38:75–81.

Feyerabend, P. K., and C. Thomas, eds. 1984. *Kunst und Wissenschaft*. Zürich: Verlag der Fachvereine.

Feynman, R. P. 1985. *QED. The Strange Theory of Light and Matter*. Princeton: Princeton University Press.

Field, H. 1973. Theory Change and the Indeterminacy of Reference. *Journal of Philosophy* 70:462–481.

Fine, A. 1967. Consistency, Derivability, and Scientific Change. *Journal of Philosophy* 64:231–240.

———. 1975. How to Compare Theories: Reference and Change. *Nous* 9:17–32.

Fine, A., M. Forbes, and L. Wessels, eds. 1990. *PSA 1990. Proceedings of the*

1990 Biennial Meeting of the Philosophy of Science Association. Vol. 1. East Lansing: Philosophy of Science Association.

Fine, A., and P. Machamer, eds. 1986. *PSA 1986. Proceedings of the 1986 Biennial Meeting of the Philosophy of Science Association.* Vol. 1. East Lansing: Philosophy of Science Association.

Finocchiaro, M. A. 1986. Judgment and Reasoning in the Evaluation of Theories. In Fine and Machamer, 1986, 227–235.

Fischer, P. 1985. *Licht and Leben. Ein Bericht über Max Delbrück, den Wegbereiter der Molekularbiologie.* Konstanz: Universitätsverlag.

Fleck, L. 1935. *Entstehung und Entwicklung einer wissenschaftlichen Tatsache. Einführung in die Lehre vom Denkstil und Denkkollektiv.* Frankfurt: Suhrkamp. (Translated as *Genesis and Development of a Scientific Fact.* Chicago: University of Chicago Press, 1979.)

Flonta, M. 1978. A "weak" and a "strong" Version of the Incommensurability Thesis. *Philosophie et Logique* 20:395–406.

Foley, L. A. 1980. Review of *Black-Body Theory and the Quantum Discontinuity, 1894–1912,* by T. S. Kuhn. *Review of Metaphysics* 33:639–641.

Franklin, A. 1984. Are Paradigms Incommensurable? *British Journal for the Philosophy of Science* 35:57–60.

Franklin, A. et al. 1989. Can a Theory-Laden Observation Test the Theory? *British Journal for the Philosophy of Science* 40:229–231.

Franzen, W. 1985. "Vernunft nach Menschenmass." Hilary Putnams neue Philosophie als mittlerer Weg zwischen Absolutheitsdenken und Relativismus. *Philosophische Rundschau* 32:161–197.

Friedmann, J. 1981. *Kritik konstruktivistischer Vernunft. Zum Anfangs- und Begründungsproblem bei der Erlanger Schule.* München: Fink.

Fuller, S. 1989. *Philosophy of Science and Its Discontents.* Boulder: Westview.

Gadamer, H.-G. 1960. *Wahrheit und Methode,* 4th ed. Tübingen: Mohr, 1975. English ed. *Truth and method.* 2d rev. ed., trans. rev. by J. Weinsheimer and D. G. Marshall. New York: Crossroads. 1989.

Galison, P. 1981. Kuhn and the Quantum Controversy. *British Journal for the Philosophy of Science* 32:71–85.

Gavroglu, K., Y. Goudaroulis, and P. Nicolacopoulos, eds. 1989. *Imre Lakatos and Theories of Scientific Change.* Dordrecht: Kluwer.

Geraets, T. F. 1979. *Rationality To-Day.* Ottawa: University of Ottawa Press.

Giedymin, J. 1970. The Paradox of Meaning Variance. *British Journal for the Philosophy of Science* 21:257–268.

Giedymin, J. 1971. Consolations for the Irrationalist? *British Journal for the Philosophy of Science* 22:39–48.

Giel, K. 1969. Studie über das Zeigen. In Bollnow, 1969, S.51–75.

Goodman, N. 1975. Words, Works, Worlds. In Goodman, 1978, 1–22.

———. 1978. *Ways of Worldmaking.* Indianapolis: Hackett.

———. 1984. *Of Mind and Other Matters.* Cambridge: Harvard University Press.

Grandy, R. E. 1983. Incommensurability: Kinds and Causes. *Philosophica* 32:7–24.

Greene, J. C. 1971. The Kuhnian Paradigm and the Darwinian Revolution in Natural History. In Roller, 1971, 3–25.

Greenwood, J. D. 1990. Two Dogmas of Neo-Empiricism: The "Theory-Informity" of Observation and the Quine-Duhem Thesis. *Philosophy of Science* 57:553–574.

Groh, D. 1971. Strukturgeschichte als "totale" Geschichte? In Schieder and Gräubig, 1977, 311–351.

Groh, D., and R. Wirtz. 1983. Vom Nutzen und Nachteil des Quantifizierens für die Historie. In Hoyningen-Huene, 1983a, pp. 175–198.

Gumin, H., and A. Mohler. 1985. *Einführung in den Konstruktivismus.* München: Oldenburg.

Gutting, G. 1979. Continental Philosophy of Science, in Asquith and Kyburg 1979: pp. 94–117.

———. ed. 1980. *Paradigms and Revolutions. Applications and Appraisals of Thomas Kuhn's Philosophy of Science.* Notre Dame: University of Notre Dame Press.

Hacking, I. 1982. Language, Truth, and Reason. In Hollis and Lukes, 1982, 48–66.

———. 1983. *Representing and Intervening. Introductory Topics in the Philosophy of Natural Science.* Cambridge: Cambridge University Press.

———. 1984. Five Parables, in Rorty, Schneewind and Skinner 1984: pp. 103–124.

———. ed. 1981. *Scientific Revolutions.* Oxford: Oxford University Press.

Hall, A. R. 1963. Commentary on The Function of Dogma in Scientific Research, by T. S. Kuhn. In Crombie, 1963, 370–375.

Hanson, N. R. 1958. *Patterns of Discovery. An Inquiry into the Conceptual Foundations of Science.* Cambridge: Cambridge University Press.

Harper, W. L. 1978. Conceptual Change, Incommensurability and Special Relativity Kinematics. *Acta Philosophica Fennica* 30:430–461.

Harvey, L. 1982. The Use and Abuse of Kuhnian Paradigms in the Sociology of Knowledge. *Sociology* 16:85–107.

Hattiangadi, J. N. 1971. Alternatives and Incommensurables: the Case of Darwin and Kelvin. *Philosophy of Science* 38:502–507.

Hautamäki, A. 1983. Scientific Change and Intensional Logic. *Philosophica* 32:25–42.

Haverkamp, A. 1987. Paradigma Metapher, Metapher Paradigma. Zur Metakinetik hermeneutischer Horizonte (Blumenberg/Derrida, Kuhn/Foucault, Black/White). In Herzog and Kosellek, 1987, 547–560.

Hegel, G. W. F. 1969. *Werke in zwanzig Bänden.* Frankfurt: Suhrkamp. English trans. cited: *Phenomenology of Spirit.* Trans. A. V. Miller. Oxford: Oxford University Press, 1977. *Hegel's Science of Logic.* Trans. A. V. Miller. London: Allen and Unwin, 1969.

Heidegger, M. 1927. *Sein und Zeit*. 14th ed. Tübingen: Niemeyer. 1977, *Being and Time*. English ed., trans. J. Macquarrie and E. Robinson. New York: Harper, 1962.

Heidegger, M. 1962. *Die Frage nach dem Ding*. 2d ed. Tübingen: Niemeyer, 1975.

Heisenberg, W. 1942. Ordnung der Wirklichkeit. In Blum, Dürr and Rechenberg, 1984, 217–306.

Hempel, C. G. 1977. Die Wissenschaftstheorie des analytischen Empirismus im Lichte zeitgenössischer Kritik. In Patzig, Scheibe, and Wieland, 1977, 20–34.

———. 1979. Scientific Rationality: Analytic vs. Pragmatic Perspectives. In Geraets, 1979, 46–58.

Hentschel, K. 1985. On Feyerabend's Version of "Mach's Theory of Research and its Relation to Einstein." *Studies in History and Philosophy of Science* 16:387–394.

Herzog, R., and R. Kosellek, eds. 1987. *Epochenschwelle und Epochenbewusstsein. Poetik und Hermeneutik XII*. München: Fink.

Hesse, M. 1983. Comment [on Kuhn's "Commensurability, Comparability, Communicability"]. In Asquith and Nickles 1983, 704–711.

Hintikka, J. 1988. On the Incommensurability of Theories. *Philosophy of Science* 55:25–38.

Hodysh, H. W. 1977. Kuhnian Paradigm and its Implications for Historiography of Curriculum Change. *Paedagogica Historica* 17:75–87.

Holcomb, H. R. 1987. Circularity and Inconsistency in Kuhn's Defense of His Relativism. *Southern Journal of Philosophy* 25:467–480.

Holcomb, H. R. 1989. Rational Progress in Science and Meta-Science. *Philosophical Forum* 20:286–310.

Holenstein, E. 1976. 'Implicational Universals' versus 'Familienähnlichkeiten.' In Holenstein, *Linguistik, Semiotik, Hermeneutik. Pladoyers für eine strukturale Phänomenologie,* by E. Holenstein, 125–133. Frankfurt: Suhrkamp.

———. 1980. Von der Hintergehbarkeit der Sprache (und der Erlanger Schule), in Holenstein, *Von der Hintergehbarkeit der Sprache. Kognitive Unterlagen der Sprache,* by E. Holenstein, 10–52. Frankfurt: Suhrkamp.

Hollinger, D. A. 1973. T. S. Kuhn's Theory of Science and its Implications for History. *American Historical Review* 78:370–393. In Gutting, 1980, 195–222.

Hollis, M., and S. Lukes, eds. 1982. *Rationality and Relativism*. Cambridge: MIT Press.

Hörz, H. 1988. *Wissenschaft als Prozess. Grundlagen einer dialektischen Theorie der Wissenschaftsentwicklung*. Berlin: Akademie-Verlag.

Hörz, H., and U. Röseberg, eds. 1981. *Materialistische Dialektik in der physikalischen und biologischen Erkenntnis*. Berlin: Akademie-Verlag.

Hoyningen-Huene, P., ed. 1983a. *Die Mathematisierung der Wissenschaften,* Zürich: Artemis.

———. 1983b. Autonome historische Prozesse—kybernetisch betrachtet. *Geschichte und Gesellschaft* 9:119–123.

———. 1984a. Der Glaube an die Wissenschaft. In *Kindlers Enzyklopädie Der Mensch,* bd. 7, 488–496. München: Kindler.

———. 1984b. Das Problemfeld "Wissenschaftlicher Fortschritt." In Feyerabend and Thomas, 1984, 201–206.

———. 1985. Levels of Dispute. Review of *Science and Values. The Aims of Science and Their Role in Scientific Debate,* by L. Laudan. *Nature* 315:781.

———. 1987a. On the Distinction Between the "Context" of Discovery and the "Context" of Justification. In *Les relations mutuelles entre la philosophie des sciences et l'histoire des sciences. Epistemologia* 10:81–88.

———. 1987b. Context of Discovery and Context of Justification. *Studies in History and Philosophy of Science* 18:501–515.

———. 1988. Comment [on J. Mittelstraß: Philosophische Grundlagen der Wissenschaften. Über wissenschaftstheoretischen Historismus, Konstruktivismus und Mythen des wissenschaftlichen Geistes]. In Hoyningen-Huene and Hirsch, 1988, 219–225.

———. 1989. Naturbegriff—Wissensideal—Experiment. Warum ist die neuzeitliche Wissenschaft technisch verwertbar? *Zeitschrift für Wissenschaftsforschung.* 5:43–55.

Hoyningen-Huene, P., and G. Hirsch, eds. 1988. *Wozu Wissenschaftsphilosophie? Positionen und Fragen zur heutigen Wissenschaftsphilosophie.* Berlin: de Gruyter.

Hronszky, I., M. Fehér, and B. Dajka, eds. 1988. *Scientific Knowledge Socialized. Selected Proceedings of the 5th International Conference on the History and Philosophy of Science Organized by the IUHPS.* Dordrecht: Kluwer.

Hübner, K. 1978. *Kritik der wissenschaftlichen Vernunft.* Freiburg: Alber.

Hung, H.-C. 1987. Incommensurability and Inconsistency of Language. *Erkenntnis* 27:323–352.

Husserl, E. 1992. *Ideen zu einer reinen Phänomenologie und phänomenologischen Philosophie. Allgemeine Einführung in die reine Phänomenologie.* 4th ed. Tübingen: Max Niemeyer. 1980. *Ideas: general introduction to pure phenomenology.* English ed., trans. W. R. B. Gibson. New York: Humanities Press, 1967.

Jauss, H. R. 1969. Paradigmawechsel in der Literaturwissenschaft. *Linguistische Berichte* 3:44–56.

Jonas, H. 1973. *Organismus und Freiheit. Ansätze zu einer philosophischen Biologie.* Göttingen: Vandenhoeck and Ruprecht.

Jones, G. E. 1981. Kuhn, Popper, and Theory Comparison. *Dialectica* 35:389–397.

Jones, K. 1986. Is Kuhn a Sociologist? *British Journal for the Philosophy of Science* 37:443–452.

Kambartel, F. 1973. Struktur. In Krings, Baumgartner and Wild, 1973, 1430–1439.

Kamlah, W., and P. Lorenzen. 1967. *Logische Propädeutik. Vorschule des vernünftigen Redens.* Mannheim: Bibliographisches Institut. *Logical Propaedeutic. Pre-School of Reasonable Speech.* English ed., trans. H. Robinson. Lanham, MD: University Press of America, 1984.

Kant, I. 1960. *Werke in zehn Bänden,* ed. W. Weischedel, Darmstadt: Wissenschaftliche Buchgesellschaft. English citations from *Critique of Pure Reason.* 2d ed., trans. N. K Smith. New York: Macmillan, 1965.

Katz, J. J. 1979. Semantics and Conceptual Change. *Philosophical Review* 88:327–365.

Kay, L. E. 1985. Conceptual Models and Analytical Tools: The Biology of Physicist Max Delbrück. *Journal of the History of Biology* 18:207–247.

Keita, L. 1988. "Theory Incommensurability" and Kuhn's History of Science. A Critical Analysis. *Diogenes* 143:41–65.

King, M. D. 1971. Reason, Tradition, and the Progressiveness of Science. *History and Theory* 10:3–32.

Kisiel, T., and G. Johnson. 1974. New Philosophies of Science in the USA: A Selective Survey. *Zeitschrift für Allgemeine Wissenschaftstheorie* 5:138–191.

Kitcher, P. 1978. Theories, Theorists, and Theoretical Change. *Philosophical Review* 87:519–547.

———. 1983. Implications of Incommensurability. In Asquith and Nickles, 1983, 689–703.

Klein, M. J. 1979. Essay Review of *Black-Body Theory and the Quantum Discontinuity, 1894–1912,* by T. S. Kuhn. *Isis* 70:430–434.

Kobi, E. E. 1977. Modelle und Paradigmen in der heilpädagogischen Theoriebildung. In A. Bührli (ed.): *Sonderpädagogische Theoriebildung. Vergleichende Sonderpädagogik,* ed. A. Bührli, 11–24. Luzern: Schweizerische Zentralstelle für Heilpädagogik.

Koertge, N. 1983. Theoretical Pluralism and Incommensurability: Implications for Science and Education. *Philosophica* 31:85–108.

———. 1988. Is Reductionism the Best Way to Unify Science? In Radnitzky, 1988a, 19–44.

Kordig, C. R. 1970. Objectivity, Scientific Change, and Self-Reference. In Buck and Cohen, 1971, 519–523.

———. 1971a. *The Justification of Scientific Change.* Dordrecht: Reidel.

———. 1971b. Scientific Transitions, Meaning Invariance, and Derivability. *Southern Journal of Philosophy* 9:119–125.

———. 1971c. Stipulative Invariance. *Southern Journal of Philosophy* 9:129.

———. 1971d. The Comparability of Scientific Theories. *Philosophy of Science* 38:467–485.

Kosellek, R. 1973. Ereignis and Struktur. In *Geschichte—Ereignis und Erzählung,* R. Kosellek and W. D. Stempel, 560–570. München: Fink.

Kragh, H. 1987. *An Introduction to the Historiography of Science*. Cambridge: Cambridge University Press.

Krajewski, W., ed. 1982. Polish Essays in Philosophy of the Natural Sciences. Dordrecht: Reidel.

Krausz, M., ed. 1989. *Relativism: Interpretation and Confrontation*. Notre Dame: Notre Dame University Press.

Krige, J. 1980. *Science, Revolution and Discontinuity*. Sussex: Harvester.

Krings, H., H. M. Baumgartner, and C. Wild, eds. 1973. *Handbuch philosophischer Grundbegriffe*. München: Kösel.

Krüger, L. 1974. Die systematische Bedeutung wissenschaftlicher Revolutionen, pro und contra Thomas Kuhn. In Diederich, 1974, pp. 210–246.

Künne, W. 1983. "Im übertragenen Sinne." Zur Theorie der Metapher. *Conceptus* 17(nos. 40,41): 181–200.

Laitko, H. 1981. Thomas S. Kuhn und das Prolem der Entstehung neuen Wissens. In Bayertz, 1981, 174–191.

Lakatos, I. 1970. Falsification and the Methodology of Scientific Research Programmes. In Lakatos and Musgrave 1970: pp. 91–196. In Lakatos, 1978, 8–101.

———. 1971. History of Science and its Rational Reconstructions. In Buck and Cohen, 1971, 91–136. Also in Lakatos, 1978, 102–138.

———. 1978. The methodology of Scientific Research Programmes. In *Philosophical Papers*. Vol. 1, ed. J. Worral and G. Currie. Cambridge: Cambridge University Press.

Lakatos, I., and A. Musgrave, eds. 1970. *Criticism and the Growth of Knowledge*. London: Cambridge University Press.

La Mer, V. 1954. Some Current Misinterpretations of N. L. Sadi Carnot's Memoir and Cycle. *American Journal of Physics* 22:20–27.

———. 1955. Some Current Misinterpretations of N. L. Sadi Carnot's Memoir and Cycle. II. *American Journal of Physics* 23:95–102.

Lane N. R., and S. A. Lane. 1981. Paradigms and Perception. *Studies in History and Philosophy of Science* 12:47–60.

Laudan, L. 1976. Two Dogmas of Methodology. *Philosophy of Science* 43:585–597.

———. 1977: *Progress and Its Problems. Towards a Theory of Scientific Growth*, Berkeley: University of California Press.

———. 1984. *Science and Values. The Aims of Science and Their Role in Scientific Debate*, Berkeley: University of California Press.

———. 1990. *Science and Relativism. Some Key Controversies in the Philosophy of Science*. Chicago: University of Chicago Press.

Laudan, L., A. Donovan, R. Laudan, R., P. Barker, H. Brown, J. Leplin, P. Thagard, and S. Wykstra. 1986. Scientific Change: Philosophical Models and Historical Research. *Synthese* 69:141–223.

Lee, K. 1984. Kuhn—a Re-appraisal. *Explorations in Knowledge* 1:33–88.

Leibniz, G. W. 1684. *Meditations de Cognitione, Veritate et Ideis.* In *Die philosophischen Scjriften von Gottfried Wilhelm Leibniz,* ed. C. J. Gerhardt, Bd. 4, 422–426. Berlin: Weidman, 1880.

Leplin, J., ed. 1984. *Scientific Realism.* Berkeley: University of California Press.

Levi, I. 1985. Messianic vs. Myopic Realism. In Asquith and Kitcher, 1985, 617–636.

Levin, M. E. 1979. On Theory-Change and Meaning-Change. *Philosophy of Science* 46:407–424.

Levinson, P., ed. 1982. *In Pursuit of Truth. Essays in Honour of Karl Popper's 80th Birthday.* Atlantic Highlands, NJ: Humanities Press.

Lewis, D. 1986. *On the Plurality of Worlds.* Oxford: Basil Blackwell.

Lindholm, L. M. 1981. Is Realistic History of Science Possible? A Hidden Inadequacy in the New History of Science. In Agassi and Cohen, 1981, 159–186.

Lodynski, A. 1982. Some Remarks in Defense of the Incommensurability Thesis. In Krajewski, 1982, 91–102.

Loeck, G. 1987. Wissenserzeugung durch Beobachteränderung, *Erkenntnis* 26:195–229.

Lorenz, K., and F. M. Wuketits, ed. 1983. *Die Evolution des Denkens.* München: Piper.

Lübbe, H. 1977. *Geschichtsbegriff und Geschichtsinteresse. Analytik und Pragmatik der Historie.* Basel: Schwabe.

Lugg, A. 1987. "The Priority of Paradigms" Revisited. *Zeitschrift für allgemeine Wissenschaftstheorie* 18:175–182.

MacCormac, E. R. 1971. Meaning Variance and Metaphor, *British Journal for the Philosophy of Science* 22:145–169.

MacIntyre, A. 1977. Epistemological Crises, Dramatic Narrative, and the Philosophy of Science. *The Monist* 60:453–471. Also in Gutting, 1980, 54–74.

Malpas, J. E. 1989. The Intertranslatability of Natural Languages. *Synthese* 78:233–264.

Mandelbaum, M. 1977. A Note on Thomas S. Kuhn's *The Structure of Scientific Revolutions. The Monist* 60:445–452.

Mandelbaum, M. 1979. Subjective, Objective, and Conceptual Relativism. *The Monist* 62:403–428. In Meiland and Krausz, 1982, 34–61.

Marquard, O. 1978. Kompensation. Überlegungen zu einer Verlaufsfigur geschichtlicher Prozess. In Faber and Meier, 1978, 330–362.

Martin, G. 1969. *Immanuel Kant. Ontologie und Wissenschaftstheorie.* 4th ed. Berlin: de Gruyter.

Martin, M. 1971. Referential Variance and Scientific Objectivity. *British Journal for the Philosophy of Science* 22:17–26.

———. 1972. Ontological Variance and Scientific Objectivity. *British Journal for the Philosophy of Science* 23:252–256.

———. 1984. How To Be a Good Philosopher of Science: A Plea for Empiricism in Matters Methodological. In Cohen and Wartofsky, 1984, 33–42.

Mastermann, M. 1970. The Nature of a Paradigm. In Lakatos and Musgrave, 1970, 59–89.

Maudgil, A. 1989. World-Pictures and Paradigms: Wittgenstein and Kuhn. In Weingartner and Schurz 1989, 285–290.

Maxwell, N. 1984. *From Knowledge to Wisdom. A Revolution in the Aims and Methods of Science.* Oxford: Basil Blackwell.

Mayr, E. 1971. The Nature of the Darwinian Revolution. In Mayr, 1976, 277–296.

———. 1976. *Evolution and the Diversity of Life. Selected Essays.* Cambridge: Harvard University Press.

———. 1990. When is Historiography Whiggish? *Journal of the History of Ideas* 51:301–309.

McMullin, E. 1983. Values in Science. In Asquith and Nickles, 1983, 3–28.

Meiland, J. W. 1974. Kuhn, Scheffler and Objectivity in Science. *Philosophy of Science* 41:179–187.

Meiland, J. W., and M. Krausz, eds. 1982. *Relativism. Cognitive and Moral.* Notre Dame: University of Notre Dame Press.

Merton, R. K. 1942. Science and Democratic Social Structure. In Merton, 1968, 604–615. Also in Merton, 1973, 267–278.

———. 1945. Paradigm for the Sociology of Knowledge. In Merton, 1973, 7–40 (original title: Sociology of Knowledge).

———. 1968. *Social Theory and Social Structure.* New York: Free Press.

———. 1973. *The Sociology of Science. Theoretical and Empirical Investigations.* Chicago: University of Chicago Press.

———. 1977. The Sociology of Science. An Episodic Memoir. In Merton and Gaston, 1977, 3–141.

Merton, R. K., and J. Gaston, eds. 1977. *The Sociology of Science in Europe.* Carbondale: Southern Illinois University Press.

Mey, M. de 1982. *The Cognitive Paradigm.* Dordrecht: Reidel.

Meyer-Abich, K. M. 1986. Peace with Nature, or Plants as Indicators to the Loss of Humanity. *Experientia* 42:115–120.

Mittelstraß, J. 1984. Forschung, Begründung, Rekonstruktion. Wege aus dem Begründungsstreit. In Schnädelbach, 1984, 117–140.

———. 1988. Philosophische Grundlagen der Wissenschaften. Über wissenschaftstheoretischen Historismus, Konstruktivismus und Mythen des wissenschaftlichen Geistes. In Hoyningen-Huene and Hirsch, 1988, 179–212.

Moberg, D. W. 1979. Are There Rival, Incommensurable Theories? *Philosophy of Science* 46:244–262.

Mocek, R. 1988. *Neugier und Nutzen. Blicke in die Wissenschaftsgeschichte.* Berlin: Dietz.

Munz, P. 1985. *Our Knowledge of the Growth of Knowledge. Popper or Wittgenstein?* London: Routledge & Kegan Paul.

Musgrave, A. 1971. Kuhn's Second Thoughts. *British Journal for the Philosophy of Science* 22:287–297.

———. 1978. How to Avoid Incommensurability. *Acta Philosophica Fennica* 30:336–346.

Naumann, H., ed. 1973. *Der moderne Strukturbegriff. Materialien zu seiner Entwicklung.* Darmstadt: Wissenschaftliche Buchgesellschaft.

Nersessian, N. J. 1982. Why is "Incommensurability" a Problem? *Acta Biotheoretica* 31:205–218.

———. 1984. *Faraday to Einstein: Constructing Meaning in Scientific Theories.* Dordrecht: Kluwer.

———. 1989a. Conceptual Change in Science and in Science Education. *Synthese* 80:163–183.

———. 1989b. Scientific Discovery and Commensurability of Meaning. In Gavroglu, Goudaroulis, and Nicolacopoulos, 1989, 323–334.

Neumaier, O., ed. 1986. *Wissen und Gewissen. Arbeiten zur Verantwortungsproblematik.* Vienna: VWGÖ.

Neurath, O. 1935. Pseudorationalismus der Falsifikation. *Erkenntnis* 5:353–365.

Newton-Smith, W. H. 1981. *The Rationality of Science.* London: Routledge.

Nicholas, J. 1982. Review of *Black-Body Theory and the Quantum Discontinuity, 1894–1912,* by T. S. Kuhn. *Philosophy of Science* 49:295–297.

Niiniluoto, I., 1985. The Significance of Verisimilitude. In Asquith and Kitcher, 1985, 591–613.

———. 1987. *Truthlikeness.* Dordrecht: Kluwer.

Nola, R., ed. 1988. *Relativism and Realism in Science.* Dordrecht: Kluwer.

Nordmann, A. 1986. Comparing Incommensurable Theories. *Studies in History and Philosophy of Science* 17:231–246.

Oddie, G. 1986. The Poverty of the Popperian Program for Truthlikeness. *Philosophy of Science* 53:163–178.

Oddie, G. 1988. On a Dogma Concerning Realism and Incommensurability. In Nola, 1988, 169–203.

———. 1989. Partial Interpretation, Meaning Variance, and Incommensurability. In Gavroglu, Goudaroulis, and Nicolacopoulos, 1989, 305–322.

Ortony, A., ed. 1979. *Metaphor and Thought.* Cambridge: Cambridge University Press.

Parsons, K. P. 1971a. On Criteria of Meaning Change. *British Journal for the Philosophy of Science* 22:131–144.

———. 1971b. A Note on Meaning Invariance. *Southern Journal of Philosophy* 9:126–128.

Patzig, G., E. Scheibe, and W. Wieland, eds. 1977. *Logik, Ethik, Theorie der Geisteswissenschaften*. Hamburg: Meiner.

Pearce, D. 1982. Stegmüller on Kuhn and Incommensurability. *British Journal for the Philosophy of Science* 33:389–396.

———. 1986. Incommensurability and Reduction Reconsidered. *Erkenntnis* 24:293–308.

———. 1987. *Roads to Commensurability*. Dordrecht: Reidel.

———. 1988. The Problem of Incommensurability: A Critique of Two Instrumentalist Approaches. In Hronszky, Fehér and Dajka, 1988, 385–398.

Pearce, G., and P. Maynard, eds. 1973. *Conceptual Change*. Dordrecht: Reidel.

Pegg, D. T. 1975. Absorber Theory of Radiation. *Reports on Progress in Physics* 38:1339–1383.

Percival, W. K. 1976. Applicability of Kuhn's Paradigms to History of Linguistics. *Language* 52:285–294.

———. 1979. Applicability of Kuhn's Paradigms to the Social Sciences. *American Sociologist* 14:28–31.

Perelman, C., and L. Olbrechts-Tyteca. 1958. *La Nouvelle Rhétorique: Traité de l'Argumentation*. Paris: Presses Universitaires de France.

Perry, N. 1977. A Comparative Analysis of 'Paradigm' Proliferation. *British Journal of Sociology* 28:38–50.

Peterson-Falshöft, G. 1980. Die Erfahrung Des Neuen, *Philosophische Rundschau* 27:101–117.

Phillips, D. L. 1975. Paradigms and Incommensurability. *Theory and Society* 2:37–61.

Pinch, T. J. 1979. Essay Review of *Black-Body Theory and the Quantum Discontinuity, 1894–1912,* by T. S. Kuhn. *Isis* 70:437–440.

Plato. 1963. *Plato: The Collected Dialogues,* ed. E. Hamilton and H. Cairns. Princeton: Princeton University Press.

Poldrack, H. 1983. Historische Hintergründe des "new approach" in der bürgerlichen Wissenschaftsforschung. *Deutsche Zeitschrift für Philosophie* 31:856–860.

Polikarov, A. 1981. Strukturmodelle der Wissenschaftsentwicklung. In Rapp, 1981, 111–128.

Popp, J. A. 1975. Paradigms in Educational Inquiry. *Educational Theory* 25:28–39.

Popper, K. R. 1973. *Logik der Forschung*. 5th ed. Tübingen: Mohr. *The Logic of Scientific Discovery*. English ed. New York: Harper & Row, 1968.

———. 1972. *Conjectures and Refutations. The Growth of Scientific Knowledge*. 4th ed. London: Routledge.

———. 1970. Normal Science and its Dangers. In Lakatos and Musgrave, 1970, 51–58.

———. 1972. *Objective Knowledge. An Evolutionary Approach.* Oxford: Clarendon.

———. 1979. *Objective Knowledge. An Evolutionary Approach* rev. ed. Oxford: Clarendon.

Porus, N. L. 1988. Incommensurability, Scientific Realism and Rationalism. In Hronszky, Fehér and Dajka, 1988, 375–384.

Postiglione, G. A., and J. A. Scimecca. 1983. The Poverty of Paradigmaticism: A Symptom of the Crisis in Sociological Explanation. *The Journal of Mind and Behavior* 4:179–190.

Prauss, G. 1980. *Einführung in die Erkenntnistheorie.* Darmstadt: Wissenschaftliche Buchgesellschaft.

Przelecki, M. 1978. Commensurable Referents of Incommensurable Terms. *Acta Philosophica Fennica* 30:347–365.

Purtill, R. L. 1967. Kuhn on Scientific Revolutions. *Philosophy of Science* 34:53–58.

Putnam, H. 1974. The 'Corroboration' of Theories. In Putnam, 1975, 250–269. Originally in Schilpp, 1974.

———. 1975. *Mathematics, Matter and Method. Philosophical Papers.* Vol. 1. Cambridge: Cambridge University Press.

———. 1978. *Meaning and the Moral Sciences.* London: Routledge & Kegan Paul.

———. 1981. *Reason, Truth and History.* Cambridge: Cambridge University Press.

———. 1983. *Realism and Reason. Philosophical Papers,* Vol. 3. Cambridge: Cambridge University Press.

Quine, W. v. O. 1960. *Word and Object.* Cambridge: MIT Press.

Rabb, J. D. 1975. Incommensurable Paradigms and Critical Idealism. *Studies in History and Philosophy of Science* 6:343–346.

Radnitzky, G. 1982. Popper as a Turning Point in the Philosophy of Science: Beyond Foundationalism and Relativism. In Levinson, 1982, 64–80.

———. 1983. Science, Technology, and Political Responsibility. *Minerva* 21:234–264.

———. 1986. Responsibility in Science and in the Decisions about the Use or Non-Use of Technologies. In Neumaier, 1986, 99–124.

———. 1988. Wozu Wissenschaftstheorie? Die falsifikationistische Methodologie im Lichte des Ökonomischen Ansatzes. In Hoyningen-Huene and Hirsch, 1988, 85–132.

———. ed. 1988a. *Centripetal Forces in the Sciences.* Vol. 2. New York: Paragon.

Rakover, S. S. 1989. Incommensurability: The Scaling of Mind-Body Theories as a Counterexample. *Behaviorism* 17:103–118.

Rapp, F., ed. 1981. *Naturverständnis und Naturbeherrschung. Philosophiegeschichtliche Entwicklung und gegenwärtiger Kontext.* München: Fink.

Reichenbach, H. 1938. *Experience and Prediction. An Analysis of the Foundations and the Structure of Knowledge*. Chicago: University of Chicago Press.

Reisch, G. A. 1991. Did Kuhn Kill Logical Empiricism? *Philosophy of Science* 58:264–277.

Rescher, N. 1973. *Conceptual Idealism*. Oxford: Basil Blackwell.

———. 1982. *Empirical Inquiry*. Ottowa: Rowman and Littlefield.

———. 1984. *The Limits of Science*. Berkeley: University of California Press.

Riehl, A. 1925. *Der philosophische Kritizismus. Geschichte und System. Zweiter Band: Die sinnlichen und logischen Grundlagen der Erkenntnis*. 2d ed. Leipzig: Kröner.

Rodnyj, N. I. 1973. Das Problem der wissenschaftlichen Revolution in der Konzeption der Wissenschaftsentwicklung Th.S. Kuhns. In Domin, 1978, 185–197.

Roller, D. H. D., ed. 1971. *Perspectives in the History of Science and Technology*. Norman: University of Oklahoma Press.

Rorty, R. 1979. *Philosophy and the Mirror of Nature,* Princeton: Princeton University Press.

Rorty, R., J. B. Schneewind, and Q. Skinner, eds. 1984. *Philosophy in History. Essays on the Historiography of Philosophy*. Cambridge: Cambridge University Press.

Röseberg, U. 1984. *Szenarium einer Revolution. Nichtrelativistische Quantenmechenik und philosophische Widerspruchsproblematik*. Berlin: Akademie-Verlag.

Roth, P. A. 1984. Who Needs Paradigms? *Metaphilosophy* 15:225–238.

Rouse, J. 1981. Kuhn, Heidegger and Scientific Realism. *Man and World* 14:269–290.

Rudwick, M. J. S. 1985. *The Great Devonian Controversy. The Shaping of Scientific Knowledge among Gentleman Specialists*. Chicago: University of Chicago Press.

Ruse, M. E. 1970. The Revolution in Biology. *Theoria* 36:1–22.

———. 1971. Two Biological Revolutions. *Dialectica* 25:17–38.

Rüsen. J. 1977. Der Strukturwandel der Geschichtswissenschaft und die Aufgabe der Historik. In Diemer, 1977, 110–119.

Sagal, P. T. 1972. Incommensurability, Then and Now. *Zeitschrift für allgemeine Wissenschaftstheorie* 3:298–301.

Salmon, W. C. 1990. Rationality and Objectivity in Science, or Tom Kuhn Meets Tom Bayes. In Savage, 1990, 175–204.

Sankey, H. 1990. In Defence of Untranslatability. *Australasian Journal of Philosophy* 68:1–21.

———. 1991a. Incommensurability and the Indeterminacy of Translation. *Australasian Journal of Philosophy* 69:219–223.

———. 1991b. Translation Failure Between Theories. *Studies in History and Philosophy of Science* 22:223–236.

Savage, C. W., ed. 1990. *Scientific Theories. Minnesota Studies in the Philosophy of Science 14*. Minneapolis: University of Minnesota Press.

Scheffler, I. 1967. *Science and Subjectivity*. Indianapolis: Hackett.

———. 1972. Vision and Revolution: A Postscript on Kuhn. *Philosophy of Science* 39:366–374.

Scheibe, E. 1988. The Physicists' Conception of Progress. *Studies in History and Philosophy of Science* 19:141–159.

———. 1988a. Paul Feyerabend und die rationalen Rekonstruktionen. In Hoyningen-Huene and Hirsch, 1988, 149–171.

Schieder, T., and K. Gräubig, eds. 1977. *Theorieprobleme der Geschichtswissenschaft*. Darmstadt: Wissenschaftliche Buchgesellschaft.

Schilpp, P. A., ed. 1974. *The Philosophy of Karl Popper*. La Salle: Open Court.

Schlick, M. 1925. *Allgemeine Erkenntnislehre*. 2d ed. Berlin: Julius Springer.

———. 1986. *Die Probleme der Philosophie in ihrem Zusammenhang. Vorlesung aus dem Wintersemester 1933 und 34,* ed. H. L. Mulder, A. J. Kox and R. Hegselmann. Frankfurt: Suhrkamp.

Schlobach, J. 1978. Die klassisch-humanistische Zyklentheorie und ihre Anfechtung durch das Fortschrittsbewußtsein der französischen Frühaufklärung. In Faber and Meier, 1978, 127–142.

Schlüchter, H. 1974. *Der Strukturbegriff im Verwaltungsrecht*. Würzburg: Schmitt & Meyer.

Schmidt, B. 1985. *Das Widerstandsargument in der Erkenntnistheorie. Ein Angriff auf die Automatisierung des Wissens*. Frankfurt: Suhrkamp.

Schmidt, S. J., ed. 1987. *Der Diskurs des Radikalen Konstruktivismus*. Frankfurt: Suhrkamp.

Schmidt, W. 1981. *Struktur, Bedingungen und Funktionen von Paradigmen und Paradigmenwechseln: eine wissenschafts-historisch-systematische Untersuchung am Beispiel der empirischen Psychologie*. Bern: Lang.

Schnädelbach, H. 1984. *Rationalität. Philosophische Beiträge*. Frankfurt: Suhrkamp.

Schorsch, C. 1988. *Die New-Age Bewegung. Utopie und Mythos der Neuen Zeit*. Gütersloh: Mohn.

Schramm, A. 1975. *Theoriewandel oder Theoriefortschritt? Zur Diskussion um Thomas S. Kuhns "Die Struktur wissenschaftlicher Revolutionen."* Wien: VWGÖ.

Schulz, W. 1972. *Philosophie in der veränderten Welt*. Pfullingen: Neske.

Seel, G. 1988. Die Welt: kein Ort für Laboratorien? Kritische Fragen an Karin Knorr-Cetina. In Hoyningen-Huene and Hirsch, 1988, pp. 345–358.

Seiler, S. 1980. *Wissenschaftstheorie in der Ethnologie. Zur Kritik und Weiterführung der Theorie von Thomas S. Kuhn anhand ethnographischen Materials*. Berlin: Reimer.

Shapere, D. 1964. Review of *The Structure of Scientific Revolutions,* by T. S. Kuhn. *Philosophical Review* 73:383–394. Also in Shapere, 1984.

———. 1966. Meaning and Scientific Change. In Colodny 1966, 41–85. Also in Shapere, 1984, and excerpted in Hacking, 1981, 28–59.

———. 1969. Notes Toward a Post-Positivistic Interpretation of Science. Part 1. In Shapere, 1984, 102–119.

———. 1971. The Paradigm Concept. *Science* 172:706–709. Also in Shapere, 1984.

———. 1974. [Remark in discussion.] In Suppe, 1977, 506–507.

———. 1977. What Can the Theory of Knowledge Learn from the History of Knowledge? In Shapere, 1984, 182–202. Originally in *The Monist* 60(1977):488–508.

———. 1984. *Reason and the Search for Knowledge. Investigations in the Philosophy of Science.* Dordrecht: Reidel.

———. 1984a. Interpretations of Science in America. In Shapere, 1984, 156–166.

———. 1989. Evolution and Continuity in Scientific Change. *Philosophy of Science* 56:419–437.

Shaw, G. 1969. Das Problem des Dinges an sich in der englischen Kantinterpretation. *Kantstudien:* Ergänzungshefte 97.

Shimony, A. 1976. Comments on Two Epistemological Theses of Thomas Kuhn. In Cohen, Feyerabend, and Wartofsky, 1976, 569–588.

———. 1979. Essay Review of *Black-Body Theory and the Quantum Discontinuity, 1894–1912,* by T. S. Kuhn. *Isis* 70:434–437.

Shrader-Frechette, K. 1977. Atomism in Crisis: an Analysis of the Current High Energy Paradigm. *Philosophy of Science* 44:409–440.

Siegel, H. 1976. Meiland on Scheffler, Kuhn, and Objectivity in Science. *Philosophy of Science* 43:441–448.

———. 1980. Objectivity, Rationality, Incommensurability, and more. Review of *The Essential Tension,* by T. S. Kuhn. *British Journal for the Philosophy of Science* 31:359–384.

———. 1980a. Justification, Discovery and the Naturalizing of Epistemology. *Philosophy of Science* 47:297–321.

———. 1983. Brown on Epistemology and the New Philosophy of Science. *Synthese* 56:61–89.

———. 1987. *Relativism Refuted. A Critique of Contemporary Epistemological Relativism.* Dordrecht: Reidel.

Sintonen, M. 1986, Selectivity and Theory Choice. In Fine and Machamer, 1986, 364–373.

Skousgaard, S. 1981. *Phenomenology and the Understanding of Human Destiny.* Washington: The University Press of America.

Sneed, J.D. 1971. *The Logical Structure of Mathematical Physics.* Dordrecht: Reidel.

Solla-Price, D. J. de 1963. *Little Science, Big Science.* New York: Columbia University Press.

Spaemann, R. 1983a. *Philosophische Essays*. Stuttgart: Reclam.

———. 1983b. *Die kontroverse Natur der Philosophie*. In Spaemann, 1983a, 104–129.

Spaemann, R., and R. Löw. 1981. *Die Frage Wozu? Geschichte und Wiederentdeckung des teleologischen Denkens*. München: Piper.

Spiegel-Rösing, I. S. 1973. *Wissenschaftsentwicklung und Wissenschaftssteuerung, Einführung und Material zur Wissenschaftsforschung*. Frankfurt: Athenäum.

Staudinger, H. 1984. Forschung—ein Spiel? In Ströker, 1984, 27–42.

Stegmüller, W. 1970. *Probleme und Resultate der Wissenschaftstheorie und Analytischen Philosophie. Band II. Theorie und Erfahrung*. Berlin: Springer.

———. 1973. *Probleme und Resultate der Wissenschaftstheorie und Analytischen Philosophie. Band II. Theorie und Erfahrung. 2. Halbband: Theorienstrukturen und Theoriendynamik*. Berlin: Springer.

———. 1974. Theoriendynamik und logisches Verständnis. In Diederich, 1974, 167–209.

———. 1979. *Hauptströmungen der Gegenwartsphilosophie. Eine kritische Einführung. Band II*. 6th expanded ed. Stuttgart: Kröner.

———. 1980. Accidental ('Non-Substantial') Theory Change and Theory Dislodgment. In Gutting, 1980, 75–93.

———. 1985. Thesen zur "Evolutionären Erkenntnistheorie." *Information Philosophie* 1985(Juli): 26–32.

———. 1986. *Probleme und Resultate der Wissenschaftstheorie und Analytischen Philosophie. Band II. Theorie und Erfahrung. 3. Teilband: Die Entwicklung des neuen Strukturalismus seit 1973*. Berlin: Springer.

Stephens, J. 1973. *Kuhnian Paradigm and Political Inquiry—Appraisal. American Journal of Political Science* 17:467–488.

Stock, W. G. 1980. Die Bedeutung Ludwik Flecks für die Theorie der Wissenschaftsgeschichte. *Grazer philosophische Studien* 10:105–118.

Storer, N. W. 1966. *The Social System of Science*. New York: Holt, Rinehart and Winston.

Ströker, E. 1974. Geschichte als Herausforderung. Marginalien zur jüngsten wissenschaftstheoretischen Kontroverse. *Neue Hefte für Philosophie* 7, 8:27–66.

———. ed. 1984. *Ethik der Wissenschaften? Philosophische Fragen*. München: Fink and Schöningh.

Strug, C. 1984. Kuhn's Paradigm Thesis: A Two Edged Sword for the Philosophy of Religion. *Religious Studies* 20:269–280.

Stuewer, R. H., ed. 1970. *Historical and Philosophical Perspectives of Science. Minnesota Studies in the Philosophy of Science 5*. Minneapolis: University of Minnesota Press.

Suppe, F. 1974. The Search for Philosophic Understanding of Scientific Theories. In Suppe, 1977, 1–241.

———. 1974a. Exemplars, Theories and Disciplinary Matrixes. In Suppe, 1977, 483–499.

———. 1977. *The Structure of Scientific Theories*. Urbana: University of Illinois Press.

Suppe, F., and P. D. Asquith, eds. 1976. *PSA 1976: Proceedings of the 1976 Biennial Meeting of the Philosophy of Science Association*. Vol. 1. East Lansing: Philosophy of Science Association.

Szumilewicz, I. 1977. Incommensurability and the Rationality of the Development of Science. *British Journal for the Philosophy of Science* 28:345–350.

Szumilewicz-Lachmann, I. 1985. Poincaré versus Le Roy on Incommensurability. In Andersson, 1985, 261–275.

Thackray, A. 1970. Science: Has its Present Past a Future? In Stuewer, 1970, 112–127.

Thagard, P. 1990. The Conceptual Structure of the Chemical Revolution. *Philosophy of Science* 57:183–209.

———. 1991. Concepts and Conceptual Change. In Fetzer, 1991, 101–120.

Thimm, W. 1975. Behinderung als Stigma. Überlegungen zu einer Paradigma-Alternative. *Sonderpädagogik* 5:149–157.

Tianji, J. 1985. Scientific Rationality, Formal or Informal? *British Journal for the Philosophy of Science* 36:409–423.

Toellner, R. 1977. Mechanismus—Vitalismus: ein Paradigmenwechsel? Testfall Haller. In Diemer, 1977, 61–72.

Törnebohm, H. 1978. Paradigms in Fields of Research. *Acta Philosophica Fennica* 30:62–90.

Toulmin, S. 1963. Discussion [on The Function of Dogma in Scientific Research by T. S. Kuhn], in Crombie, 1963, 382–384.

———. 1967. The Evolutionary Development of Natural Science. *American Scientist* 55:456–471.

———. 1970. Does the Distinction between Normal and Revolutionary Science Hold Water? In Lakatos and Musgrave, 1970, 39–47.

———. 1971. Rediscovering History. New Directions in the Philosophy of Science. *Encounter* 36(1): 53–64.

———. 1972. *Human Understanding. The Collective Use and Evolution of Concepts*. Princeton: Princeton University Press.

Trenckmann, U., and F. Ortmann. 1980. Psychodynamic Concept of Disease in Romanticism—Test Case for the Use of Kuhn's Term Paradigm in a Social Science. *Zeitschrift für Psychologie* 188:331–339.

Tuchanska, B. 1988. The Idea of Incommensurability and the Copernican Revolution. *The Polish Sociological Bulletin* 1:65–79.

Vaihinger, H. 1922. *Kommentar zu Kants Kritik der reinen Vernunft*. 2d ed. Stuttgart: Union deutsche Verlagsgesellschaft, 1922. Reprint. Aalen: Scientia, 1970.

Van Bendegem, J. P. 1983. Incommensurability—An Algorithmic Approach. *Philosophica* 32:97–115.

Van der Veken, W. 1983. Incommensurability in the Structuralist View. *Philosophica* 32:43–56.

Vico, G. 1968. The New Science of Giambattista Vico. 3d ed. trans. Bergin and Fisch. Ithaca: Cornell.

Vision, G. 1988. *Modern Anti-Realism and Manufactured Truth*. London: Routledge.

Vollmer, G. 1983. Mesokosmos und objektive Erkenntnis—Über Probleme, die von der evolutionären Erkenntnistheorie gelöst werden. In Lorenz and Wuketits, 1983, 29–91. Also in Vollmer, 1985, 57–115.

———. 1985. *Was können wir wissen? Band 1. Die Natur der Erkenntnis. Beiträge zur Evolutionären Erkenntnistheorie*. Stuttgart: Hirzel.

———. 1988. Metakriterien wissenschaftlicher Rationalität. *Zeitschrift für Wissenschaftsforschung* 4(2):201–213.

Watanabe, S. 1975. Needed: A Historico-Dynamical View of Theory Change. *Synthese* 32:113–134.

Watkins, J. W. 1970. Against 'Normal Science.' In Lakatos and Musgrave, 1970, 25–37.

Weimer, W., and D. Palermo. 1973. Paradigms and Normal Science in Pyschology. *Science Studies* 3:211–244.

Weingartner, P., and G. J. W. Dorn, eds. 1990. *Studies on Mario Bunge's Treatise*. Amsterdam: Rodopi.

Weingartner, P., and G. Schurz, eds. 1989. *Reports of the Thirteenth International Wittgenstein-Symposium* 1988. Vienna: Hölder-Pichler-Tempsky.

Weizsäcker, C. F. von. 1977. *Der Garten des Menschlichen. Beiträge zur geschichtlichen Anthropologie*. München: Hanser.

Wendel, H. J. 1990. *Moderner Relativismus. Zur Kritik antirealistischer Sichtweisen des Erkenntnisproblems*. Tübingen: Mohr.

Wieland, W. 1962. *Die aristotelische Physik. Untersuchungen über die Grundlegung der Naturwissenschaft und die sprachlichen Bedingungen der Prinzipienforschung bei Aristoteles*. Göttingen: Vandenhoeck & Ruprecht.

Winston, M. E. 1976. Did a (Kuhnian) Scientific Revolution Occur in Linguistics? In Suppe and Asquith, 1976, 25–33.

Wisdom, J. O. 1974. The Nature of 'Normal' Science. In Schilpp, 1974, 820–842.

Wittgenstein, L. 1960. *Philosophische Untersuchungen*. In Schriften. Vol. I. Frankfurt: Suhrkamp. English edition: *Philosophical Investigations*. Trans. G. E. M. Anscombe. Oxford, Basil Blackwell, 1968.

Wittich, D. 1978a. Eine aufschlussreiche Quelle für das Verständnis der gesellschaftlichen Rolle des Denkens Thomas S. Kuhns. *Deutsche Zeitschrift für Philosophie* 26:105–113.

———. 1978b. Die gefesselte Dialektik. Zu den philosophischen Ideen des Wissenschaftshistorikers Th.S. Kuhn. *Deutsche Zeitschrift für Philosophie* 26:785–797.

———. 1981. Die innere Logik der Philosophie Thomas S. Kuhns. In Bayertz 1981: pp. 136–159.

Wong, D. B. 1989. Three Kinds of Incommensurability. In Krausz, 1989, 140–158.

Worrall, J. 1990. Scientific Revolutions and Scientific Rationality: The Case of the "Elderly Holdout." In Savage, 1990, 319–354.

Wright, G. H. v. 1986. *Wittgenstein.* Frankfurt: Suhrkamp.

Wuchterl, K., and A. Hübner. 1979. *Ludwig Wittgenstein.* Reinbeck bei Hamburg: Rowohlt.

Young, R. W. 1979. Paradigms in Geography—Implications of Kuhn's Interpretation of Scientific Inquiry. *Australian Geographical Studies* 17:204–209.

Zheng, L. 1988. Incommensurability and Scientific Rationality. *International Studies in the Philosophy of Science* 2:227–236.

Additional Works of Thomas S. Kuhn

1991a The Road since Structure. In *PSA 1990. Proceedings of the 1990 Biennial Meeting of the Philosophy of Science Association,* vol. 2, ed. A. Fine, M. Forbes, and L. Wessels, 2–13. East Lansing: Philosophy of Science Association.

1991b The Natural and the Human Sciences. In *The Interpretive Turn,* ed. D. R. Hiley, J. E. Bohman, and R. Shusterman, 17–24.

1992 The Trouble with the Historical Philosophy of Science. Robert and Maurine Rothschild Distinguished Lecture, 19 November 1991. Cambridge: An Occasional Publication of the Department of the History of Science, Harvard University.

INDEX